The Physics of Dilute Magnetic Alloys

Available for the first time in English, this classic text by Jun Kondo describes the Kondo effect thoroughly and intuitively. Its clear and concise treatment makes this book of interest to graduate students and researchers in condensed matter physics.

The first half of the book describes the rudiments of the theory of metals at a level that is accessible for undergraduate students. The second half discusses key developments in the Kondo problem, covering topics including magnetic impurities in metals, the resistance minimum phenomenon, infrared divergence in metals and scaling theory, including Wilson's renormalization group treatment and the exact solution by the Bethe ansatz. A new chapter has been added covering advances made since the Japanese edition was published, such as the quantum dot and heavy fermion systems.

JUN KONDO is Emeritus Professor of Toho University and Special Advisor at the National Institute for Advanced Industrial Science and Technology (AIST), Japan, and a Member of the Japan Academy. He is well known for solving the problem of the resistance minimum phenomenon, now known as the Kondo effect. Theoretical and experimental studies in this field continue today, and new applications for the theory are still being found.

The Physics of Dilute Magnetic Alloys

JUN KONDO
Member of the Japan Academy, Japan

Translated by
Shigeru Koikegami, Second Lab, LLC
Kosuke Odagiri, National Institute of Advanced Industrial Science and Technology
Kunihiko Yamaji, National Institute of Advanced Industrial Science and Technology
Takashi Yanagisawa, National Institute of Advanced Industrial Science and Technology

CAMBRIDGE UNIVERSITY PRESS
Cambridge, New York, Melbourne, Madrid, Cape Town,
Singapore, São Paulo, Delhi, Mexico City

Cambridge University Press
The Edinburgh Building, Cambridge CB2 8RU, UK

Published in the United States of America by Cambridge University Press, New York

www.cambridge.org
Information on this title: www.cambridge.org/9781107024182

First published 2012

Printed and bound in the United Kingdom by the MPG Books Group

A catalog record for this publication is available from the British Library

Library of Congress Cataloging in Publication data
Kondo, Jun, 1930–
The physics of dilute magnetic alloys / Jun Kondo ; translated by
Shigeru Koikegami ... [et al.].
p. cm.
Includes bibliographical references and index.
ISBN 978-1-107-02418-2 (hardback)
1. Dilute alloys. 2. Magnetic alloys. 3. Kondo effect. 4. Free electron theory
of metals. I. Title.
QC176.8.M3K66 2012
538′.4–dc23 2012014479

ISBN 978-1-107-02418-2 Hardback

Contents

Preface

I wrote this book as an introduction to the theory of electrons in metals. There are a good many texts on this topic, and the emphasis of this present book is in discussing the physics of dilute magnetic alloys. The first half of this book is devoted to discussion of the topics that are necessary for the discussions in the later chapters. Recent activity in the theory of dilute magnetic alloys has made the field highly complex, so I have tried to describe this at a level that is suitable for those who are new to the subject area.

While metals are characterized by the presence of electrons that move about freely in them, it is most important that, unlike in semi-conductors, there is huge electron density. As a result, quantum effects become predominant and, because electrons are fermions, the phenomenon of degeneracy takes place. We may even go so far as to say that almost all of the characteristic behavior of metals is due to this phenomenon of degeneracy.

Concerning the quantum theory of electrons in metals, there have been five major developments. The first is the Sommerfeld theory, which introduced the concept of degeneracy to explain the behavior of the electronic specific heat. The second is the Bohm–Pines theory, which discusses the effect of the inter electronic Coulomb interaction, together with the many-body treatment of the problem which emerged from this theory. Next, the sensitivity of the magnetic resistance and the de Haas–van Alphen effect to the shape of the Fermi surface became the motivation for the development that occurred in the field of fermiology. As a result, the Fermi surface came to be called the 'face' of the metal.

The most important development in the field of the theory of electrons in metals has been the Bardeen–Cooper–Schrieffer (BCS) theory of superconductivity. We may say that this meant the discovery of a new state of matter, and furthermore it considerably influenced other areas of physics.

Fifth, and finally, we have the problem of magnetic alloys. This problem began with the old discovery, in the 1930s, of the electric resistance minimum effect. When metals are cooled down to liquid helium temperatures, at first, the electric resistance goes down. However, below $\sim$10 K, an increase in the resistivity was often observed. It was established that this was due to the mixing of a small number of magnetic atoms. However, the relationship between the magnetism due to the impurity atoms and the increase in the electric resistivity was unclear.

In the end, we came to the understanding of this problem as being due to the magnetic moment which affects the degenerate electrons, but the effort of many was necessary for the complete solution to this problem. A system composed of electrons in metals and magnetic moments may seem simple at first, but the theoretical treatment of this problem is quite involved, and commonplace methods of approximation are not the best way of tackling the core of the problem. Many approaches blossomed in this flowerbed, supported by the stem of the combined intellect of many, nurturing an especially fruitful area of physics which also affected the development of other areas.

The aim and goal of this book is to showcase this theoretical fruit, which will be found in Chapters 5 and beyond. The first four chapters can be thought of as preparatory introductory material. Chapters 1 and 2 discuss the more elementary topics in the description of many-electron systems. Chapters 3 and 4 move on to the elementary topics in the description of metals. We start the discussion of magnetic alloys in Chapter 5. In my opinion, the discussions up to this chapter are fairly simple. Chapters 6 and beyond turn to a more complex content, but I took care to keep the discussion thorough. Thus I believe that those who are new to the subject area may also understand the full content of this book upon careful study. Nonetheless, the author's incapabilities may have encumbered his intentions. I very much hope that readers will be forgiving in this respect.

Finally, my sincere gratitude goes to Professor Yukihito Tanabe of the University of Tokyo, who encouraged me to write this book; Mr. Kyohei Endo and Mr. Saneatsu Makiya of Shokabo; and Ms. Masako Kobayashi who helped me especially in the proofreading.

Jun Kondo
October 1983

Translators' foreword

This present volume is a translation of *"kinzoku denshi-ron – jisei-gokin wo chushin to shite"* ("Theory of electrons in metals – with emphasis on magnetic alloys"), written by Professor J. Kondo in Japanese and published by Shokabo in 1983. The translation contains an additional chapter which discusses some of the developments that have taken place since the original publication of the book. The title of the book could very well have been "The Kondo effect", had the author not been Professor Kondo himself. The discussion of the Kondo effect takes the prime position in this book, though the author never refers to it as such!

The author and his theory need no introduction. Suffice it to say that his work has been a milestone in condensed matter physics, with far-reaching consequences such as those in the study of many-body problems in general. But not only that. The Kondo effect also marked the beginning of the concept of asymptotic freedom, where the relevant coupling strength increases logarithmically with decreasing energy/temperature scale. This phenomenon is of central importance in the physics of strong interaction, in particle and nuclear physics, which is now believed to be described by quantum chromo-dynamics (QCD). We should add that there has been renewed interest in the study of the Kondo effect in the context of heavy electron systems and quantum dots.

Professor Kondo's famous work was carried out in the 1960s, in what has now become a central block of the National Institute of Advanced Industrial Science and Technology (AIST) in Tsukuba, Japan, and was then called the Electro-technical Laboratory or ETL and situated in Tanashi, Japan. This is the laboratory in which the translators are presently based, and Professor Kondo has been a mentor to all four of us.

In the preparation of the initial manuscript, the workload was split as follows. K. Y. produced the initial manuscript of the first four chapters. S. K. took responsibility for Chapters 5, 7, 8 and the Appendix. T. Y. worked on

Chapter 6. The new chapter on recent developments was written by Professor Kondo, and this was translated by K. O. The manuscript was then wholly and thoroughly revised by T. Y. and K. O. for the language and physics; on the whole we believe that we have been able to maintain the original style of exposition.

As stated by the author in the Preface, the book is written at a level that is suitable for newcomers to the field, and the Japanese language version of the book indeed exemplifies his lucid style of exposition. Nonetheless, we benefited from numerous discussions with Professor Kondo, who would willingly, but ever unassumingly, respond to our queries concerning the more intricate points.

We would like to thank Tatsuya Ono of Shokabo, Lindsay Barnes, Simon Capelin, Graham Hart, Caroline Mowatt, Mairi Sutherland and Antoaneta Ouzounova of Cambridge University Press, for their help and generosity during the preparation of the manuscript and the publication of the book. We would also like to thank Drs. Izumi Hase, Naoki Shirakawa and all of our other colleagues at AIST for discussions, help and encouragement.

S. K.
K. O.
K. Y.
T. Y.
October 2011

1
Atoms

We present here an overview of the electronic structure of atoms. We begin with the mean-field approximation. This scheme is sometimes also called the Hartree approximation, and is the most basic starting point when discussing many-electron systems. In this approach, the atomic states are distinguished from one another by their electronic configuration. An electronic configuration is, in general, degenerate with a number of other configurations. However, when we take into account the corrections due to the deviation of the Coulomb interaction away from the mean field, the energy levels are split into a number of distinct levels, and each of these split energy levels is called a multiplet. In order to demonstrate this point, we introduce the Slater determinant. After this, we discuss the Coulomb integral and the exchange integral. In particular, because of the Pauli principle, the exchange integral exists only between electrons with the same spin orientations. This allows us to explain Hund's rule, that is, the multiplet that has the largest value of composite spin has the lowest energy.

1.1 Mean-field approximation and electronic configurations

The usual starting point for discussing the electronic structure of atoms is the mean-field approximation.

The motion of an electron is affected by attractive Coulomb interaction due to the positive charge *Ze* of the nucleus and repulsive Coulomb interaction due to the other electrons. The latter is time dependent owing to the motion of the other electrons, but we may, as an approximation, replace these electrons by an appropriate charge distribution and consider the Coulomb force due to it. This is called the mean-field approximation. In this way, we regard each electron as moving in the mean field, independently of the other electrons.

To determine the charge distribution which replaces the other electrons, we make use of the self-consistency condition. This point will be discussed later. Here, we consider the electronic wavefunction for a given mean field.

The potential energy of the electron in the mean field is denoted by $V(\boldsymbol{r})$. The vector $\boldsymbol{r}$ gives the spatial coordinates of the electron. The nucleus is assumed to remain stationary at the origin. Insofar as the problem concerns single atoms, we may almost always regard $V(\boldsymbol{r})$ to be a function of $r = |\boldsymbol{r}|$ only. $V(r)$ is then called the central field. When r is sufficiently large, $V(r)$ approaches $-e^2/r$. This is because when an electron is sufficiently far away from the origin, the Ze positive charge at the origin and the $Z-1$ electrons around it exert a combined force, to the distant electron, which is asymptotically equivalent to that of a single positive charge at the origin. On the other hand, when r is sufficiently small, we see immediately that $V(r)$ approaches $-Ze^2/r$+const.

Now, in the mean-field approximation, the atomic Hamiltonian is expressed as

$$H_0 = -\frac{\hbar^2}{2m}\sum_{i=1}^{Z}\Delta_i + \sum_{i=1}^{Z}V(r_i). \tag{1.1}$$

Δ_i is the Laplacian with respect to the spatial coordinates of the ith electron. The Schrödinger equation,

$$H_0\Psi = E\Psi, \tag{1.2}$$

is reduced, upon the substitution

$$\Psi = \prod_i \psi_i(\boldsymbol{r}_i), \tag{1.3}$$

to Z one-electron Schrödinger equations given by

$$\left[-\frac{\hbar^2}{2m}\Delta_i + V(r_i)\right]\psi_i(\boldsymbol{r}_i) = \varepsilon_i\psi_i(\boldsymbol{r}_i), \tag{1.4}$$

$$E = \sum_i \varepsilon_i. \tag{1.5}$$

Since $V(r_i)$ is spherically symmetric, the solution to eq. (1.4) is specified by quantum numbers n, l and m.

$$\psi_{nlm}(\boldsymbol{r}_i) = R_{nl}(r_i)Y_{lm}(\theta_i,\varphi_i). \tag{1.6}$$

Y_{lm} are the spherical harmonic functions, and $R_{nl}(r)$ satisfy the following equations:

$$\left[-\frac{d^2}{dr^2}-\frac{2}{r}\frac{d}{dr}+U(r)+\frac{l(l+1)}{r^2}\right]R_{nl}(r)=\frac{2m\varepsilon_{nl}}{\hbar^2}R_{nl}(r), \tag{1.7}$$

$$U(r)=\frac{2m}{\hbar^2}V(r). \tag{1.7a}$$

By solving eq. (1.7) with the boundary condition $R_{nl}(r) \to 0$ as $r \to \infty$, we obtain negative eigenvalues ε_{nl}. For a given l, n is assigned the values $n = l+1, l+2, \ldots$. The eigenvalues increase in this order. The quantum number l represents the magnitude of the orbital angular momentum, and the quantum number m represents the z-component of l. Eigenvalues ε_{nl} are $(2l + 1)$-fold degenerate since they are independent of m.

A solution, given by eq. (1.6), to the one-electron Schrödinger equation is called an 'atomic orbital', or simply 'orbital'. As stated earlier, an orbital is specified by the three quantum numbers n, l and m. By convention, $l = 0, 1, 2, 3, \ldots$ are assigned the letters s, p, d, f, ... For example, the '2p orbital' refers to the wavefunction of eq. (1.6) with the quantum numbers $n = 2$ and $l = 1$.

For most atoms, writing the orbitals in the ascending order of the corresponding energy leads to 1s, 2s, 2p, 3s, 3p, 4s, 3d, 4p, 5s, 4d, 5p, 6s, 4f, 5d, 6p, ... To obtain the ground state, we then place Z electrons into the orbitals starting from the lowest state, in a way that respects the Pauli exclusion principle. According to this principle, only one spin-up electron and one spin-down electron can enter each orbital given by eq. (1.6). Each s orbital therefore accommodates only two electrons, and the full set of p orbitals accommodates six electrons in total.

For example, in the case of Si, the ground state is denoted by $(1s)^2(2s)^2(2p)^6(3s)^2(3p)^2$. This is called the electronic configuration of the atom. Orbitals that are completely filled with electrons, that is, the 'closed shell', are usually omitted in the notation. Si is therefore expressed as $(3p)^2$ or $(3s)^2(3p)^2$, while $(3s)^1(3p)^3$ denotes the electronic configuration of an excited state of Si.

1.2 Multiplets

Let us consider an electronic configuration. The sum of the energies of all of the orbitals contained in the configuration, given by eq. (1.5), is called the configuration energy, and is a first approximation for the value of the atomic energy. In general, when the shell is not closed, the configuration energy has a large degree of degeneracy.

For example, the $(2p)^2$ configuration is ${}_6C_2 = 15$-fold degenerate. This is due to the mean-field approximation and, as the approximation is improved, the level should, in principle, split into several levels. To obtain such a splitting, we take the following Hamiltonian instead of H_0:

$$H = -\frac{\hbar^2}{2m}\sum_i \Delta_i - \sum_i \frac{Ze^2}{r_i} + \sum_{i>j} \frac{e^2}{r_{ij}}. \tag{1.8}$$

The second term here is the potential energy due to the attractive Coulomb interaction between each electron and the nucleus. The third term is the potential energy of the Coulomb interaction between electrons ($r_{ij} = |\boldsymbol{r}_i - \boldsymbol{r}_j|$). For example, in the case of $(2p)^2$, we can improve the approximation using perturbation theory if we diagonalize H, using the 15 wavefunctions that are the eigenstates of H_0. In order to do so, we have to explicitly take into account the spin-coordinate dependence of the wavefunctions, and furthermore make the wavefunctions satisfy the Pauli principle.

Spin is an intrinsically quantum mechanical concept, but as a classical analogy, we can imagine the electrons to be spinning, and the corresponding angular momentum is the spin. When we equate its magnitude with $\sqrt{s(s+1)}\hbar$, we find that $s = 1/2$ in the case of the electron. Its z-component is only allowed to take either of the two values $+\frac{1}{2}\hbar$ and $-\frac{1}{2}\hbar$.

This can be quantified by introducing the spin coordinate. In addition to the spatial coordinates, the electron has a spin coordinate which takes the value $+\frac{1}{2}$ or $-\frac{1}{2}$. The state in which the probability of the former (latter) is equal to unity is designated as the state whose z-component of spin is equal to $+\frac{1}{2}\hbar$ $(-\frac{1}{2}\hbar)$. This state is denoted as $\alpha(\zeta)$ $(\beta(\zeta))$. ζ is the spin coordinate. Let us write the spin angular momentum divided by $\hbar$ as $\boldsymbol{s} = (s_x, s_y, s_z)$. This operator acts on functions of the spin coordinate. We then have

$$s_z\alpha(\zeta) = \frac{1}{2}\alpha(\zeta), \quad s_z\beta(\zeta) = -\frac{1}{2}\beta(\zeta). \tag{1.9a}$$

We obtain the following from the generic properties of angular momentum operators:

$$s_x\alpha(\zeta) = \frac{1}{2}\beta(\zeta), \quad s_x\beta(\zeta) = \frac{1}{2}\alpha(\zeta), \tag{1.9b}$$

$$s_y\alpha(\zeta) = \frac{i}{2}\beta(\zeta), \quad s_y\beta(\zeta) = -\frac{i}{2}\alpha(\zeta). \tag{1.9c}$$

Furthermore, because of the orthonormality between α and β, we have

$$\left.\begin{aligned}\langle\alpha|\alpha\rangle &= \sum_{\zeta=1/2,-1/2} \alpha(\zeta)^2 = 1\\ \langle\beta|\beta\rangle &= \sum_{\zeta=1/2,-1/2} \beta(\zeta)^2 = 1\\ \langle\alpha|\beta\rangle &= \sum \alpha(\zeta)\beta(\zeta) = 0\\ \langle\beta|\alpha\rangle &= \sum \beta(\zeta)\alpha(\zeta) = 0\end{aligned}\right\}. \tag{1.10}$$

In this way, the wavefunction of an electron is specified by four quantum numbers $nlms$, that is, the wavefunction is given by $\psi_{nlm}(\boldsymbol{r}_i)\alpha(\zeta_i)$ or $\psi_{nlm}(\boldsymbol{r}_i)\beta(\zeta_i)$. Here, $s = 1/2$ corresponds to α and $s = -1/2$ to β. From now on, we write the set of three quantum numbers nlm as λ and the set of λ and s as k. We then write the wavefunction of an electron as $\phi_k(i)$. The argument i refers to the spatial and spin coordinates. In the general case where there are Z electrons with quantum numbers $k_1, k_2, \ldots, k_Z$, the wavefunctions are written as

$$\Psi(1, 2, \ldots, Z) = \phi_{k_1}(1)\phi_{k_2}(2)\cdots\phi_{k_Z}(Z).$$

This is eq. (1.3) modified by including the spin functions.

Before going on to diagonalizing eq. (1.8) using this result, we have to rewrite Ψ so that it satisfies the Pauli principle. According to this principle, the wavefunction $\Psi(1, 2, \ldots)$ for a many-electron system must change its sign when any pair of electrons exchange their coordinates. For example,

$$\Psi(2, 1, \ldots) = -\Psi(1, 2, \ldots). \tag{1.11}$$

All of the arguments, other than the first two, are identical on both sides.

Ψ as written above by itself does not satisfy the Pauli principle, but one can construct such a function as follows. Let us denote the operation of the permutation of coordinates as P. PΨ then refers to Ψ with the order of the coordinates rearranged according to a permutation P. It is also an eigenfunction of H_0 with the same eigenvalue as Ψ. $Z!$ such permutations are linearly combined as follows:

$$\begin{aligned}\Phi(1, 2, \ldots) &= \frac{1}{\sqrt{Z!}} \sum_{\mathrm{P}} (-)^{\mathrm{P}} \mathrm{P}\Psi\\ &= \frac{1}{\sqrt{Z!}} \begin{vmatrix} \phi_{k_1}(1) & \phi_{k_2}(1) & \cdots \\ \phi_{k_1}(2) & \phi_{k_2}(2) & \\ \vdots & & \ddots \end{vmatrix}.\end{aligned} \tag{1.12}$$

Here, $(-)^{\rm P}$ is equal to 1 when P is an even permutation, and -1 when it is an odd one. This is called the Slater determinant and is easily seen to satisfy the Pauli principle. We also see that eq. (1.12) vanishes if there is a pair of identical labels, no matter which, from among $k_1, k_2, \ldots, k_Z$. This is nothing other than the exclusion principle. The overall factor $1/\sqrt{Z!}$ is needed to normalize eq. (1.12).

Now we take $(2p)^2$ as an example. As we mentioned earlier, this electronic configuration is 15-fold degenerate, which is found by counting the number of ways of choosing k_1, k_2. We list them explicitly in the following. Out of the four components of k, we omitted n and l and wrote out only m and s explicitly. We have defined $M_L = \sum m_i$ and $M_S = \sum s_i$.

	k_1		k_2		M_L	M_S
Φ_1	1	$\frac{1}{2}$	1	$-\frac{1}{2}$	2	0
Φ_2	1	$\frac{1}{2}$	0	$\frac{1}{2}$	1	1
Φ_3	1	$\frac{1}{2}$	0	$-\frac{1}{2}$	1	0
Φ_4	1	$-\frac{1}{2}$	0	$\frac{1}{2}$	1	0
Φ_5	1	$-\frac{1}{2}$	0	$-\frac{1}{2}$	1	-1
Φ_6	1	$\frac{1}{2}$	-1	$\frac{1}{2}$	0	1
Φ_7	1	$\frac{1}{2}$	-1	$-\frac{1}{2}$	0	0
Φ_8	1	$-\frac{1}{2}$	-1	$\frac{1}{2}$	0	0
Φ_9	0	$\frac{1}{2}$	0	$-\frac{1}{2}$	0	0
Φ_{10}	1	$-\frac{1}{2}$	-1	$-\frac{1}{2}$	0	-1
Φ_{11}	0	$\frac{1}{2}$	-1	$\frac{1}{2}$	-1	1
Φ_{12}	0	$\frac{1}{2}$	-1	$-\frac{1}{2}$	-1	0
Φ_{13}	0	$-\frac{1}{2}$	-1	$\frac{1}{2}$	-1	0
Φ_{14}	0	$-\frac{1}{2}$	-1	$-\frac{1}{2}$	-1	-1
Φ_{15}	-1	$\frac{1}{2}$	-1	$-\frac{1}{2}$	-2	0

In order to diagonalize H using these functions as the bases, it is convenient to classify the wavefunctions according to the total angular momentum. For the total orbital angular momentum $\boldsymbol{L} = (L_x, L_y, L_z)$ and the total spin angular momentum $\boldsymbol{S} = (S_x, S_y, S_z)$, their x-components, for example, are defined by

$$L_x = \sum_{i=1}^{Z} l_{ix}, \quad S_x = \sum_{i=1}^{Z} s_{ix}.$$

The five operators $\boldsymbol{L}^2, L_z, \boldsymbol{S}^2, S_z$ and H can be shown to commute with each other. Hence if we choose the eigenfunctions of $\boldsymbol{L}^2, L_z, \boldsymbol{S}^2$ and S_z as the bases, there are no matrix elements of H between the basis wavefunctions that have different eigenvalues.

The diagonalization therefore needs to be carried out only among the sets of wavefunctions with identical $\boldsymbol{L}^2, L_z, \boldsymbol{S}^2$ and S_z eigenvalues. Let us take $L(L+1)$, $M_L, S(S+1)$ and M_S as the eigenvalues of $\boldsymbol{L}^2, L_z, \boldsymbol{S}^2$ and S_z. The 15 Φs tabulated above are already eigenfunctions of L_z and S_z, and the values of M_L and M_S are written therein. They are the sums of the corresponding values for individual electrons. Let us first consider Φ_1. Since it has $M_L = 2$ and $M_S = 0$, and no other states have the same set of eigenvalues, Φ_1 is already an eigenfunction of $\boldsymbol{L}^2$ and $\boldsymbol{S}^2$ at the same time. In fact, since we can easily show that $\boldsymbol{L}^2\Phi_1 = 2(2+1)\Phi_1$ and $\boldsymbol{S}^2\Phi_1 = 0$, we conclude that $L = 2$ and $S = 0$. On the other hand, since, in general, an eigenfunction of the angular momentum has the following property:

$$(L_x - iL_y)\Phi(L, M_L) = \sqrt{L(L+1) - M_L(M_L - 1)}\Phi(L, M_L - 1), \quad (1.13)$$

we obtain another eigenfunction, with M_L reduced by 1, by applying $L_x - iL_y = (l_{1x} - il_{1y}) + (l_{2x} - il_{2y})$ to Φ_1 and expressing the result as a linear combination of the states Φ_1 to Φ_{15}. Repeating this procedure, we obtain the following five functions for $L = 2$ and $M_L = 2, 1, \ldots, -2$:

${}^1\mathrm{D}(L = 2 \quad S = 0)$	M_L	M_S
Φ_1	2	0
$\frac{1}{\sqrt{2}}(\Phi_3 - \Phi_4)$	1	0
$\frac{1}{\sqrt{6}}(\Phi_7 - \Phi_8 + 2\Phi_9)$	0	0
$\frac{1}{\sqrt{2}}(\Phi_{12} - \Phi_{13})$	−1	0
Φ_{15}	−2	0

Next, we examine Φ_2. Since it has eigenvalues $M_L = 1$ and $M_S = 1$, and no other state has the same set of eigenvalues, it must be an eigenfunction of $\boldsymbol{L}^2$ and $\boldsymbol{S}^2$. We find that it has $L = 1$ and $S = 1$. By applying $L_x - iL_y$ and $S_x - iS_y$, respectively, we obtain functions that have M_L and M_S reduced by unity:

$^3\mathrm{P}(L = 1 \quad S = 1)$	M_L	M_S
Φ_2	1	1
$\frac{1}{\sqrt{2}}(\Phi_3 + \Phi_4)$	1	0
Φ_5	1	−1
Φ_6	0	1
$\frac{1}{\sqrt{2}}(\Phi_7 + \Phi_8)$	0	0
Φ_{10}	0	−1
Φ_{11}	−1	1
$\frac{1}{\sqrt{2}}(\Phi_{12} + \Phi_{13})$	−1	0
Φ_{14}	−1	−1

We have now obtained 14 functions. The one remaining function should be a linear combination of Φ_7, Φ_8 and Φ_9, because these three have appeared only in two eigenfunctions. From the orthogonalization condition involving these two, the remaining function is determined as

$$\frac{1}{\sqrt{3}}(-\Phi_7 + \Phi_8 + \Phi_9). \tag{1.14}$$

This is found to have $L = 0$ and $S = 0$, and we label it as $^1\mathrm{S}$. In general, $L = 0, 1, 2, \ldots$ are labeled S, P, D, ... and the value of $2S + 1$ is written as the superfix on the left shoulder. The $L = 1, S = 1$ state is denoted as $^3\mathrm{P}$.

In the above examples, each function has a different set of L, S, M_L, M_S values, and each of the 15 functions diagonalizes H. However, the 15 eigenvalues of H do not split into 15 levels. All functions with the same L and S but different values of M_L and M_S have the same eigenvalue. In order to demonstrate this point, we use eq. (1.13) to obtain

$$\begin{aligned}
&\int \Phi(L, M_L - 1)^* H \Phi(L, M_L - 1) d\tau \\
&\quad = [L(L+1) - M_L(M_L - 1)]^{-1} \\
&\qquad \times \int \left[(L_x - iL_y)\Phi(L, M_L)\right]^* H (L_x - iL_y)\Phi(L, M_L) d\tau \\
&\quad = [L(L+1) - M_L(M_L - 1)]^{-1} \\
&\qquad \times \int \Phi(L, M_L)^* H (L_x + iL_y)(L_x - iL_y)\Phi(L, M_L) d\tau .
\end{aligned}$$

Here, we made use of the fact that H commutes with L_x and L_y. Furthermore, using

$$\begin{aligned}
(L_x + iL_y)(L_x - iL_y)\Phi(L, M_L) &= (\boldsymbol{L}^2 - L_z^2 + L_z)\Phi(L, M_L) \\
&= [L(L+1) - M_L(M_L - 1)]\Phi(L, M_L),
\end{aligned}$$

we find that the above integral is independent of M_L. We can derive a similar result for M_S. Therefore, in the above example, the eigenvalues split into three levels: ^{1}D (quintet or fivefold), ^{3}P (ninefold) and ^{1}S (singlet). Each level is called a multiplet. In general, the configuration energies obtained in the mean-field approximation split into several multiplets when we take into account the deviation of the Coulomb energy from the mean value. Each multiplet is $(2L + 1)(2S + 1)$-fold degenerate.

The degeneracy of the multiplet is broken further when we take into account the magnetic interaction between the electrons. Dirac's relativistic equation of the electron leads to a term in the Hamiltonian that is proportional to $\sum \boldsymbol{l}_i \cdot \boldsymbol{s}_i$, where $\boldsymbol{l}_i$ is the orbital angular momentum of the ith electron. This term is proportional to $\boldsymbol{L} \cdot \boldsymbol{S}$ in a particular multiplet, and is therefore called the LS coupling term. It gives rise to an energy that depends on the angle between L and S. The vector sum of $\boldsymbol{L}$ and $\boldsymbol{S}$ is the total angular momentum $\boldsymbol{J}$, and we obtain

$$J(J+1) = \boldsymbol{J}^2 = \boldsymbol{L}^2 + \boldsymbol{S}^2 + 2\boldsymbol{L} \cdot \boldsymbol{S} = L(L+1) + S(S+1) + 2\boldsymbol{L} \cdot \boldsymbol{S},$$

so that the energy depends on the value of J. J takes a value between $L + S$ and $|L - S|$ inclusive, and each level is $(2J + 1)$-fold degenerate corresponding to $J_z = J, \ldots, -J$.

Coming back to the problem of $(np)^2$, we obtain, from the above discussion,

$$E(^1\mathrm{D}) = \int \Phi_1^* H \Phi_1 d\tau = H_{11}, \tag{1.15}$$

$$E(^3\mathrm{P}) = \int \Phi_2^* H \Phi_2 d\tau = H_{22}. \tag{1.16}$$

$E(^1\mathrm{S})$ can be derived using eq. (1.14). However, we can derive it more simply as follows. If we diagonalize the matrix elements between the states Φ_7, Φ_8 and Φ_9, that is:

$$\begin{pmatrix} H_{77} & H_{78} & H_{79} \\ H_{87} & H_{88} & H_{89} \\ H_{97} & H_{98} & H_{99} \end{pmatrix},$$

one solution should coincide with eq. (1.14), and the other two should be the $M_L = 0$ and $M_S = 0$ components of the $^3\mathrm{P}$ and $^1\mathrm{D}$ states. The sum of the eigenvalues of this matrix is therefore equal to $E(^1\mathrm{S}) + E(^3\mathrm{P}) + E(^1\mathrm{D})$. On the other hand, since the sum of the eigenvalues is equal to the sum of the diagonal elements, we obtain

$$E(^1\mathrm{S}) = H_{77} + H_{88} + H_{99} - H_{11} - H_{22}. \tag{1.17}$$

Thus, in the present example, all of the energy levels of the multiplets can be expressed in terms of the diagonal elements of H.

1.3 Coulomb and exchange integrals

Next, we need to calculate the matrix elements of H such as that given by eq. (1.15). In general, let us write

$$H = \sum_i h(i) + \sum_{i>j} V(i,j). \tag{1.18}$$

Here, $h(i)$ includes only the ith coordinates (which may include the spin coordinate) and, in general, the derivatives with respect to them. $V(i,j)$ is the interaction term between the ith and jth electrons. When Φ is a Slater determinant consisting of $\phi_{k_i} (i = 1, \ldots, Z)$ as shown in eq. (1.12) (ϕ_{k_i}s being orthonormal), we obtain, according to Appendix A,

$$\int \Phi^* H \Phi d\tau = \sum_i h(k_i, k_i) + \sum_{i>j} [V(k_i, k_j; k_i, k_j) - V(k_i, k_j; k_j, k_i)], \tag{1.19}$$

where

$$h(k, k') = \int \phi_k(1)^* h(1) \phi_{k'}(1) d\tau_1, \tag{1.20}$$

$$V(k, k'; k'', k''') = \iint \phi_k(1)^* \phi_{k'}(2)^* V(1,2) \phi_{k''}(1) \phi_{k'''}(2) d\tau_1 d\tau_2. \tag{1.21}$$

Here, the integration with respect to τ stands for both the integration over the space coordinates and the summation over the spin coordinates.

When $\phi_{k_i}(1)$ is of the form $\psi_{\lambda_i}(\boldsymbol{r}_1)\alpha(\zeta_1)$ or $\psi_{\lambda_i}(\boldsymbol{r}_1)\beta(\zeta_1)$, and at the same time $h(1)$ and $V(1,2)$ are functions of the space coordinates only, the summation over the spin coordinates is given by eq. (1.10). Equation (1.20) therefore remains finite only when both k and k' contain only α or only β spin types. Likewise, eq. (1.21) remains finite only when both k and k'' contain the same spin and moreover both k' and k''' have the same spin. Therefore

$$\int \Phi^* H \Phi d\tau = \sum_i h(\lambda_i, \lambda_i) + \sum_{i>j} Q(\lambda_i, \lambda_j) - {\sum_{i>j}}' J(\lambda_i, \lambda_j), \tag{1.22}$$

where

$$Q(\lambda_i, \lambda_j) = \iint |\psi_{\lambda_i}(\boldsymbol{r}_1)|^2 |\psi_{\lambda_j}(\boldsymbol{r}_2)|^2 V(1,2) dv_1 dv_2, \tag{1.23}$$

$$J(\lambda_i, \lambda_j) = \iint \psi_{\lambda_i}(\boldsymbol{r}_1)^* \psi_{\lambda_j}(\boldsymbol{r}_1) \psi_{\lambda_j}(\boldsymbol{r}_2)^* \psi_{\lambda_i}(\boldsymbol{r}_2) V(1,2) dv_1 dv_2. \tag{1.24}$$

$Q(\lambda_i, \lambda_j)$ is the Coulomb interaction energy between the charge distribution given by ψ_{λ_i} and that given by ψ_{λ_j}. This is called the Coulomb integral. $J(\lambda_i, \lambda_j)$ has no classical analogue, and is called the exchange integral. It is non-zero only when both i and j have the same spin. $\sum'$ in the last term of eq. (1.22) denotes summation with a restriction that i and j have the same spin.

Regarding the integrals that appear in the above expressions, the angular integration can be carried out by using eq. (1.6) for ψ_λ. After this, $h(nlm, nlm)$ is independent of m, and so is reduced to an integral over r which involves R_{nl}. Let us denote this integral as $h(nl)$.

The Coulomb and exchange integrals consist of a coefficient which depends on lm and $l'm'$, and a double integral which involves R_{nl}s. The latter integration can, in general, only be carried out numerically. Using eq. (1.22), we obtain from eqs. (1.15) and (1.16)

$$E(^1\mathrm{D}) = 2h(2\mathrm{p}) + Q(2\mathrm{p}1, 2\mathrm{p}1),$$

$$E(^3\mathrm{P}) = 2h(2\mathrm{p}) + Q(2\mathrm{p}1, 2\mathrm{p}0) - J(2\mathrm{p}1, 2\mathrm{p}0).$$

We obtain $E(^1\mathrm{S})$ in the same way using eq. (1.17). When we express these in terms of integrals that involve R_{nl}, we arrive at the relation

$$\frac{E(^1\mathrm{S}) - E(^1\mathrm{D})}{E(^1\mathrm{D}) - E(^3\mathrm{P})} = \frac{3}{2}.$$

This is independent of the values of the integrals. This relation is known to hold approximately in atoms with configurations of the type $(\mathrm{np})^2$. The absolute values of the energies of the multiplets are obtained only by integrations which involve R_{nl}, and this is determined by $V(r)$ through eq. (1.7). In general, $V(r)$ is determined by the self-consistent field method (see §1.4).

When we compare the energies of the multiplets, we encounter an empirical rule which is called Hund's rule. This states, firstly, that the multiplet with the largest value of S has the lowest energy. When there is more than one multiplet with the largest value of S, it then states that the one with the largest value of L has the lowest energy.

Large S implies that the electrons tend to have spin aligned in the same direction. In this case, as can be deduced from eq. (1.22), a large number of exchange integrals survive. Since the exchange integrals are positive, they provide a negative contribution to the energy. On the other hand, when L is large, the electrons tend to revolve in the same direction, because if the electrons revolve in opposite directions, they will have greater probability of coming close to each other, and this would increase the energy. In the above example, it follows that $E(^3\mathrm{P}) < E(^1\mathrm{D}) < E(^1\mathrm{S})$.

As to why the energy is lowered by the exchange integrals when the electrons' spins are aligned in the same direction, we can explain this as follows.

Let us consider a Slater determinant for the case where the two orbitals ψ_λ and $\psi_{\lambda'}$ are occupied by two electrons with parallel spin orientations, and another one for the case with antiparallel spin, as follows:

$$\Psi_{\uparrow\uparrow} = \frac{1}{\sqrt{2}} \begin{vmatrix} \psi_\lambda(\boldsymbol{r}_1)\alpha(\zeta_1) & \psi_{\lambda'}(\boldsymbol{r}_1)\alpha(\zeta_1) \\ \psi_\lambda(\boldsymbol{r}_2)\alpha(\zeta_2) & \psi_{\lambda'}(\boldsymbol{r}_2)\alpha(\zeta_2) \end{vmatrix}, \tag{1.25}$$

$$\Psi_{\uparrow\downarrow} = \frac{1}{\sqrt{2}} \begin{vmatrix} \psi_\lambda(\boldsymbol{r}_1)\alpha(\zeta_1) & \psi_{\lambda'}(\boldsymbol{r}_1)\beta(\zeta_1) \\ \psi_\lambda(\boldsymbol{r}_2)\alpha(\zeta_2) & \psi_{\lambda'}(\boldsymbol{r}_2)\beta(\zeta_2) \end{vmatrix}. \tag{1.26}$$

The probability $\rho_{\uparrow\uparrow}(\boldsymbol{r}_1, \boldsymbol{r}_2)$ or $\rho_{\uparrow\downarrow}(\boldsymbol{r}_1, \boldsymbol{r}_2)$, for the electrons of any spin direction to be at $\boldsymbol{r}_1$ and $\boldsymbol{r}_2$, are given by integrating either $|\Psi_{\uparrow\uparrow}|^2$ or $|\Psi_{\uparrow\downarrow}|^2$ with respect to the spin coordinates. Using eq. (1.10), we obtain

$$\rho_{\uparrow\uparrow}(\boldsymbol{r}_1, \boldsymbol{r}_2) = \frac{1}{2}|\psi_\lambda(\boldsymbol{r}_1)\psi_{\lambda'}(\boldsymbol{r}_2) - \psi_\lambda(\boldsymbol{r}_2)\psi_{\lambda'}(\boldsymbol{r}_1)|^2,$$

$$\rho_{\uparrow\downarrow}(\boldsymbol{r}_1,\boldsymbol{r}_2) = \frac{1}{2}(|\psi_\lambda(\boldsymbol{r}_1)\psi_{\lambda'}(\boldsymbol{r}_2)|^2 + |\psi_\lambda(\boldsymbol{r}_2)\psi_{\lambda'}(\boldsymbol{r}_1)|^2).$$

The average of the Coulomb interaction $V(\boldsymbol{r}_1,\boldsymbol{r}_2)$ with either $\Psi_{\uparrow\uparrow}$ or $\Psi_{\uparrow\downarrow}$ is nothing but

$$\int \rho_{\uparrow\uparrow}(\boldsymbol{r}_1,\boldsymbol{r}_2)V(\boldsymbol{r}_1,\boldsymbol{r}_2)dv_1dv_2 \quad \text{or} \quad \int \rho_{\uparrow\downarrow}(\boldsymbol{r}_1,\boldsymbol{r}_2)V(\boldsymbol{r}_1,\boldsymbol{r}_2)dv_1dv_2.$$

We see easily that the former is equivalent to $Q(\lambda,\lambda') - J(\lambda,\lambda')$ and the latter is equivalent to $Q(\lambda,\lambda')$. $\rho_{\uparrow\uparrow}$ vanishes when $\boldsymbol{r}_1 = \boldsymbol{r}_2$, so that the probability of two electrons coming close to one another is small. Therefore the Coulomb interaction energy is smaller. In general, the probability of two electrons having the same spin and coming to the same position is zero. This follows from the fact that Ψ must be zero by eq. (1.11) when the coordinates of electron 1 are exactly equal to those of electron 2. In other words, the Pauli principle makes electrons with parallel spin avoid each other. The decrease in the Coulomb energy due to this effect is given by the exchange integral.

1.4 Hartree's method

In the previous section, the energy of an atom was expressed in terms of integrals which involve the radial function R_{nl}. Now we come to the problem of evaluating the radial function. For doing so, there is one method, which uses hydrogenic wavefunctions based on the wavefunctions of the hydrogen atom. However, the best radial function is obtained by the self-consistent field method. In this method, we first introduce an appropriate approximate form of $V(r)$. Then R_{nl} is evaluated temporarily by eq. (1.7). $V(r)$ is then calculated using the following equation:

$$V(r) = \sum_{j\neq i} \int \frac{e^2|\psi_{\lambda_j}(\boldsymbol{r}_1)|^2}{|\boldsymbol{r}-\boldsymbol{r}_1|}dv_1. \tag{1.27}$$

The summation over j is over all $(Z-1)$ orbitals excluding the orbital i under consideration. Since this is, in general, different from the $V(r)$ that was adopted at first, we substitute this into eq. (1.7). We then evaluate R_{nl} again, and substitute it into eq. (1.27) to obtain $V(r)$. This procedure is repeated until the input $V(r)$ and the evaluated $V(r)$ agree with each other. This was first performed by Hartree, and hence it is called the Hartree approximation.

This approximation has the following meaning. Say that there is a general system, not limited to atoms, with Hamiltonian given by eq. (1.18). Writing the wavefunction of the system as

$$\Psi = \prod_i \psi_{\lambda_i}(r_i), \tag{1.28}$$

the average energy is given by

$$E = \sum_i h(\lambda_i, \lambda_i) + \sum_{i>j} Q(\lambda_i, \lambda_j). \tag{1.29}$$

When we minimize this by varying with respect to ψ_{λ_i} under the constraint of the normalization condition, we immediately obtain

$$[h(r) + V(r)]\psi_{\lambda_i}(r) = \varepsilon\psi_{\lambda_i}(r). \tag{1.30}$$

$V(r)$ is given by eq. (1.27) and ε is the Lagrange multiplier associated with the normalization. That is to say, the determination of $V(r)$ by the above self-consistency method corresponds to nothing other than finding the best (i.e., one that minimizes the energy) wavefunction out of those of the form eq. (1.28).

This potential term $V(r)$ depends on i, so the ψ_{λ_i}s are not orthogonal to one another. If we had made use of eq. (1.12) instead of eq. (1.28) as the wavefunction, the energy would be given by eq. (1.22). The minimization condition would then lead to the Hartree–Fock equation. In this case, the potential term becomes non-local, so that it is more difficult to solve the equation. However, the potential becomes universal, i.e., independent of i. The Hartree–Fock scheme can be said to be a better approximation than the Hartree scheme. Calculations in this scheme have been carried out on many atoms and ions, and a good level of agreement with experiments has been obtained regarding the levels of multiplets and so on, especially for light atoms.

References and further reading

There are many textbooks on atomic structure, a representative one being:

J. C. Slater (1960) *Quantum Theory of Atomic Structure*, Vols. I, II (New York: McGraw–Hill).

The following is a convenient review of atoms and molecules:

H. Eyring, J. Walter and G. E. Kimball (1944) *Quantum Chemistry* (New York: John Wiley).

A considerable part of this chapter is based on this textbook.

Note added by the translators:

Both of these books unfortunately seem to be out of print at the time of writing (2011). However, copies may be found in large libraries and second-hand stores. While this present book itself is largely self-contained, its conciseness may motivate some readers to search for further information on related topics. We regret that our expertise is limited when it comes to suggesting a suitable alternative, though we are sure that there are plenty of good books on the market. For now, we can only hope that both these classic books will become available again in the near future.

2
Molecules

We consider the H_2 molecule. We introduce the molecular orbitals from the viewpoint of mean-field approximation, and these are classified into bonding and antibonding orbitals. We discuss the molecular bonds also from the Heitler–London viewpoint, which is based on atomic orbitals. In both of these cases, the state with zero total spin, or the spin singlet state, is found to form a stable molecule. We discuss the relationship between the two viewpoints, and introduce the configuration interaction as an improvement to both. In order to facilitate the treatment of complex molecules, a model is proposed, and a second-quantization procedure which is convenient for its description is introduced.

2.1 The H_2^+ molecule

Let us, in this chapter, define the molecule as a system that consists of more than one nucleus and one or more electrons. Its wavefunction is a function of the coordinates of the nuclei and the electrons. An intuitive picture is as follows. Electrons are lighter by far than nuclei and are moving around fast. It is therefore reasonable to consider the nuclei as being instantaneously fixed, and to solve for the wavefunction of the electrons. The wavefunction and energy thus obtained are functions of the positions of the nuclei, and the energy can be considered to play the role of the potential energy with respect to the motion of the nuclei. This approximation is called the adiabatic approximation, or the Born–Oppenheimer approximation, and is valid when the ratio m/M of the masses of the electron and the nuclei is sufficiently small. This chapter is devoted to the consideration of the wavefunction of the electronic system when the nuclei are fixed.

When studying the electronic structure of molecules, the mean-field approximation is still at the core of our approach. The wavefunction of an electron that moves in the mean field is called the molecular orbital. The placement of electrons into these molecular orbitals in accordance with the Pauli principle is called, again, the electronic configuration. In the discussion of molecules, unlike in the discussion of atoms, the mean field is not spherically symmetric. This makes it difficult to solve the problem, or to carry out calculations that aim at the same degree of accuracy as has been achieved in the case of atoms using Hartree's method. Even so, each molecule has its own symmetries, thanks to which the problem can be simplified.

Let us consider H_2^+ as the simplest example. The system consists of two protons that are separated by distance R and an electron that moves around them. As a trial function for the lowest-energy molecular orbital, we consider a linear combination of the 1s orbitals of the two hydrogen atoms. We label the two nuclei a and b, and the two 1s orbitals are correspondingly named ψ_a and ψ_b. Writing the molecular orbital as

$$\psi = c_a\psi_a + c_b\psi_b, \tag{2.1}$$

we obtain, by multiplying the Schrödinger equation

$$H\psi = E\psi$$

by ψ_a and ψ_b, and integrating both sides, the following secular equation:

$$\begin{vmatrix} H_{aa} - E & H_{ba} - SE \\ H_{ab} - SE & H_{bb} - E \end{vmatrix} = 0, \tag{2.2}$$

where

$$S = \int \psi_a\psi_b dv, \quad H_{ij} = \int \psi_i H\psi_j dv.$$

Note that the overlap integral S is non-zero. Obviously, $H_{aa} = H_{bb}$ and $H_{ab} = H_{ba}$. By solving eq. (2.2), we obtain the two solutions:

$$\left.\begin{aligned} \psi_1 &= \frac{\psi_a + \psi_b}{\sqrt{2(1+S)}}, \quad E_1 = \frac{H_{aa} + H_{ab}}{1+S} \\ \psi_2 &= \frac{\psi_a - \psi_b}{\sqrt{2(1-S)}}, \quad E_2 = \frac{H_{aa} - H_{ab}}{1-S} \end{aligned}\right\}. \tag{2.3}$$

In the present case, the Hamiltonian H is written as

$$H = -\frac{\hbar^2}{2m}\Delta - \frac{e^2}{|\boldsymbol{r} - \boldsymbol{R}_a|} - \frac{e^2}{|\boldsymbol{r} - \boldsymbol{R}_b|}. \tag{2.4}$$

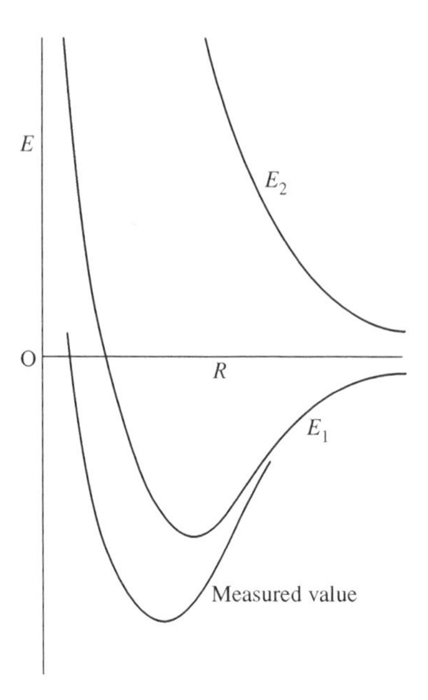

Figure 2.1 Binding energies of H_2^+ as functions of the interatomic distance.

S, H_{aa} and H_{ab} etc. are easily evaluated. This yields E_1 and E_2 as functions of R, and these are shown in Fig. 2.1. Note that the repulsive interaction e^2/R between the nuclei is included in this figure. When R is large, the overlap between ψ_a and ψ_b becomes small, and so S and H_{ab} tend to zero, whereas H_{aa} tends to the energy E_{1s} of the 1s orbital of the hydrogen atom. Hence both E_1 and E_2 tend to E_{1s}. This corresponds to the transition to a system that consists of a hydrogen atom and a proton. This behavior of the energy is correct, even though, no matter how large R is, ψ_1 and ψ_2 refer to states where an electron is coming and going between the two protons. ψ_1 corresponds to the ground state, which is, according to the actual calculations, the most stable at $R = 1.32\,\text{Å}$. The difference between the minimum energy and the energy at $R = \infty$ is called the dissociation energy. In the present case, this energy is 1.76 eV. In contrast, ψ_2 cannot be an orbital that makes up a stable molecule. ψ_1 is called the bonding orbital, and ψ_2 is called the antibonding orbital.

From the symmetry of the system, we can say, generally, that the electronic wavefunction must be either symmetric or antisymmetric with respect to the vertical plane that divides the line segment between the two protons at the midpoint. ψ_1 belongs to the former case, while ψ_2 belongs to the latter. Hence ψ_2 vanishes at the midpoint of the line segment, while ψ_1 has a considerable density at that point. The presence of an attractive Coulomb interaction between this charge density and the protons can be thought of as the reason why ψ_1 is stabilized. In general, we can say that an orbital that has a high electron density in between two nuclei contributes to molecular bonding.

When $R = 0$, ψ_1 reduces to the 1s orbital of the hydrogen atom, whereas eq. (2.4) becomes the Hamiltonian for He^+. Because of this, we cannot obtain the correct energy by using eq. (2.3). On the other hand, if we take ψ_a and ψ_b as the 1s orbitals captured by a nucleus of charge Ze, and then minimize the energy by optimizing Z as a variational parameter, we should obtain the correct energies at the two extremes, i.e., $Z = 1$ at $R = \infty$ and $Z = 2$ at $R = 0$. In between these two extremes, the description of the energy is also improved considerably. We obtain the dissociation energy of 2.25 eV with distance 1.06 Å between the two nuclei, and these are close to the observed values 2.791 eV and 1.06 Å. In any case, the method in which the molecular orbital is approximated by a linear combination of atomic orbitals is called the LCAO (linear combination of atomic orbitals) method and is in wide use.

2.2 The H_2 molecule

As the symmetry of the H_2 molecule is the same as that of the H_2^+ molecule, the symmetry of its molecular orbitals is also the same. Let us again adopt eq. (2.3) for the molecular orbitals. The ground state is given by filling two electrons in the bonding orbital ψ_1. From eq. (1.12), the wavefunction is given by

$$\begin{aligned}\Phi_S &= \frac{1}{\sqrt{2}}\begin{vmatrix}\psi_1(1)\alpha(1) & \psi_1(1)\beta(1)\\ \psi_1(2)\alpha(2) & \psi_1(2)\beta(2)\end{vmatrix}\\ &= \psi_1(1)\psi_1(2)\frac{1}{\sqrt{2}}[\alpha(1)\beta(2) - \beta(1)\alpha(2)].\end{aligned} \tag{2.5}$$

The last factor is an expression of the (singlet) state, where the two spins are combined with each other to give zero total spin (see Appendix B). This factor is unnecessary for the evaluation of the energy of Φ_S. The mean energy is given by

$$E_S = 2E_1 + \int \psi_1(r_1)^2\psi_1(r_2)^2\frac{e^2}{r_{12}}dv_1dv_2 + \frac{e^2}{R}. \tag{2.6}$$

Here, the Hamiltonian includes, along with the sum of the Hamiltonians of the type given by eq. (2.4) for the two electrons, the repulsive energy e^2/r_{12} between the two electrons. When this energy is evaluated as a function of R, a minimum, which is similar to that in the ψ_1 curve in Fig. 2.1, is obtained. However, its dissociation energy 2.65 eV and internuclear distance 0.85 Å are in poor agreement with the observed 4.72 eV and 0.74 Å. This indicates that the use of the wavefunction of eq. (2.5) is not a good approximation. This is further clarified by the following discussion. From eq. (2.3), we obtain

$$\psi_1(1)\psi_1(2) \propto \psi_a(1)\psi_a(2)+\psi_b(1)\psi_b(2)+\psi_a(1)\psi_b(2)+\psi_b(1)\psi_a(2). \quad (2.7)$$

The first and second terms represent the H^+H^- state, where both electrons gather around a single nucleus. When $R \to \infty$, having two hydrogen atoms is more stable than the H^+H^- state, and so eq. (2.5) does not give the correct limiting behavior. The presence of these unstable terms can be thought of as the main cause of the discrepancy in the calculated value of the energy for finite R.

In order to improve on this, we generally employ a method that is called the configuration interaction.

Equation (2.5) is an electronic configuration in which two electrons fill ψ_1. Other than this configuration, we have $\psi_2(1)\psi_2(2)$, $\psi_1(1)\psi_2(2)$ and $\psi_2(1)\psi_1(2)$. The configuration interaction method corresponds to taking the wavefunction as a linear combination of these four. Here, a proviso is that, in the case of zero composite spin such as in eq. (2.5), the last two terms must always appear in the combination $\psi_2(1)\psi_1(2) + \psi_1(1)\psi_2(2)$, since otherwise the Pauli principle is violated. However, taking these four as the basis set is equivalent to taking the four terms on the right-hand side of eq. (2.7). If we make a linear combination of the latter four and determine their coefficients by minimizing the energy, the coefficients of $\psi_a(1)\psi_a(2)$ and $\psi_b(1)\psi_b(2)$ should vanish naturally in the limit $R\to\infty$, and therefore the results that are derived by using the former linear combination should reduce to the same result. In this way, for finite R, the coefficients of $\psi_a(1)\psi_a(2)$ and $\psi_b(1)\psi_b(2)$ are variational parameters; if these terms raise the energy, the coefficients should become small.

As a matter of fact, before the above molecular orbital method was introduced, the following function had been employed by Heitler and London. This was the first instance of the application of quantum mechanics to the problem of molecules.

$$\Phi_{\mathrm{HL}} = \frac{1}{\sqrt{2(1+S^2)}}[\psi_a(1)\psi_b(2) + \psi_b(1)\psi_a(2)]. \quad (2.8)$$

The factor at the front is for the normalization. The expectation value of energy is then given as

$$E_{\mathrm{HL}} = \frac{1}{1+S^2}\iint \psi_a(1)\psi_b(2)H[\psi_a(1)\psi_b(2) + \psi_b(1)\psi_a(2)]dv_1dv_2. \quad (2.9)$$

As ψ_a and ψ_b represent the 1s orbital of the hydrogen atom, this may be expressed as follows:

$$E_{\mathrm{HL}} + \frac{e^2}{R} = 2E_{1\mathrm{s}} + \frac{Q}{1+S^2} + \frac{J}{1+S^2}, \tag{2.10}$$

where

$$Q = \iint \psi_a(1)^2 \psi_b(2)^2 \left(-\frac{e^2}{|\boldsymbol{r}_1 - \boldsymbol{R}_b|} - \frac{e^2}{|\boldsymbol{r}_2 - \boldsymbol{R}_a|} + \frac{e^2}{r_{12}} + \frac{e^2}{R} \right) dv_1 dv_2, \tag{2.11}$$

and

$$J = \iint \psi_a(1)\psi_b(1)\psi_a(2)\psi_b(2) \times \left(-\frac{e^2}{|\boldsymbol{r}_1 - \boldsymbol{R}_b|} - \frac{e^2}{|\boldsymbol{r}_2 - \boldsymbol{R}_a|} + \frac{e^2}{r_{12}} + \frac{e^2}{R} \right) dv_1 dv_2. \tag{2.12}$$

By computing eq. (2.10) as a function of R, we obtain the dissociation energy 3.14 eV and the internuclear distance 0.87 Å, and so the function is slightly improved from eq. (2.5).

In the above, we only considered the zero-spin (singlet) state. If two electrons have identical spin, one must fill ψ_1 and the other must fill ψ_2. The wavefunction then becomes

$$\begin{aligned} \Phi_T &= \frac{1}{\sqrt{2}} \begin{vmatrix} \psi_1(1)\alpha(1) & \psi_2(1)\alpha(1) \\ \psi_1(2)\alpha(2) & \psi_2(2)\alpha(2) \end{vmatrix} \\ &= \frac{1}{\sqrt{2(1-S^2)}} [\psi_b(1)\psi_a(2) - \psi_a(1)\psi_b(2)]\alpha(1)\alpha(2). \end{aligned} \tag{2.13}$$

The spin part can be either $\beta(1)\beta(2)$ or $[\alpha(1)\beta(2) + \beta(1)\alpha(2)]/\sqrt{2}$. These are components of the (triplet) state whose composite spin is equal to 1 (see Appendix B). For calculating the energy, the spin part is not necessary. The energy is given, in the place of eq. (2.10), by

$$2E_{1\mathrm{s}} + \frac{Q}{1-S^2} - \frac{J}{1-S^2}. \tag{2.14}$$

This does not have a minimum as a function of R, and so the state of eq. (2.13) is unstable. The difference between eq. (2.13) and eq. (2.8) is merely in the sign of the second term: this causes the negative sign before J and, as J is negative, renders the state unstable. This sign difference is ultimately due to the Pauli principle. The singlet spin function changes its sign under the exchange of 1 and 2, and so the part of the wavefunction that depends on the spatial position must remain constant under the exchange of 1 and 2. In the case of the triplet, there is an opposite relation, and this is the cause of the negative sign of the second term in eq. (2.13).

As for the value of J, as is seen from eq. (2.12), its absolute value goes up when the overlap between the two 1s orbitals becomes greater, so that the singlet state becomes more stable. In general, when the overlap between the orbitals of two atoms is large, the spin singlet state of the two electrons belonging to the two orbitals makes a large contribution to the bonding between the atoms. This is called the covalent bond. Because of the ease in drawing an intuitive picture, the Heitler–London scheme, explained above, has been applied even to more complex molecules, and is powerful for the qualitative treatment of covalent bonds.

2.3 The configuration interaction

When the Heitler–London scheme is applied to complicated molecules, or when the configuration interaction is used in the molecular orbital method, simple models are often employed in order to make the problems tractable and to obtain qualitatively useful results. Let us illustrate this using the case of the H_2 molecule. First, we assume the orthogonality of the atomic orbitals ($S = 0$) in this model. We take the Hamiltonian to be the sum of $h(1)+h(2)$, which relates only to single electrons, and $V(1,2)$, which relates to both electrons. For the matrix elements, we use the following notation:

$$\varepsilon \equiv \int \psi_a(1)h(1)\psi_a(1)dv_1 = \int \psi_b(1)h(1)\psi_b(1)dv_1, \tag{2.15}$$

$$t \equiv \int \psi_a(1)h(1)\psi_b(1)dv_1 = \int \psi_b(1)h(1)\psi_a(1)dv_1, \tag{2.16}$$

$$U_0 \equiv \int \psi_a(1)^2\psi_a(2)^2V(1,2)dv_1dv_2 = \int \psi_b(1)^2\psi_b(2)^2V(1,2)dv_1dv_2, \tag{2.17}$$

$$U_1 \equiv \int \psi_a(1)^2\psi_b(2)^2V(1,2)dv_1dv_2 = \int \psi_b(1)^2\psi_a(2)^2V(1,2)dv_1dv_2, \tag{2.18}$$

$$J' \equiv \int \psi_a(1)\psi_b(1)\psi_a(2)\psi_b(2)V(1,2)dv_1dv_2. \tag{2.19}$$

We have used no simplifications for h, but for V we adopt the simplification that the matrix elements other than those listed above vanish. Simplifications in this spirit can easily be extended to more complicated molecules, and on the basis of this, a treatment that utilizes the molecular orbital method or the Heitler–London scheme can be adopted. In the case of the H_2 molecule, the molecular orbitals are given by

$$\left.\begin{aligned}\psi_1 &= \frac{\psi_a + \psi_b}{\sqrt{2}}\\ \psi_2 &= \frac{\psi_a - \psi_b}{\sqrt{2}}\end{aligned}\right\}. \tag{2.20}$$

The wavefunctions in the molecular orbital method, in the Heitler–London scheme, and the triplet wavefunction are given, respectively, by

$$\Phi_S = \psi_1(1)\psi_1(2), \tag{2.21}$$

$$\Phi_{\mathrm{HL}} = \frac{\psi_a(1)\psi_b(2) + \psi_b(1)\psi_a(2)}{\sqrt{2}}, \tag{2.22}$$

$$\Phi_T = \frac{\psi_a(1)\psi_b(2) - \psi_b(1)\psi_a(2)}{\sqrt{2}}. \tag{2.23}$$

The corresponding energies are given by

$$E_S = 2(\varepsilon + t) + \tfrac{1}{2}U_0 + \tfrac{1}{2}U_1 + J', \tag{2.24}$$

$$E_{\mathrm{HL}} = 2\varepsilon + U_1 + J', \tag{2.25}$$

$$E_T = 2\varepsilon + U_1 - J', \tag{2.26}$$

in the same order, and they, in turn, correspond to eqs. (2.6), (2.10) and (2.14), respectively.

Let us now consider the influence of the configuration interaction on E_{HL}. For this purpose, we take account of $\psi_a(1)\psi_a(2)$ and $\psi_b(1)\psi_b(2)$. From the symmetry of the problem, we immediately see that they enter the problem in the form

$$\Phi' = \frac{1}{\sqrt{2}}[\psi_a(1)\psi_a(2) + \psi_b(1)\psi_b(2)].$$

Let us therefore adopt a wavefunction of the form $c_1\Phi_{\mathrm{HL}} + c_2\Phi'$. From eqs. (2.15)–(2.19), the energy of Φ' is equal to $2\varepsilon + U_0 + J'$, and the matrix element between Φ_{HL} and Φ' is equal to $2t$. The problem is therefore reduced to the following secular equation:

$$\begin{vmatrix} 2\varepsilon + U_1 + J' - E & 2t \\ 2t & 2\varepsilon + U_0 + J' - E \end{vmatrix} = 0. \tag{2.27}$$

If $U_0 = U_1$, then one of the solutions is equal to eq. (2.24), and eq. (2.21) is the correct wavefunction. However, in contrast to U_0, which is the Coulomb energy in the state where two electrons fill a 1s orbital, U_1 is the Coulomb energy of the state in which the electrons stay in two different 1s orbitals separated by the internuclear distance. In the case where $U_0 \gg U_1$, the energy of eq. (2.24), which includes U_0, is much larger and so eq. (2.21) no longer remains as good an approximation as eqs. (2.22) and (2.23). When $U_0 - U_1 \gg |t|$, the influence of Φ' can be taken into Φ_{HL} as a perturbation. This is equivalent to expanding the solution to eq. (2.27) in powers of $t/(U_0 - U_1)$. As a result, eq. (2.25) is modified to

$$E_{\mathrm{HL}} = 2\varepsilon + U_1 + J' - \frac{4t^2}{U_0 - U_1}. \tag{2.28}$$

On the other hand, for the triplet state, there are no electronic configurations to take into account, because, in the case of two electrons with the same spin, two electrons are forbidden from entering a single orbital by the Pauli principle. There are therefore no corrections to eq. (2.26). Thus, the energy difference between the singlet and triplet states is equal to

$$2J_{\mathrm{eff}} \equiv 2J' - \frac{4t^2}{U_0 - U_1}. \tag{2.29}$$

As seen in the above discussions, when $U_0 - U_1$ is sufficiently large, we obtain a picture regarding the spatial motion of electrons that each electron is localized around a separate nucleus. However, there remains the spin degree of freedom, which may be expressed by the following effective Hamiltonian:

$$-2J_{\mathrm{eff}}\boldsymbol{s}_a \cdot \boldsymbol{s}_b. \tag{2.30}$$

Here, $\boldsymbol{s}_a$ and $\boldsymbol{s}_b$ are the spin operators pertinent to electrons localized around the nucleus a and b, respectively. Expression (2.30) is proportional to their scalar product. If we define $\boldsymbol{S} = \boldsymbol{s}_a + \boldsymbol{s}_b$, we obtain $S(S+1) = \boldsymbol{S}^2 = {\boldsymbol{s}_a}^2 + {\boldsymbol{s}_b}^2 + 2\boldsymbol{s}_a \cdot \boldsymbol{s}_b = 3/2 + 2\boldsymbol{s}_a \cdot \boldsymbol{s}_b$. In the case of the singlet ($S = 0$), we obtain $\boldsymbol{s}_a \cdot \boldsymbol{s}_b = -3/4$. In the case of the triplet ($S = 1$), we obtain $\boldsymbol{s}_a \cdot \boldsymbol{s}_b = 1/4$. The energy difference is therefore just equal to eq. (2.29). The Hamiltonian of the form of eq. (2.30) is easily extended to the case of many electrons, and this is called the Heisenberg Hamiltonian.

2.4 Second quantization

Rather than expressing the wavefunction of the many-electron system by a Slater determinant as in eq. (1.12), it is often more convenient to employ the method of second quantization. Here, we pay attention to the orbitals that are included in the wavefunction, and do not explicitly incorporate the spatial coordinates. Let us now adopt the electronic orbitals ϕ_k which form a complete orthonormal set. We define the creation and annihilation operators ${a_k}^\dagger$ and a_k corresponding to an orbital k. These are the Hermitian conjugates of each other. We define $|0\rangle$ to be the state with zero electrons. The operation of ${a_k}^\dagger$ on this state generates the state

$$|k\rangle = {a_k}^\dagger|0\rangle, \tag{2.31}$$

which denotes that an electron is present in the orbital k. When the annihilation operator a_k is applied to the zero-electron state $|0\rangle$, we obtain zero:

$$a_k|0\rangle = 0. \tag{2.32}$$

This may be regarded as the definition of $|0\rangle$. The operator

$$n_k \equiv {a_k}^\dagger a_k$$

is regarded as expressing the number of electrons in orbital k. Since the electron is a fermion, we impose the following anticommutation relation:

$${a_k}^\dagger a_k + a_k {a_k}^\dagger = 1. \tag{2.33}$$

Since, from these relations, we obtain

$$n_k|0\rangle = 0, \quad n_k|k\rangle = |k\rangle,$$

$|0\rangle$ is the state with no electrons and $|k\rangle$ is the state where one electron is in orbital k. Note that

$$a_k {a_k}^\dagger|0\rangle = |0\rangle. \tag{2.34}$$

Since two electrons cannot be copresent in an orbital k, we impose

$${a_k}^\dagger {a_k}^\dagger = 0, \quad a_k a_k = 0. \tag{2.35}$$

$|0\rangle$ and $|k\rangle$ are defined with the normalization,

$$\langle 0|0\rangle = 1, \quad \langle 0|a_k {a_k}^\dagger|0\rangle = 1. \tag{2.36}$$

Next, when there are a large number of electrons, we take

$$|k_1 k_2 \cdots\rangle = a^{\dagger}_{k_1} a^{\dagger}_{k_2} \cdots |0\rangle \tag{2.37}$$

as the wavefunction corresponding to eq. (1.12). When the order of k_i is modified, and the parity of the permutation is odd, the sign changes as shown in eq. (1.12). In order to take account of this property, we impose the following anticommutation relations:

$$a_k{}^{\dagger} a_{k'}{}^{\dagger} + a_{k'}{}^{\dagger} a_k{}^{\dagger} = 0, \quad a_k a_{k'} + a_{k'} a_k = 0. \tag{2.38}$$

Equations (2.35) are included in these as a special case. We also impose

$$a_k{}^{\dagger} a_{k'} + a_{k'} a_k{}^{\dagger} = \delta_{kk'}. \tag{2.39}$$

This includes eq. (2.33). The conjugate of eq. (2.37) is

$$\langle k_1 k_2 \cdots | = \langle 0 | \cdots a_{k_2} a_{k_1}. \tag{2.40}$$

Equations (2.37) and (2.40) are also normalized by eq. (2.36).

In this way, we have obtained a method to describe the state that is in one-to-one correspondence with the Slater-determinant method. In this method, if we take the Hamiltonian as follows, we obtain identical matrix elements to the case when we adopt eq. (1.18) as the Hamiltonian in the Slater-determinant method:

$$H = \sum_{kk'} h(k, k') a_k{}^{\dagger} a_{k'} + \frac{1}{2} \sum_{kk'k''k'''} V(k, k'; k''', k'') a_k{}^{\dagger} a_{k'}{}^{\dagger} a_{k''} a_{k'''}. \tag{2.41}$$

The summation over k is over all of the ks with which ϕ_k span a complete set. To demonstrate this, we adopt two states corresponding to eqs. (A. 1) and (A. 2) in Appendix A:

$$|\Phi\rangle = a_{k_1}{}^{\dagger} a_{k_2}{}^{\dagger} \cdots |0\rangle, \tag{2.42}$$

$$|\Phi'\rangle = a_{k'_1}{}^{\dagger} a_{k'_2}{}^{\dagger} \cdots |0\rangle. \tag{2.43}$$

We need to obtain $\langle \Phi' | H | \Phi \rangle$. Following the example of Appendix A, we first consider the first term of eq. (2.41).

(A) Case where $k_l = k'_l \quad (l = 1, \ldots, Z)$

That is, $|\Phi\rangle = |\Phi'\rangle$. We need $\langle \Phi | a^{\dagger}_k a_{k'} | \Phi \rangle$. $a^{\dagger}_k a_{k'} | \Phi \rangle$ vanishes if k' is not equal to any of $k_1, \ldots, k_Z$. If $k' = k_l$, we obtain $a_k{}^{\dagger} a_{k'} | \Phi \rangle = a^{\dagger}_{k_1} a^{\dagger}_{k_2} \cdots a_k{}^{\dagger} \cdots |0\rangle$.

Here, $a_k^\dagger$ is at the position where $a_{k_l}^\dagger$ was originally. In the inner product of this and $\langle\Phi|$, all factors except those associated with l yield unity by eq. (2.36), and $\langle 0|a_{k_l}a_k{}^\dagger|0\rangle$ remains. This is finite and equal to unity only when $k = k_l$. Therefore $\langle\Phi|a_k{}^\dagger a_{k'}|\Phi\rangle$ is equal to unity when $k = k'$ is equal to any of $k_1, \ldots, k_Z$, and zero otherwise. We thus obtain as the matrix element

$$\sum_{l=1}^{Z} h(k_l, k_l).$$

(B) Case where $k_1 \neq k_1'$ and $k_l = k_l'$ $(l = 2, \ldots, Z)$

As we saw above, in order to pick up non-zero matrix elements, it suffices to know the values of k and k' that make $a_k{}^\dagger a_{k'}|\Phi\rangle$ identical to $|\Phi'\rangle$. Using eq. (2.34), we then find that $k' = k_1$ and $k = k_1'$ is the only case. We thus obtain $h(k_1', k_1)$ as the matrix element. In the general case of $k_i \neq k_i'$ and $k_l = k_l'$ $(l \neq i)$, we obtain $h(k_i', k_i)$ for the matrix element in the same way.

Moreover, for the cases that are reducible to either of the above two cases by changing the order of operation of $a_{k_l}^\dagger$, we can use the above results after taking care of the sign. In other cases, that is, when $|\Phi\rangle$ and $|\Phi'\rangle$ differ for two or more orbitals, the matrix element vanishes.

Next, we consider the second term of eq. (2.41). First, consider case (A) above. For $a_k{}^\dagger a_{k'}{}^\dagger a_{k''}a_{k'''}|\Phi\rangle$ to be equal to $|\Phi\rangle$, we have the following four combinations:

$$\left.\begin{array}{l} k = k_l \\ k' = k_m \end{array}\right\} \quad \left\{\begin{array}{l} k'' = k_l \\ k''' = k_m \end{array}\right.$$

$$\left.\begin{array}{l} k = k_m \\ k' = k_l \end{array}\right\} \quad \left\{\begin{array}{l} k'' = k_m \\ k''' = k_l \end{array}\right.$$

(each pair on the left connected to each pair on the right)

Adding together the four contributions, we obtain $\sum_{l>m}[V(k_l, k_m; k_l, k_m) - V(k_m, k_l; k_l, k_m)]$. In case (B), for $a_k{}^\dagger a_{k'}{}^\dagger a_{k''}a_{k'''}|\Phi\rangle$ to be equal to $|\Phi'\rangle$, it is necessary to cancel $a_{k_1}^\dagger$ inside $|\Phi\rangle$ by using $a_{k''}$ or $a_{k'''}$, and install $a_{k_1'}{}^\dagger$ by using $a_k{}^\dagger$ or $a_{k'}{}^\dagger$. Therefore there are four combinations. Out of the two annihilation operators, the other one must cancel a certain $a_{k_l}^\dagger$, other than $a_{k_1}^\dagger$, which is included in $|\Phi\rangle$. The operator that has been so removed must be reinstated by the other creation operator. In this way, we obtain $\sum_{l\neq 1}[V(k_1', k_l; k_1, k_l) - V(k_l, k_1'; k_1, k_l)]$. We may proceed in the same way when $|\Phi\rangle$ and $|\Phi'\rangle$ are different, for orbitals other than $l = 1$.

(C) Case where $k_1 \neq k_1', k_2 \neq k_2'$, and $k_l = k_l' (l = 3, \ldots, Z)$.

There are again four combinations that make $a_k{}^\dagger a_{k'}{}^\dagger a_{k''} a_{k'''} |\Phi\rangle$ equal to $|\Phi'\rangle$. Adding together these contributions yields $V(k_1', k_2'; k_1, k_2) - V(k_2', k_1'; k_1, k_2)$. In general, when $|\Phi\rangle$ and $|\Phi'\rangle$ are different only for the ith and jth orbitals, we may proceed in the same way, and this yields $V(k_i', k_j'; k_i, k_j) - V(k_j', k_i'; k_i, k_j)$.

For the cases that are reducible to any of the above three cases by changing the order of the operation of $a_{k_l}^\dagger$, we can use the above results after taking care of the sign. In other cases, that is, where $|\Phi\rangle$ and $|\Phi'\rangle$ differ for three or more orbitals, all matrix elements vanish.

In this way, the method of second quantization is found to be completely equivalent to the Slater-determinant method, which uses the spatial wavefunctions. The method of second quantization has the following advantages. Consider the Hamiltonian, which is written in the form of eq. (2.41). By regarding $a_k{}^\dagger a_{k'}$ $(k \neq k')$ as the operator that transfers an electron from k' to k and $a_k{}^\dagger a_k$ as that which expresses the electron number in orbital k, we can directly read off the electronic processes from the Hamiltonian. Conversely, when we set up a model that involves certain electronic processes and energy terms, the corresponding Hamiltonian can be written down immediately. Note that the Hamiltonian does not express how many electrons are present in the system under consideration. In other words, eq. (2.41) encodes no information on Z.

Up to this point, k has been taken as denoting both λ, which specifies the orbital, and s, which designates the direction of spin. When $h(l)$ and $V(l, m)$ do not involve spin, $h(k, k')$ vanishes unless k and k' have the same direction of spin, and $V(k, k'; k''', k'')$ vanishes unless k and k''', as well as k' and k'', have the same direction of spin. The modification of eq. (2.41) to take this into account leads to

$$H = \sum_{\lambda\lambda' s} h(\lambda, \lambda') a_{\lambda s}{}^\dagger a_{\lambda' s} + \frac{1}{2} \sum_{\lambda\lambda'\lambda''\lambda''' s s'} V(\lambda, \lambda'; \lambda''', \lambda'') a_{\lambda s}{}^\dagger a_{\lambda' s'}{}^\dagger a_{\lambda'' s'} a_{\lambda''' s}. \tag{2.44}$$

Although we have used the same symbols for h and V as in eq. (1.20) and (1.21), respectively, they are defined here by

$$h(\lambda, \lambda') = \int \psi_\lambda{}^*(1) h(1) \psi_{\lambda'}(1) dv_1, \tag{2.45}$$

$$V(\lambda, \lambda'; \lambda'', \lambda''') = \int \psi_\lambda{}^*(1) \psi_{\lambda'}{}^*(2) V(1, 2) \psi_{\lambda''}(1) \psi_{\lambda'''}(2) dv_1 dv_2. \tag{2.46}$$

The above is a general discussion. If we retain some particular set of terms and drop the others, this would be an approximation and corresponds to setting up a model. For V, the terms to be considered first are the $\lambda = \lambda' = \lambda'' = \lambda'''$ terms. They give rise to the following contribution to the Hamiltonian:

$$\sum_{\lambda} V(\lambda, \lambda; \lambda, \lambda) a_{\lambda\uparrow}{}^{\dagger} a_{\lambda\uparrow} a_{\lambda\downarrow}{}^{\dagger} a_{\lambda\downarrow}, \tag{2.47}$$

where $\uparrow$ and $\downarrow$ denote $s = 1/2$ and $-1/2$, respectively. This contribution implies that the occupation of the orbital λ by a spin $\uparrow$ electron and a spin $\downarrow$ electron gives rise to an energy of $V(\lambda, \lambda; \lambda, \lambda)$. Next, for the case where $\lambda = \lambda'''$ and $\lambda' = \lambda''$ ($\lambda \neq \lambda'$), we obtain the Hamiltonian

$$\sum_{\lambda > \lambda'} V(\lambda, \lambda'; \lambda, \lambda')(a_{\lambda\uparrow}{}^{\dagger} a_{\lambda\uparrow} + a_{\lambda\downarrow}{}^{\dagger} a_{\lambda\downarrow})(a_{\lambda'\uparrow}{}^{\dagger} a_{\lambda'\uparrow} + a_{\lambda'\downarrow}{}^{\dagger} a_{\lambda'\downarrow}). \tag{2.48}$$

This expresses the Coulomb interaction energy between the electron in the λ orbital and that in the λ' orbital. Likewise, for the case where $\lambda = \lambda''$ and $\lambda' = \lambda'''$ ($\lambda \neq \lambda'$), we obtain the Hamiltonian

$$\begin{aligned} -\sum_{\lambda > \lambda'} V(\lambda, \lambda'; \lambda', \lambda)(& a_{\lambda\uparrow}{}^{\dagger} a_{\lambda\uparrow} a_{\lambda'\uparrow}{}^{\dagger} a_{\lambda'\uparrow} + a_{\lambda\downarrow}{}^{\dagger} a_{\lambda\downarrow} a_{\lambda'\downarrow}{}^{\dagger} a_{\lambda'\downarrow} \\ & + a_{\lambda\uparrow}{}^{\dagger} a_{\lambda\downarrow} a_{\lambda'\downarrow}{}^{\dagger} a_{\lambda'\uparrow} + a_{\lambda\downarrow}{}^{\dagger} a_{\lambda\uparrow} a_{\lambda'\uparrow}{}^{\dagger} a_{\lambda'\downarrow}). \end{aligned} \tag{2.49}$$

The integral that appears here is called the exchange integral, and has no classical analogy.

As an example, let us consider the model of eqs. (2.15)–(2.19). We first choose the two orbitals ψ_a and ψ_b for ψ_λ. We denote these as $\lambda = \mathrm{A}$ and B. Now, if we set

$$\begin{aligned} \varepsilon &= h(\mathrm{A}, \mathrm{A}) = h(\mathrm{B}, \mathrm{B}), \\ t &= h(\mathrm{A}, \mathrm{B}) = h(\mathrm{B}, \mathrm{A}), \\ U_0 &= V(\mathrm{A}, \mathrm{A}; \mathrm{A}, \mathrm{A}) = V(\mathrm{B}, \mathrm{B}; \mathrm{B}, \mathrm{B}), \\ U_1 &= V(\mathrm{A}, \mathrm{B}; \mathrm{A}, \mathrm{B}), \\ J' &= V(\mathrm{A}, \mathrm{B}; \mathrm{B}, \mathrm{A}), \end{aligned}$$

then the Hamiltonian of this model, expressed in the second-quantization formalism, is given by the sum of eqs. (2.47)–(2.49) and

$$\begin{aligned} &\varepsilon(a_{\mathrm{A}\uparrow}{}^{\dagger} a_{\mathrm{A}\uparrow} + a_{\mathrm{A}\downarrow}{}^{\dagger} a_{\mathrm{A}\downarrow} + a_{\mathrm{B}\uparrow}{}^{\dagger} a_{\mathrm{B}\uparrow} + a_{\mathrm{B}\downarrow}{}^{\dagger} a_{\mathrm{B}\downarrow}) \\ &\quad + t(a_{\mathrm{A}\uparrow}{}^{\dagger} a_{\mathrm{B}\uparrow} + a_{\mathrm{B}\uparrow}{}^{\dagger} a_{\mathrm{A}\uparrow} + a_{\mathrm{A}\downarrow}{}^{\dagger} a_{\mathrm{B}\downarrow} + a_{\mathrm{B}\downarrow}{}^{\dagger} a_{\mathrm{A}\downarrow}). \end{aligned}$$

The wavefunctions are given by $a_{A\uparrow}{}^{\dagger}|0\rangle$, $a_{A\uparrow}{}^{\dagger}a_{B\downarrow}{}^{\dagger}|0\rangle$, etc. We can look for the eigenstates by using the linear combination of the above wavefunctions, in a way that is exactly parallel to our discussion from eq. (2.21) onwards.

Returning to eq. (2.41), let us consider the case where all *V*s are zero. That is, we neglect all electron–electron interactions. The Hamiltonian is then called the one-electron Hamiltonian. We try to diagonalize this into the form

$$\sum_n \varepsilon_n a_n{}^{\dagger} a_n. \tag{2.50}$$

This problem is equivalent to diagonalization in the case where there is a Hamiltonian $h(\boldsymbol{r})$ which has matrix elements $\langle\phi_k|h|\phi_{k'}\rangle = h(k,k')$ among an orthonormal set ϕ_k. If the latter problem is solved, and the eigenfunctions are given by

$$\phi_n = \sum_k c_{kn}\phi_k, \tag{2.51}$$

that is, the coefficients c_{kn} satisfying

$$\langle\phi_n|h|\phi_{n'}\rangle = \sum_{kk'} c_{kn}{}^{*} c_{k'n'} h(k,k') = \delta_{nn'}\varepsilon_n \tag{2.52}$$

are found, the former problem can be solved by diagonalizing in terms of them using

$$a_n = \sum_k c_{kn}{}^{*} a_k. \tag{2.53}$$

This is because, by inverting eq. (2.53), we obtain

$$a_k = \sum_n c_{kn} a_n, \tag{2.54}$$

and, substituting this into the first term of eq. (2.41), we obtain eq. (2.50).

References and further reading

The following textbooks are related to this chapter:

J. C. Slater (1963, 1965, 1967) *Quantum Theory of Molecules and Solids*, Vols. I, II, III (New York: McGraw-Hill).

From the reference section of Chapter 1, the following textbook is also useful:

H. Eyring, J. Walter and G. E. Kimball (1944) *Quantum Chemistry* (New York: John Wiley).

Note added by the translators:

With great regret, we must report again that all of these books unfortunately seem to be out of print at the time of writing (2011). Please see the comments in the reference section of Chapter 1 for our standpoint with respect to this problem.

3
The Sommerfeld theory of metals

Solids are classified according to the qualitative nature of the interaction by which the atoms in solids attract each other. In metals, in particular, there is an important contribution due to the itinerant electrons. We then discuss the Sommerfeld theory, which describes such electrons. Almost all electrons are dormant because of the Fermi–Dirac statistics, and so they do not contribute to either specific heat or susceptibility. Such an electron system is said to be Fermi degenerate, and exhibits a number of singular properties. These are especially marked when the electrons interact with localized spin, and this point will be the main theme of this book in Chapters 5 and beyond.

3.1 Classification of solids

Thanks to recent quantum mechanical treatment, thorough understanding has been obtained not only on the problem of why molecular bonds appear but also on the problem of why it is more energetically favorable for atoms to get together and form solids than to be separated into individual atoms and molecules. While some aspects of these problems can be understood to some extent in terms of classical or phenomenological analysis, for the problem of the binding in metals, in particular, a quantum mechanical treatment is indispensable. In this section, let us classify solids on the basis of their condensation mechanism.

3.1.1 Molecular crystals

Since rare-gas atoms such as He, Ne and so on, which have closed-shell structures, and also molecules such as H_2, Cl_2 and so on, do not interact strongly with one another, it is difficult for them to form solids. However, any atoms and molecules, when they are separated by some distance, interact with each other through a weak attractive force which is called the van der Waals force.

The strength of this force is inversely proportional to the seventh power of the distance in the cases of atoms and non-polar molecules such as those mentioned above.

The origin of this force can only be understood through quantum mechanics. An intuitive explanation is that if the electron cloud of an atom or a molecule is temporarily polarized so that an electronic polarization is induced in the atom, it induces polarization in the other atoms, and this, altogether, brings about the decrease of the energy. This decrease in the energy is inversely proportional to the sixth power of the distance, and therefore the force is inversely proportional to the seventh power.

Because of this force, these atoms and molecules attract one another and form solids at sufficiently low temperatures. When the distance between the atoms or molecules becomes sufficiently small, the electron clouds start to overlap, and a strong repulsive force comes into play. This force is also of quantum mechanical origin, as it is caused by the same-spin electrons avoiding each other due to the Pauli principle. The crystals that are formed by this force exhibit a closest-packed structure in general, similarly to the packing of rigid spheres. These crystals are called molecular crystals, as the atoms or molecules that compose the crystals retain their original form almost without alteration.

3.1.2 Ionic crystals

The ionization energies (i.e., the energy required for an electron to be taken to infinity) of metallic atoms are generally small, while the electron affinity (i.e., the energy that is released when an electron that is located infinitely far away is attached to the atom) of halogen and chalcogen (i.e., the oxygen family) atoms is generally large.

Let us take the case of NaCl as an example. To ionize Na via $\mathrm{Na} \to \mathrm{Na}^+ + (-e)$, we require 5.14 eV. On the other hand, 3.7 eV is released in the reaction $(-e) + \mathrm{Cl} \to \mathrm{Cl}^-$. When N Na atoms and N Cl atoms are placed far enough away from one another, the energy that is required to change all atoms to Na^+ and Cl^- is $1.44N$ eV. This is not negative but is small. When these ions are placed close together, the Coulomb energy is released. When the mutual distance becomes sufficiently small, the latter overcomes the former, and this leads to a stable crystal. In this case, the greater in number the negative (positive) ions approaching the positive (negative) ions are, the lower the energy is, so that the NaCl structure or the CsCl structure is favored (see Fig. 3.1). When the electron clouds of ions overlap with one another, a strong repulsive force comes into play, similarly to the case of molecular crystals. This type of crystal is called the ionic crystal.

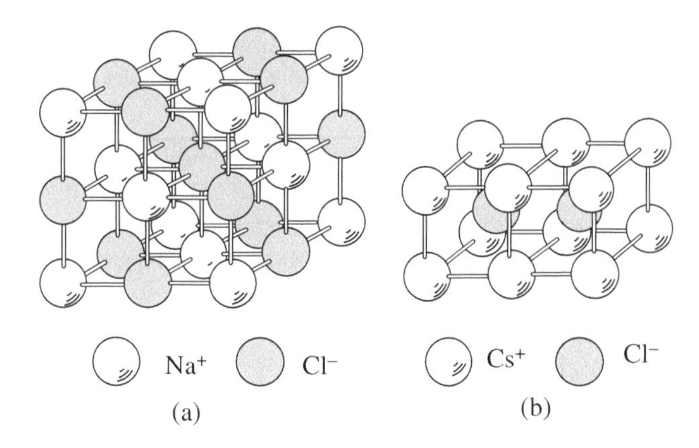

Figure 3.1 (a) The crystal structure of NaCl, (b) the crystal structure of CsCl.

3.1.3 Covalent-bond crystals

As we saw in §2.2 when discussing the H_2 molecule, in the covalent bond, the large overlap between the orbitals belonging to the two atoms makes the covalent bond stronger. The overlap certainly increases as the distance between atoms is decreased, but when this distance becomes too small, the structure is energetically unfavorable because of the repulsive force between the nuclei. Here, by overlap, we mean the amount by which an orbital of a certain atom extends to that of another atom.

When each of the orbitals of the two atoms extends towards the other atom and, further, the two electrons in the two orbitals form a spin singlet state, a strong covalent bond is formed. The ground state of the C atom takes the electronic configuration $(2s)^2(2p)^2$, but when this is excited to $(2s)^1(2p)^3$, the electron cloud extends to four directions as shown in Fig. 3.2. Therefore, in the diamond structure shown in Fig. 3.3, each C atom is bound with four other C atoms by covalent bonds. The energy gain due to this compensates the energy that is needed for the excitation $(2s)^2(2p)^2 \rightarrow (2s)^1(2p)^3$, and a stable crystal is formed. The group IV elements Si and Ge also adopt this structure. Moreover, many of the compounds of the group III and V elements, for example InSb, adopt a similar structure (zinc-blende structure, Fig. 3.4), in which half of the sites of the diamond structure are occupied by In and the other half by Sb, in such a way that four Sb(In) atoms are placed around the In(Sb) atom. This can be understood as being due to the system first going over to In^-Sb^+ resulting in the same electronic structure as the group IV elements, and then forming covalent bonds.

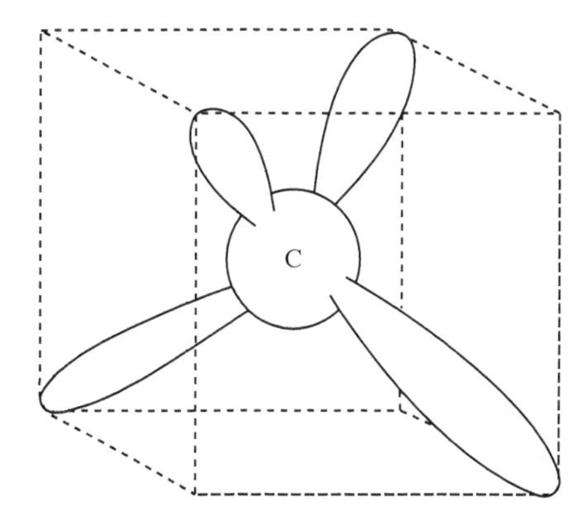

Figure 3.2 Regular tetrahedral structure made of four orbitals with electronic configuration $(2s)^1(2p)^3$.

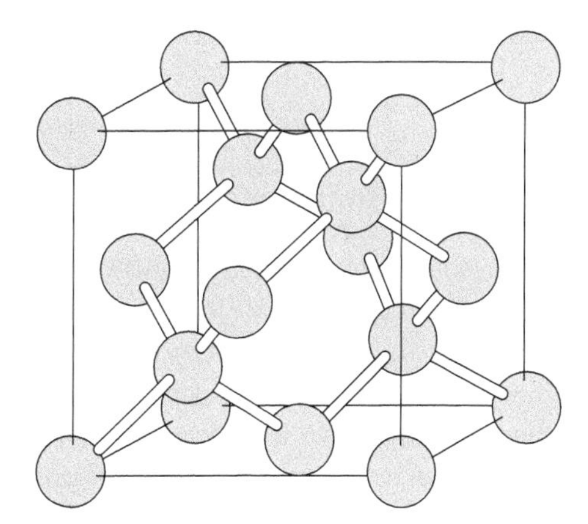

Figure 3.3 The crystal structure of diamond.

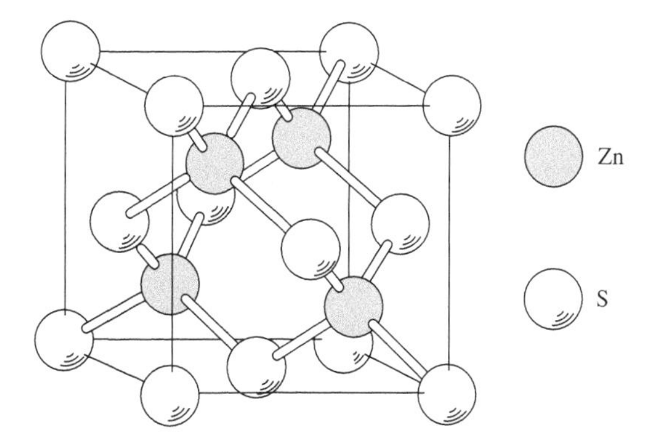

Figure 3.4 The crystal structure of zinc blende.

3.1.4 Metals

In the metallic state of Na, for example, we can consider Na^+ with a closed-shell structure to be lined up in uniform order, and the electrons originating from the 3s electron of the Na atom to be moving around amongst them. The energy gain in this state compared with the state with individually separated Na atoms can be explained as due to the 3s electrons being able to move around the whole crystal, not being bound to a single atom. This is based on the uncertainty principle. When an electron is confined to be within a range of width Δx, it moves around with a momentum of the order of $\hbar/\Delta x$, and this forces the system to have an energy of about $\hbar^2/m(\Delta x)^2$. The larger Δx is, the smaller the latter energy is, and so the spreading out of electrons over the whole crystal leads to the lowering of the energy. This is considered to be the origin of the metallic bond. More detailed discussion will be given later. Alkaline metals and noble metals release an electron per atom, while group II metals release two and group III metals three.

3.2 The Sommerfeld theory

As discussed in the previous section, generally in metals, closed-shell ions are placed in regular order and a large number of electrons move around in the space around them. These electrons are called the conduction electrons, and most of the characteristic properties of metals are due to them. When we discuss the electronic structure of metals, we usually only consider the conduction electrons, and pay no attention to those electrons that compose the closed shell.

The movement of conduction electrons is subject to the influence of ions and other conduction electrons. Here, again, we start with a mean field, which is an average over them. The simplest, and a rather bold, assumption is to set the force of the mean field equal to zero. However, at the surface of the metal, a high potential barrier rebounds the electrons.

These assumptions were adopted by Drude, who, immediately after the discovery of the electron, tried to explain the various properties of metals as originating from the electrons that move around in them. He was able to explain many of the properties by treating the electrons analogously to molecules in the kinetic theory of gases, namely by assuming that the electrons move about freely in the metal just like gas molecules in a container, colliding at times with something and bouncing off the wall. Drude introduced the relaxation time τ, which is due to the collisions, as a parameter, and succeeded in describing electrical conductivity including alternating current (AC) conductivity, thermal conductivity, the Hall effect, and so on. In particular, he provided a beautiful explanation for the Wiedemann–Frantz law, which states that the ratio of thermal conductivity to electrical conductivity divided by T is equal to a constant which depends on neither temperature nor the material.

However, there was one problem that was difficult to explain in terms of the Drude theory. This problem concerned specific heat. If there are N electrons moving about freely, there should be a contribution of $(3/2)Nk_B$ according to the equipartition law. However, the specific heat of metals is around 6.2 kcal/deg, which is similar to that of the insulators. This indicates, apparently, that there is no contribution due to the electrons. This problem was solved, after the discovery of quantum mechanics, by applying quantum statistics to electrons. This important theory, which was put forward by Sommerfeld, was to become the basis for the subsequent development of the electronic theory of metals.

As a model for electrons in metals, let us consider a cube of edge length L. The potential for electrons is constant inside it. Outside the cube, the potential energy is raised by V_0 (see Fig. 3.5). We take the zero of the energy to be equal to the potential energy inside the metal. This yields the following Schrödinger equation:

$$-\frac{\hbar^2}{2m}\Delta\psi = E\psi \quad (0 \leqq x, y, z \leqq L). \tag{3.1}$$

Its solutions are given by

$$\psi_k = \Omega^{-\frac{1}{2}} e^{i\boldsymbol{k}\cdot\boldsymbol{r}} \quad (\Omega = L^3). \tag{3.2}$$

Here, we have the following relationship between $\boldsymbol{k}$ and E:

$$E = E_k \equiv \frac{\hbar^2 k^2}{2m}. \tag{3.3}$$

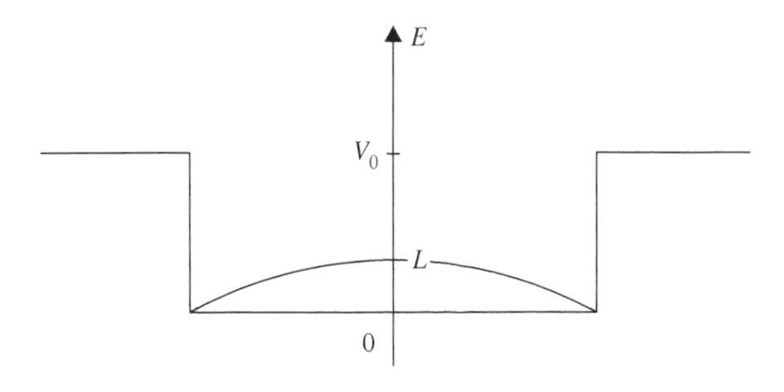

Figure 3.5 A model of the potential energy of electrons in metals. The potential energy is lower by V_0 inside the cube.

The values of $\boldsymbol{k}$, and hence the eigenvalues E, are determined upon fixing the boundary conditions. Our boundary condition is that ψ tends to zero as the electron is moved away from the cube. However, the macroscopic properties of the material, other than the surface properties, are presumably independent of the boundary condition. We thus employ a simpler periodic boundary condition. This is defined by $\psi(x+L, y, z) = \psi(x, y, z)$ and similarly for y and z. This yields the following values of $\boldsymbol{k}$:

$$k_x = \frac{2\pi}{L} n_x, \quad k_y = \frac{2\pi}{L} n_y, \quad k_z = \frac{2\pi}{L} n_z. \tag{3.4}$$

Here, n_x, n_y and n_z are integers, and they can be considered to be the quantum numbers that specify the solution. $k = |\boldsymbol{k}|$ has the meaning of the wavenumber of the electron wave. $\boldsymbol{k}$ is called the wavenumber vector. From eq. (3.3), the energy is proportional to ${n_x}^2 + {n_y}^2 + {n_z}^2$.

The ground state is obtained by filling these energy levels from the lowest level upward with electrons. Figure 3.6 illustrates the sets of (k_x, k_y) that are allowed by eq. (3.4) in the case of two dimensions. There is an obvious extension to three dimensions. In this $\boldsymbol{k}$-space, the equienergy surface is a sphere centered at the origin. Now consider a sphere in the $\boldsymbol{k}$-space that contains points whose number is equal to a half of the total number of electrons. The ground state is

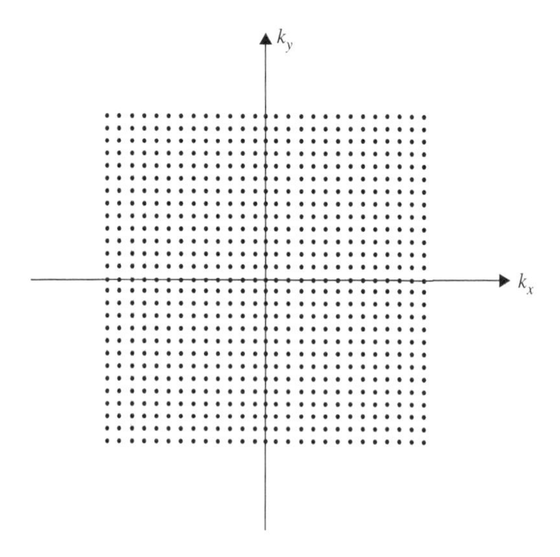

Figure 3.6 An illustration of the points specified by eq. (3.4) (in the case of two dimensions). The distance between neighboring points is equal to $2\pi/L$. The value of $E_{\boldsymbol{k}}$ in eq. (3.3) is constant over the surface of a sphere in the $\boldsymbol{k}$-space.

the state in which every point inside this sphere is filled by two electrons. This sphere is called the Fermi sphere, and its surface is called the Fermi surface. The energy of a state that lies on the Fermi surface is the Fermi energy, and its wavenumber is the Fermi wavenumber. The value of the Fermi energy must be smaller than V_0 of Fig. 3.5 as a matter of course. Let us calculate the values of these quantities as functions of the electron number.

For later convenience, we first define the density of states $\rho(E)$. This is defined by equating $\rho(E)dE$ with the number of points allowed by eq. (3.4) in between two spheres that correspond to E and $E + dE$. Since the volume per point is equal to $(2\pi/L)^3$, we obtain, by elementary algebra,

$$\rho(E) = \frac{L^3}{\sqrt{2}\pi^2} \frac{m^{3/2}}{\hbar^3} \sqrt{E}. \tag{3.5}$$

The Fermi energy, written as ε_F, is therefore given by

$$\int_0^{\varepsilon_F} \rho(E)dE = \frac{N}{2}, \tag{3.6}$$

so that we obtain

$$\varepsilon_F = \frac{\hbar^2}{2m} \left(3\pi^2 \frac{N}{L^3} \right)^{2/3}. \tag{3.7}$$

From this, it is easy to find the Fermi wavenumber k_F, which equals $(3\pi^2 N/L^3)^{1/3}$. These quantities are functions of the electron density N/L^3. In terms of ε_F, $\rho(E)$ may be expressed as

$$\rho(E) = \frac{3}{4} \frac{N}{\varepsilon_F} \sqrt{\frac{E}{\varepsilon_F}}. \tag{3.8}$$

The sum of the electron energies is then given by

$$E_{\text{tot}} = 2 \int_0^{\varepsilon_F} E\rho(E)dE = \frac{3}{5} N\varepsilon_F. \tag{3.8a}$$

What is very different from the classical theory is the fact that the system has finite, and large, energy even at absolute zero. The velocity v_F, which corresponds to the Fermi wavenumber k_F, is given by $\hbar k_F/m$. From the known values of the electron density N/L^3 in metals, we obtain rather large values for v_F: $v_F = 1.1 \times 10^8$ cm/s in Na and $v_F = 1.6 \times 10^8$ cm/s in Cu. Corresponding to these, $T_F \equiv \varepsilon_F/k_B$, which is the Fermi energy ε_F expressed in units of temperature, takes extremely large values: 38 000 K in Na and 82 000 K in Cu. In other words, the electrons in metals have a very large kinetic energy at absolute zero, of the order of the energy that is attained only at such high temperatures in the classical theory. This arises from the Pauli principle, according to which at most two electrons (of opposite spins) can occupy a state.

Next, let us think about the situation at finite temperatures. In this case, excited states appear with a probability that is proportional to the Boltzmann factor. In the excited state, electrons inside the Fermi sphere are excited to points outside the sphere. Excitations are most likely to occur with excitation energy of up to about $k_B T$. As mentioned above, since the Fermi energy is of the order of magnitude of ten thousands of degrees on the temperature scale, excitations are limited to electrons in close vicinity of the Fermi surface, and the majority of the electrons remain unaffected. This immediately resolves the difficulty associated with the specific heat. The equipartition law in classical theory says that, with increase ΔT of temperature, the energy of all electrons goes up by $\frac{3}{2}k_B \Delta T$. However, according to quantum theory, most electrons are not excited because of the Pauli principle, and they consequently give no contribution to the specific heat.

Let us formalize the above discussion by applying quantum statistics to our system. Here, we shall do so by using the second-quantization method. Following the results of eqs. (3.2)–(3.4), we can write our Hamiltonian in the second-quantization formalism as follows:

$$H = \sum_{k_x k_y k_z} E_{\boldsymbol{k}}(a^{\dagger}_{\boldsymbol{k}\uparrow} a_{\boldsymbol{k}\uparrow} + a^{\dagger}_{\boldsymbol{k}\downarrow} a_{\boldsymbol{k}\downarrow}), \tag{3.9}$$

where the summation over k_x, k_y and k_z refers to the summation over the integers n_x, n_y and n_z in eq. (3.4). Here, to take account of the Boltzmann factor, it is convenient to employ the grand canonical ensemble. To do so, we denote the chemical potential as μ and the operator for the total number of electrons as N, given by

$$N = \sum_{\boldsymbol{k}} (a^{\dagger}_{\boldsymbol{k}\uparrow} a_{\boldsymbol{k}\uparrow} + a^{\dagger}_{\boldsymbol{k}\downarrow} a_{\boldsymbol{k}\downarrow}), \tag{3.10}$$

and define

$$\left.\begin{aligned} \mathcal{H} &= H - \mu N = \textstyle\sum_{\boldsymbol{k}} \varepsilon_{\boldsymbol{k}} (a_{\boldsymbol{k}\uparrow}{}^{\dagger} a_{\boldsymbol{k}\uparrow} + a_{\boldsymbol{k}\downarrow}{}^{\dagger} a_{\boldsymbol{k}\downarrow}) \\ \varepsilon_{\boldsymbol{k}} &= E_{\boldsymbol{k}} - \mu \end{aligned}\right\}. \tag{3.11}$$

$e^{-\beta\mathcal{H}}$ is then the probability factor, where $\beta = 1/k_B T$. We need to obtain the average number of electrons that occupy $\psi_{\boldsymbol{k}}$. Since the operator that expresses the electron number is given by $a_{\boldsymbol{k}\sigma}{}^{\dagger} a_{\boldsymbol{k}\sigma}$ where $\sigma = \uparrow$ or $\downarrow$, the average number is given by

$$f_{k\sigma} = \frac{\mathrm{Tr}(a_{k\sigma}{}^{\dagger} a_{k\sigma} e^{-\beta\mathcal{H}})}{\mathrm{Tr}(e^{-\beta\mathcal{H}})}, \tag{3.12}$$

where Tr denotes the sum over the diagonal matrix elements. In the second-quantization formalism, the electron occupation number of a state is restricted to only 0 or 1 for any set of $\boldsymbol{k}\sigma$. The total combinations of 0 and 1 for all $\boldsymbol{k}\sigma$ span the states in second quantization. The states include the state where all occupation numbers are zero, and also the state where all occupation numbers are 1. Hence, when we employ the grand canonical ensemble, we have an advantage that it is not necessary to keep the electron number fixed. Since the commutability of each term in eq. (3.11) with all of the other terms is easily seen from eq. (2.38), we can perform the following decomposition into factors:

$$e^{-\beta\mathcal{H}} = \prod_{k\sigma} \exp(-\beta\varepsilon_k \, a_{k\sigma}{}^{\dagger} a_{k\sigma}).$$

As a result, in eq. (3.12), all factors with labels other than $\boldsymbol{k}\sigma$ cancel between the denominator and the numerator, and we obtain

$$\begin{aligned} f_{k\sigma} &= \frac{\sum_{n=0,1} n \exp(-\beta\varepsilon_k n)}{\sum_{n=0,1} \exp(-\beta\varepsilon_k n)} \\ &= \frac{1}{\exp(\beta\varepsilon_k) + 1}. \end{aligned} \tag{3.13}$$

This function is depicted in Fig. 3.7. In the limit $T \to 0$, it reduces to the step function, whose value changes at $E_k = \mu$. This corresponds to the situation where the states up to the Fermi surface are filled with electrons at $T = 0$. Thus $\mu = \varepsilon_F$ at $T = 0$. Moreover, we see from the figure that, at finite temperatures, excitations occur in the neighborhood of μ within a range of about $k_B T$.

At finite temperatures, μ is determined by the condition that the total number of electrons is equal to N, and is a function of temperature in general:

$$N = \sum_{k\sigma} f_{k\sigma}. \tag{3.14}$$

Correspondingly, the total energy is given by

$$E_{\mathrm{tot}} = \sum_{k\sigma} E_k f_{k\sigma}. \tag{3.15}$$

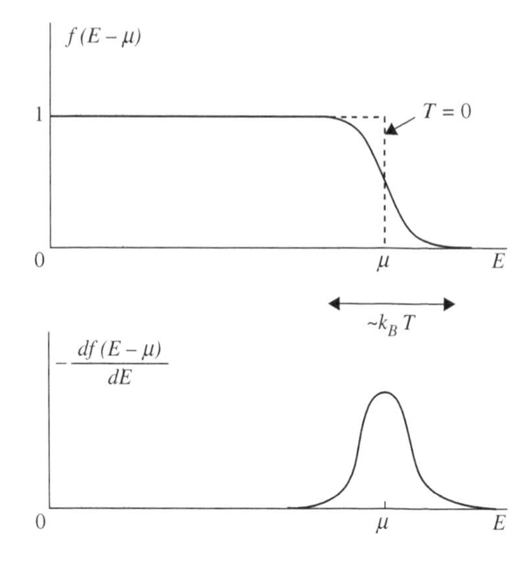

Figure 3.7 The Fermi distribution function $f(E-\mu)$. At $T=0$, it reduces to the step function whose value changes at μ, while, at finite T, it gradually moves over from 1 to 0 in the neighborhood of μ in an energy interval of order k_BT. The function $-df(E-\mu)/dE$ peaks at μ, and the area under its curve is equal to unity.

Rewriting these using the density of states $\rho(E)$ yields

$$N = 2\int_0^\infty \rho(E)f(E-\mu)dE, \tag{3.16}$$

$$E_{\text{tot}} = 2\int_0^\infty E\rho(E)f(E-\mu)dE, \tag{3.17}$$

where $f(x)$ is defined by

$$f(x) \equiv \frac{1}{e^{\beta x}+1} \tag{3.18}$$

and is called the Fermi distribution function.

For given T, μ is determined by eq. (3.16) and, by putting this into eq. (3.17), E_{tot} is obtained. Differentiating this with respect to T yields the specific heat. When $\mu \gg k_BT$, there is a general method which allows us to obtain eqs. (3.16) and (3.17) in terms of a power series expansion. For a relatively smooth function $g(x)$, let us calculate

$$I = \int_0^\infty g(E)f(E-\mu)dE. \tag{3.19}$$

Defining $G(E)$ by

$$G(E) = \int_0^E g(E)dE,$$

we obtain

$$I = G(E)f(E-\mu)\Big|_0^\infty + \int_0^\infty G(E)\left[-\frac{df(E-\mu)}{dE}\right]dE. \tag{3.20}$$

The first term vanishes. The function $-df/dE$ is, as shown in Fig. 3.7, centered at $E = \mu$ and has a width of about k_BT. The area underneath its curve is equal to 1. Therefore, if $G(E)$ is almost constant in the energy range of width $\sim k_BT$ around μ, $I \cong G(\mu)$. As a result, when we expand $G(E)$ as

$$G(E) = G(\mu) + (E-\mu)G'(\mu) + \tfrac{1}{2}(E-\mu)^2 G''(\mu) + \ldots,$$

and put this into eq. (3.20), the second term vanishes, and we obtain

$$I = G(\mu) + \frac{1}{2}g'(\mu)\int_0^\infty (E-\mu)^2\left[-\frac{df(E-\mu)}{dE}\right]dE + \ldots,$$

where we used $G''(\mu) = g'(\mu)$. The integral that appears on the right-hand side is evaluated as

$$\beta\int_{-\infty}^\infty \frac{e^{\beta\varepsilon}\varepsilon^2 d\varepsilon}{(e^{\beta\varepsilon}+1)^2} = \frac{\pi^2}{3\beta^2}.$$

Here, we changed the lower limit of integration to $-\infty$ using the condition $\mu \gg k_BT$. Thus we are led to the result

$$I = \int_0^\mu g(E)dE + \frac{\pi^2}{6}g'(\mu)(k_BT)^2. \tag{3.21}$$

Applying this to eq. (3.16) yields

$$\frac{N}{2} = \int_0^\mu \rho(E)dE + \frac{\pi^2}{6}\rho'(\mu)(k_BT)^2.$$

We take the difference between this and eq. (3.6). For sufficiently small $\varepsilon_F - \mu$, we obtain

$$\mu = \varepsilon_F - \frac{\pi^2}{6}\left[\frac{\rho'(\varepsilon_F)}{\rho(\varepsilon_F)}\right](k_BT)^2. \tag{3.22}$$

From this, we see that the second term is small, as it is suppressed by $(k_BT/\varepsilon_F)^2$ compared with the first term. So long as $g(E)$ is not singular, we find that the expansion in eq. (3.21) is also in powers of $(k_BT/\varepsilon_F)^2$. Next, from eq. (3.17), we have

$$\frac{E_{\text{tot}}}{2} = \int_0^{\varepsilon_F} E\rho(E)dE + \int_{\varepsilon_F}^{\mu} E\rho(E)dE + \frac{\pi^2}{6}\frac{d}{dE}(E\rho(E))_\mu (k_BT)^2.$$

Here, again, we make use of the fact that $\varepsilon_F - \mu$ is small to employ eq. (3.22) in the second term and use $E = \varepsilon_F$ in place of $E = \mu$ in the third term. We then obtain

$$\frac{E_{\text{tot}}}{2} = \int_0^{\varepsilon_F} E\rho(E)dE + \frac{\pi^2}{6}\rho(\varepsilon_F)(k_BT)^2. \tag{3.23}$$

The first term is the value at absolute zero. From this equation, the specific heat is given by

$$c = \frac{dE_{\text{tot}}}{dT} = \frac{2\pi^2}{3}{k_B}^2\rho(\varepsilon_F)T. \tag{3.24}$$

In the above calculation, we did not use the expression of $\rho(E)$ given by eq. (3.8). Note that eq. (3.8) was derived using the energy spectrum defined by eqs. (3.3) and (3.4). The above result is general and is for any choice of the energy spectrum as long as we use the corresponding density of states $\rho(E)$. If we use eq. (3.8), we obtain

$$c = \frac{\pi^2}{2}Nk_B\frac{k_BT}{\varepsilon_F}. \tag{3.25}$$

This has a clear implication that, rather than all electrons carrying a specific heat of $(3/2)k_B$, only those electrons in the vicinity of the Fermi surface, which amounts to about k_BT/ε_F of the whole, carry a specific heat of about k_B. Such behavior of the specific heat, being proportional to the temperature, has actually been observed in metals. The value of the specific heat, which is evaluated by substituting the value given above of ε_F into eq. (3.25), is usually smaller by a few tens of percentage than the experimentally measured value. The mechanism of this deviation is already known, but we omit its explanation here.

The result given by eq. (3.24) or by eq. (3.25) is valid for $k_BT \ll \varepsilon_F$. For later convenience, let us derive the general expression for the specific heat. Abbreviating $f(E-\mu)$ simply as f, we obtain

$$\frac{df}{dE} = -\frac{1}{k_BT}f(1-f),$$

$$\frac{df}{dT} = -\frac{df}{dE}\left(\frac{E-\mu}{T} + \frac{d\mu}{dT}\right).$$

By differentiating eq. (3.16) with respect to T and setting the left-hand side equal to zero, we obtain

$$\int_0^\infty \rho(E)\frac{df}{dE}\left(\frac{E-\mu}{T}+\frac{d\mu}{dT}\right)dE = 0. \tag{3.26}$$

Differentiating eq. (3.17) with respect to E yields

$$c = 2\int_0^\infty E\rho(E)\left(-\frac{df}{dE}\right)\left(\frac{E-\mu}{T}+\frac{d\mu}{dT}\right)dE.$$

We then make use of eq. (3.26) to obtain

$$\begin{aligned} c &= 2\int_0^\infty (E-\mu)\rho(E)\left(-\frac{df}{dE}\right)\left(\frac{E-\mu}{T}+\frac{d\mu}{dT}\right)dE \\ &= \frac{2}{T}\int_0^\infty (E-\mu)^2\rho(E)\left(-\frac{df}{dE}\right)dE \\ &\quad + 2\frac{d\mu}{dT}\int_0^\infty (E-\mu)\rho(E)\left(-\frac{df}{dE}\right)dE. \end{aligned} \tag{3.27}$$

If $\rho(E)$ can be regarded as a constant here in the neighborhood of $E = \mu$ to within a width of about $k_B T$, the second term of eq. (3.27) vanishes and the first term coincides with eq. (3.24).

The above discussion shows that the fact that the electron observes the Fermi–Dirac statistics leads to a large deviation in the specific heat with respect to the results expected on the basis of classical statistics. A large deviation is found to occur also in the susceptibility. Since the electron has a magnetic moment that accompanies the spin, according to classical statistics, its susceptibility should obey the Curie law. However, metals have only a weak and temperature-independent susceptibility. As an explanation for this, the Pauli theory, which is based on the Fermi–Dirac statistics, will be discussed next.

When a magnetic field H is applied along the direction of the z-axis, its interaction with the spin magnetic moment $-g\mu_B \boldsymbol{s}$ brings about a Zeeman energy that is equal to $g\mu_B s_z H$, where g is about 2. The energy of the spin $\uparrow$ electron therefore becomes $E_k + \mu_B H$ and that of the spin $\downarrow$ electron becomes $E_k - \mu_B H$.

As shown in Fig. 3.8, the energy spectrum for the spin $\uparrow$ electrons is raised as a whole by $\mu_B H$ and that for the spin $\downarrow$ electrons is lowered as a whole by $\mu_B H$. As the ground state is given by filling the states with electrons from the lowest state upwards, the number of spin $\downarrow$ electrons becomes larger. As is apparent from the figure, the number of electrons of each spin under a magnetic

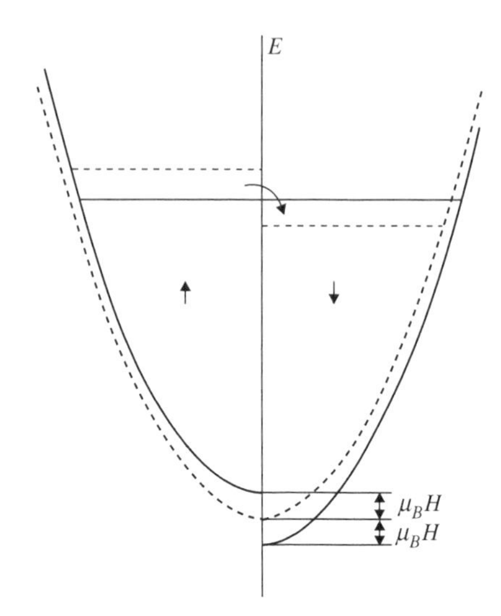

Figure 3.8 External field H raises the spin ↑ level by $\mu_B H$ and lowers the spin ↓ level by $\mu_B H$. The total energy is minimized when the number of spin ↑ electrons that turn to spin ↓ is given by $\mu_B H \rho(\varepsilon_F)$.

field H is obtained, starting from the same number of electrons of the two spin orientations, by moving $\mu_B H \rho(\varepsilon_F)$ electrons from spin ↑ to spin ↓. The difference between the numbers of electrons of the two spin orientations is equal to $2\mu_B H \rho(\varepsilon_F)$, and so a total magnetic moment of $2{\mu_B}^2 H \rho(\varepsilon_F)$ is generated. The susceptibility is therefore given by

$$\chi = 2{\mu_B}^2 \rho(\varepsilon_F). \tag{3.28}$$

Here, $\mu_B H$ is very small compared with ε_F, and so, in the above discussion, the terms that are higher order in $\mu_B H / \varepsilon_F$ are neglected.

The above is for the susceptibility at absolute zero, and similar considerations can be made for the case of finite temperatures, where the correction to eq. (3.28) is of order $(k_B T / \varepsilon_F)^2$ and is therefore small. Equation (3.28) is of order $k_B T / \varepsilon_F$ compared with the Curie susceptibility $N {\mu_B}^2 / k_B T$. The Curie susceptibility was initially obtained by calculating the probability of an electron of arbitrary spin direction to have its spin pointed in the direction of the applied magnetic field. In the case of the metallic electrons, only a fraction, of order $k_B T / \varepsilon_F$, are free to change the orientation of spin. Thus almost all electrons are dormant, so to speak, and this situation is described by the term 'degeneracy'. The degeneracy of metallic electrons is an essential factor in the discussions of Chapters 5 onward, not merely restricted to the above two matters.

For later convenience, let us derive the general expression for susceptibility. In general, let us say that when an external magnetic field H is applied to a system described by the Hamiltonian H_0, the term $-\mu_z H$ is added to the Hamiltonian. The operator μ_z expresses the z-component of the magnetic moment. The partition function of this system is given by

$$Z = \sum_n \langle n| \exp\{-\beta(H_0 - \mu_z H)\}|n\rangle \quad \left(\beta = \frac{1}{k_B T}\right). \tag{3.29}$$

Here, $|n\rangle$ denotes an arbitrary complete system. In the following, we restrict ourselves to the case where H_0 and μ_z commute with each other. The system of free electrons in metals belongs to this category. We may then choose as $|n\rangle$ the eigenstates of H_0 and μ_z, and obtain

$$Z = \sum_n \exp(-\beta E_n + \beta H \mu_{zn}),$$

where E_n and μ_{zn} are the eigenvalues of H_0 and μ_z, respectively. By expanding this up to the second power of H, we obtain

$$Z = \sum_n \exp(-\beta E_n)\left(1 + \beta H \langle\mu_z\rangle + \frac{1}{2}\beta^2 H^2 \langle{\mu_z}^2\rangle\right).$$

Here,

$$\langle A\rangle \equiv \frac{\sum_n A_n \exp(-\beta E_n)}{\sum_n \exp(-\beta E_n)}$$

denotes the average of the quantity A, which is diagonalized with respect to $|n\rangle$, in the absence of the magnetic field. $\langle\mu_z\rangle$ vanishes since the magnetic moment does not have a preferred direction in the absence of the external magnetic field. By expanding the free energy $F = -k_B T \log Z$ in powers of H up to the second order, we obtain

$$F = F_0 - \frac{1}{2}\beta\langle{\mu_z}^2\rangle H^2.$$

From this, the susceptibility is expressed as

$$\chi = -\left.\frac{\partial F^2}{\partial H^2}\right|_{H=0} = \frac{\langle{\mu_z}^2\rangle}{k_B T}. \tag{3.30}$$

Now, since in the system of electrons in metals, the following holds:

$$\mu_z = \mu_B \sum_k ({a_{k\downarrow}}^\dagger a_{k\downarrow} - {a_{k\uparrow}}^\dagger a_{k\uparrow}),$$

we obtain

$$\begin{aligned}\langle {\mu_z}^2 \rangle &= {\mu_B}^2 \sum_k \langle (a_{k\downarrow}{}^\dagger a_{k\downarrow} - a_{k\uparrow}{}^\dagger a_{k\uparrow})^2 \rangle \\ &= 2{\mu_B}^2 \sum_k f_k(1-f_k) \\ &= 2{\mu_B}^2 \int \rho(E) f(1-f) dE,\end{aligned}$$

and so

$$\chi = 2{\mu_B}^2 \int \rho(E) \left(-\frac{df}{dE}\right) dE. \tag{3.31}$$

When ρ can be regarded as a constant as in the case of the specific heat, eq. (3.31) coincides with eq. (3.28).

References and further reading

The content of this and the next chapters are found in almost all textbooks of solid-state physics, and there is perhaps not much difference among them. The oldest textbook, which is still valuable to the present time, is the following by Sommerfeld and Bethe:

A. Sommerfeld and H. Bethe (1933) Elektronentheorie der Metalle ("Electronic theory of metals") in H. Geiger and K. Scheel (eds.), *Handbuch der Physik*, Vol. 24, Part 2, pp. 333–622 (Berlin: Springer), later published as a separate book, *Elektronentheorie der Metalle* (Berlin: Springer, 1967).

Note added by the translators:

This book has apparently never been translated into English. The original text refers to the Japanese translation by T. Inoue, *Kinzoku-denshi-ron* ("*Electronic Theory of Metals*", Tokai University Press, 2002). While the translators hope that an English language version will become available some day, there is presumably little need for additional references here. For a standard textbook with good coverage of the contents of this chapter, see

N. W. Ashcroft and N. D. Mermin (1976) *Solid State Physics* (New York: Holt, Rinehart and Winston).

4
Band theory

The subject of this chapter is solving the Schrödinger equation for electrons moving in a periodic mean field in crystals. The solution is called the Bloch orbital. We discuss Bloch's theorem, which applies to it. As for the energy eigenvalues, there are allowed and forbidden values. The allowed values are distributed in extended regions that are called bands. The forbidden values comprise the gaps. In order to calculate the Bloch orbital, we make use of an approximation based on the free electron model, and another approximation based on the linear combination of atomic orbitals. As a more realistic approach, we discuss the Wigner–Seitz method.

4.1 The periodic structure of crystals

In the previous chapter, we showed that many experimental facts can be explained by assuming that the conduction electrons in metals move in a uniform mean field. However, in real metals, there are collisions with ions. A question arises as to why these collisions do not cause a significant effect. Moreover, there are some problems that cannot be explained in the free electron model. One such problem is the quantitative deviations from the experimental results. Also, the signs associated with the Hall effect and the thermoelectric power are often opposite to that expected in the free electron model. Another problem is why this model does not apply at all to the case of insulators.

In order to answer these questions, we have to move on to more realistic models than the free electron model. As we do so, an important point to bear in mind is that the atoms are placed with a regular order in crystals, so that the electrons are subject to a mean field that has the same type of regularity. As a preparation to solving the Schrödinger equation for such an electron, we discuss the geometry governing the periodicity of the arrangement of atoms.

Let us adopt $\boldsymbol{a}_1, \boldsymbol{a}_2, \boldsymbol{a}_3$ as the three fundamental vectors. Atoms of the same kind are placed with a regular order at positions specified by the vectors

$$\boldsymbol{R} = n_1\boldsymbol{a}_1 + n_2\boldsymbol{a}_2 + n_3\boldsymbol{a}_3 \tag{4.1}$$

for every combination of the three integers n_1, n_2, n_3. This is called the lattice structure of atoms. In the following, we cite three examples.

In the case of the simple cubic lattice, the fundamental vectors have the same length and are perpendicular to each other (see Fig. 4.1).

The body-centered cubic lattice has an atom at each vertex and at the center of cubes, and we may take the following as its fundamental vectors:

$$\boldsymbol{a}_1 = \left(\frac{a}{2}, \frac{a}{2}, \frac{a}{2}\right), \quad \boldsymbol{a}_2 = \left(-\frac{a}{2}, \frac{a}{2}, \frac{a}{2}\right), \quad \boldsymbol{a}_3 = \left(-\frac{a}{2}, -\frac{a}{2}, \frac{a}{2}\right) \tag{4.2}$$

(see Fig. 4.1).

The face-centered cubic lattice has an atom at each vertex and at each center of the face of the cube, and we may take the following as its fundamental vectors:

$$\boldsymbol{a}_1 = \left(\frac{a}{2}, \frac{a}{2}, 0\right), \boldsymbol{a}_2 = \left(\frac{a}{2}, 0, \frac{a}{2}\right), \boldsymbol{a}_3 = \left(0, \frac{a}{2}, \frac{a}{2}\right). \tag{4.3}$$

The choice of the fundamental vectors is not unique. If there is a certain three-dimensional form and if it is possible to fill the whole space perfectly by its translations according to the vectors of eq. (4.1), this form is called the unit cell. The choice of the unit cell is, again, not unique. One of the choices is to choose the parallelepiped whose edges correspond to the fundamental vectors. However, it is more convenient to adopt the following. We first take a certain central atom, and draw line segments between this atom and the neighboring atoms (nearest neighbors and, in the case of bcc, next-nearest neighbors being

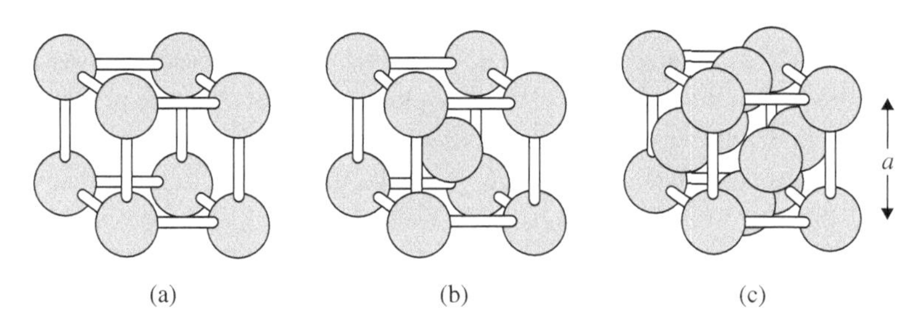

Figure 4.1 (a) Simple cubic lattice, (b) body-centered cubic lattice, (c) face-centered cubic lattice.

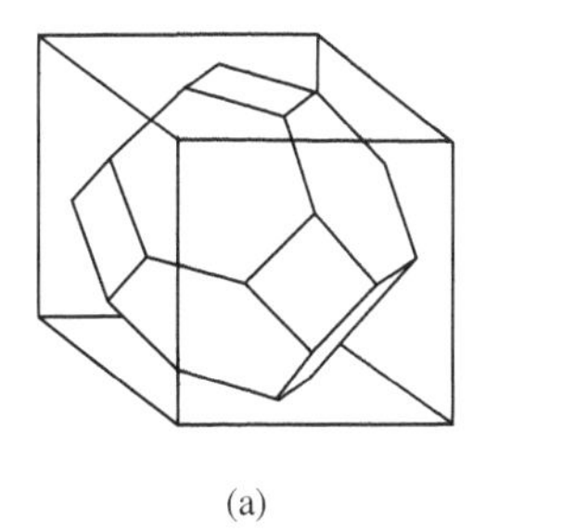

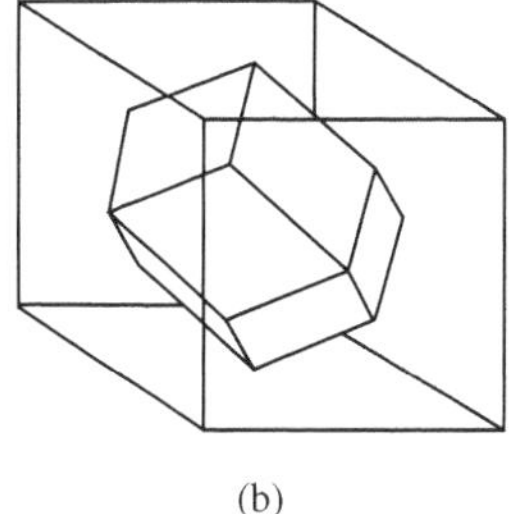

(a) (b)

Figure 4.2 Atomic polyhedra: (a) body-centered cubic lattice; (b) face-centered cubic lattice.

sufficient). We then place planes as perpendicular bisections of these line segments. The region that includes the central atom and enclosed by these planes is now the form to be taken as the unit cell.

When we take any point inside this cell, it is closer to the central atom than to any other atom. If we place this cell around every atom, the cells fill up the whole space, with no space left in between. This is called the atomic polyhedron or the Wigner–Seitz cell. Figure 4.2 shows the atomic polyhedra of the body-centered and face-centered cubic lattices.

Let us define the reciprocal lattice. In the case of a lattice that has the fundamental vectors $\boldsymbol{a}_1, \boldsymbol{a}_2, \boldsymbol{a}_3$, we define, for the reciprocal lattice, the fundamental vectors $\boldsymbol{b}_1, \boldsymbol{b}_2, \boldsymbol{b}_3$ by

$$\boldsymbol{b}_1 = \frac{2\pi}{\omega_0}\boldsymbol{a}_2 \times \boldsymbol{a}_3, \quad \boldsymbol{b}_2 = \frac{2\pi}{\omega_0}\boldsymbol{a}_3 \times \boldsymbol{a}_1, \quad \boldsymbol{b}_3 = \frac{2\pi}{\omega_0}\boldsymbol{a}_1 \times \boldsymbol{a}_2, \tag{4.4}$$

where

$$\omega_0 = |(\boldsymbol{a}_1 \times \boldsymbol{a}_2) \cdot \boldsymbol{a}_3|. \tag{4.5}$$

The set of points that is defined, for all combinations of the three integers m_1, m_2, m_3, by

$$\boldsymbol{K} = m_1\boldsymbol{b}_1 + m_2\boldsymbol{b}_2 + m_3\boldsymbol{b}_3 \tag{4.6}$$

is called the reciprocal lattice.

In the case of the body-centered cubic lattice, from eq. (4.2), we obtain $\boldsymbol{b}_1 = (2\pi/a, 0, 2\pi/a)$, $\boldsymbol{b}_2 = (-2\pi/a, 2\pi/a, 0)$ and $\boldsymbol{b}_3 = (0, -2\pi/a, 2\pi/a)$. Here, if we adopt $\boldsymbol{b}_1$, $\boldsymbol{b}_1 - \boldsymbol{b}_3$, $\boldsymbol{b}_1 + \boldsymbol{b}_2$ as the fundamental vectors, these are identical to the fundamental vectors, eq. (4.3), of the face-centered cubic lattice (a in eq. (4.3) is replaced by $4\pi/a$).

In the case of the face-centered lattice, from eq. (4.3), we obtain $\boldsymbol{b}_1 = (-2\pi/a, -2\pi/a, 2\pi/a)$, $\boldsymbol{b}_2 = (-2\pi/a, 2\pi/a, -2\pi/a)$ and $\boldsymbol{b}_3 = (2\pi/a,$

$-2\pi/a, -2\pi/a)$. Here, if we adopt $\boldsymbol{b}_1$, $-\boldsymbol{b}_3$ and $-\boldsymbol{b}_1-\boldsymbol{b}_2-\boldsymbol{b}_3$ as the fundamental vectors, they form the fundamental vectors, eq. (4.2), of the body-centered cubic lattice (a in eq. (4.2) is replaced by $4\pi/a$). Thus the body-centered cubic and face-centered cubic lattices are the reciprocal lattices of each other.

The following relationship follows from eq. (4.4):

$$\boldsymbol{a}_i \cdot \boldsymbol{b}_j = 2\pi\delta_{ij}. \tag{4.7}$$

Using this, we immediately obtain

$$\boldsymbol{K} \cdot \boldsymbol{R} = 2\pi(n_1m_1 + n_2m_2 + n_3m_3) \tag{4.8}$$

so that $e^{i\boldsymbol{K}\cdot\boldsymbol{R}} = 1$. We shall use this frequently in later discussions.

When a three-dimensional function $f(\boldsymbol{r})$ satisfies

$$f(\boldsymbol{r}) = f(\boldsymbol{r} + \boldsymbol{R}) \tag{4.9}$$

for all $\boldsymbol{R}$, we say that $f(\boldsymbol{r})$ has the periodicity of the lattice that is presently under consideration. In this case, the following expansion is possible (see Appendix C):

$$f(\boldsymbol{r}) = \sum_{\boldsymbol{K}} f_{\boldsymbol{K}} e^{i\boldsymbol{K}\cdot\boldsymbol{r}}, \tag{4.10}$$

where the summation over $\boldsymbol{K}$ refers to the summation over all of (m_1, m_2, m_3). The function $f_{\boldsymbol{K}}$ is given by

$$f_{\boldsymbol{K}} = \omega_0{}^{-1} \int_{\Omega_0} e^{-i\boldsymbol{K}\cdot\boldsymbol{r}} f(\boldsymbol{r}) dv. \tag{4.11}$$

Ω_0 denotes that the region of integration is a unit cell. Since the integrand is invariant under $\boldsymbol{r} \to \boldsymbol{r} + \boldsymbol{R}$, we may take any unit cell; ω_0 is the volume of the unit cell.

4.2 Bloch's theorem

Now, turning our attention to the mean field $V(\boldsymbol{r})$ that an electron in metals is subject to, we expect that the field satisfies the periodicity of eq. (4.9). We consider a crystal that consists of $N_1 \times N_2 \times N_3$ atoms located at positions specified by eq. (4.1) with $0 \leqq n_1 \leqq N_1 - 1, 0 \leqq n_2 \leqq N_2 - 1, 0 \leqq n_3 \leqq N_3 - 1$. We then consider solving

$$H\psi \equiv \left[-\frac{\hbar^2}{2m}\Delta + V(\boldsymbol{r}) \right] \psi = E\psi \tag{4.12}$$

with a boundary condition at the surface; we adopt the following periodic boundary condition:

$$\psi(\boldsymbol{r} + N_i \boldsymbol{a}_i) = \psi(\boldsymbol{r}) \quad (i = 1, 2, 3). \tag{4.13}$$

From the fact that $V(\boldsymbol{r})$ satisfies the periodicity of eq. (4.9), we can derive an important theorem which governs the solutions of eq. (4.12). Let us consider $\psi(\boldsymbol{r} + \boldsymbol{R})$ in relation to $\psi(\boldsymbol{r})$, where $\boldsymbol{R}$ is an element of eq. (4.1). When $\psi(\boldsymbol{r})$ is defined for all points inside the crystal, and $\boldsymbol{r}$ is such a point inside the crystal, there are some regions of $\boldsymbol{r}$ in which $\psi(\boldsymbol{r} + \boldsymbol{R})$ is undefined. On the other hand, one may extend $\psi(\boldsymbol{r})$ over to the whole space using eq. (4.13) and, in this sense, unambiguously define $\psi(\boldsymbol{r} + \boldsymbol{R})$ by moving $\psi(\boldsymbol{r})$ by $\boldsymbol{R}$. This operation of translation to form $\psi(\boldsymbol{r} + \boldsymbol{R})$ is denoted as $T_{\boldsymbol{R}}\psi$. Then, since H is invariant under translation by $\boldsymbol{R}$, we obtain

$$T_{\boldsymbol{R}} H \psi = H(\boldsymbol{r} + \boldsymbol{R})\psi(\boldsymbol{r} + \boldsymbol{R}) = H(\boldsymbol{r})\psi(\boldsymbol{r} + \boldsymbol{R}) = H T_{\boldsymbol{R}} \psi.$$

Moreover, since

$$T_{\boldsymbol{R}} T_{\boldsymbol{R}'} = T_{\boldsymbol{R}'} T_{\boldsymbol{R}} = T_{\boldsymbol{R}+\boldsymbol{R}'}, \tag{4.14}$$

all of $T_{\boldsymbol{R}}$ and H commute with one another. We can therefore regard ψ as an eigenfunction of the two operators. Hence the following relation holds when eq. (4.12) is satisfied:

$$T_{\boldsymbol{R}} \psi = c(\boldsymbol{R})\psi. \tag{4.15}$$

By eq. (4.14),

$$c(\boldsymbol{R})c(\boldsymbol{R}') = c(\boldsymbol{R} + \boldsymbol{R}') \tag{4.16}$$

needs to be satisfied.

We now come to the periodic boundary condition. From eq. (4.13), we find $c(N_i \boldsymbol{a}_i) = 1$ and, using eq. (4.16) repeatedly, we obtain $c(\boldsymbol{a}_i)^{N_i} = 1$. Hence,

$$c(\boldsymbol{a}_i) = e^{2\pi i l_i / N_i} \quad (i = 1, 2, 3), \tag{4.17}$$

where l_i are integers. Since cs for the fundamental vectors have now been determined, we obtain, for general $\boldsymbol{R}$,

$$c(\boldsymbol{R}) = c(\boldsymbol{a}_1)^{n_1} c(\boldsymbol{a}_2)^{n_2} c(\boldsymbol{a}_3)^{n_3} = \exp 2\pi i \left(\frac{n_1 l_1}{N_1} + \frac{n_2 l_2}{N_2} + \frac{n_3 l_3}{N_3} \right)$$

by eqs. (4.16) and (4.1). We then define

$$\boldsymbol{k} = \frac{l_1 \boldsymbol{b}_1}{N_1} + \frac{l_2 \boldsymbol{b}_2}{N_2} + \frac{l_3 \boldsymbol{b}_3}{N_3} \tag{4.18}$$

to obtain, from eq. (4.7),

$$c(\boldsymbol{R}) = e^{i\boldsymbol{k}\cdot\boldsymbol{R}}.$$

To summarize, when we solve the Schrödinger equation of eq. (4.12) with the periodic boundary condition of eq. (4.13), and if $V(\boldsymbol{r})$ satisfies the periodicity of eq. (4.9), we have

$$\psi(\boldsymbol{r}+\boldsymbol{R}) = e^{i\boldsymbol{k}\cdot\boldsymbol{R}}\psi(\boldsymbol{r}), \tag{4.19}$$

where $\boldsymbol{R}$ is an arbitrary vector given by eq. (4.1). This is called Bloch's theorem, and is an important theorem of solid-state physics. Wavefunctions that satisfy this property are called Bloch orbitals.

Here, $\boldsymbol{k}$ is given by eq. (4.18) in terms of the fundamental vectors of the reciprocal lattice. We can regard $\boldsymbol{k}$, or l_1, l_2, l_3, as the quantum numbers that specify the solution. The choice of $\boldsymbol{k}$ is not unique, since using $\boldsymbol{k}+\boldsymbol{K}$ in place of $\boldsymbol{k}$ (with $\boldsymbol{K}$ being an element of eq. (4.6)) yields, by eq. (4.8),

$$e^{i(\boldsymbol{k}+\boldsymbol{K})\cdot\boldsymbol{R}} = e^{i\boldsymbol{k}\cdot\boldsymbol{R}}. \tag{4.20}$$

As far as eq. (4.19) is concerned, we may therefore regard $\boldsymbol{k}$ and $\boldsymbol{k}+\boldsymbol{K}$ as being the same. Then how many $\boldsymbol{k}$s are there that are substantially different? We find that they are given by the points inside a unit cell in the reciprocal space, or the k-space. This is because it is impossible to link any two points inside this region by any choice of the vector $\boldsymbol{K}$ of eq. (4.6), and any points outside of this region may be brought inside by the addition of one of the vectors $\boldsymbol{K}$. This scheme, in which we consider only the points $\boldsymbol{k}$ that are inside the unit cell, is called the reduced zone scheme.

We ordinarily employ the atomic polyhedron in the k-space as the unit cell. This is called the first Brillouin zone. The solution to eq. (4.12) may thus be specified by the values of $\boldsymbol{k}$ taken to be inside the first Brillouin zone. However, in general, there are a large number of solutions for the same value of $\boldsymbol{k}$, and we usually discriminate between them by introducing a number n, which is sometimes called the band index. The solution is then denoted as $\psi_{n\boldsymbol{k}}(\boldsymbol{r})$.

The number of points in the unit cell of the reciprocal lattice that satisfy eq. (4.18) is given by $N_1N_2N_3$, and so the number of points in the first Brillouin zone is also equal to $N_1N_2N_3$. The volume of the unit cell in the reciprocal lattice is equal to $\boldsymbol{b}_1\cdot(\boldsymbol{b}_2\times\boldsymbol{b}_3)$, which is, from eq. (4.4), equal to $(2\pi)^3/\omega_0$. The density of the points $\boldsymbol{k}$ in the reciprocal space is therefore given by $\Omega/(2\pi)^3$, where Ω is the volume of the crystal. This result will be used frequently.

The following quantity also appears frequently:

$$\sum_{\boldsymbol{k}} e^{i\boldsymbol{k}\cdot\boldsymbol{R}}.$$

Here, $\boldsymbol{R}$ lies in the crystal, and the summation over $\boldsymbol{k}$ is carried out in the first Brillouin zone. On the other hand, from eq. (4.20), the choice of the unit cell does not affect our results. If we adopt a parallelepiped defined by $\boldsymbol{b}_1, \boldsymbol{b}_2, \boldsymbol{b}_3$ as three edges, we obtain the product of

$$\sum_{l_1=0}^{N_1-1} e^{i2\pi l_1 n_1/N_1} \quad (0 \leqq n_1 \leqq N_1 - 1)$$

with two analogous sums for 2 and 3. This vanishes unless $n_1 = n_2 = n_3 = 0$, and therefore we obtain

$$\sum_{\boldsymbol{k}} e^{i\boldsymbol{k}\cdot\boldsymbol{R}} = \delta_{\boldsymbol{R}0} N_1 N_2 N_3. \tag{4.21}$$

Similarly, we obtain

$$\sum_{\boldsymbol{R}} e^{i\boldsymbol{k}\cdot\boldsymbol{R}} = \begin{cases} N_1 N_2 N_3 & \text{if } \boldsymbol{k} = 0 \text{ or one of the } \boldsymbol{K} \text{ vectors,} \\ 0 & \text{otherwise,} \end{cases} \tag{4.22}$$

where the summation over $\boldsymbol{R}$ is for the $N_1 N_2 N_3$ lattice points in the crystal.

4.3 An approach starting from the free electron picture

When $V(\boldsymbol{r})$ can be considered to be small in eq. (4.12), let us try to treat the plane waves of free electrons as an unperturbed system and to take the effect of $V(\boldsymbol{r})$ into account as a perturbation. The plane waves are given by eq. (3.2). The form of the crystal is not always cubic. We consider the parallelepiped with edges $N_1\boldsymbol{a}_1, N_2\boldsymbol{a}_2$ and $N_3\boldsymbol{a}_3$. The values of $\boldsymbol{k}$ are then given by eq. (4.18). There is no reason to restrict it to the first Brillouin zone, and we consider all values of $\boldsymbol{k}$. The matrix elements of $V(\boldsymbol{r})$ between plane waves are given by

$$\begin{aligned} \Omega^{-1} \int e^{-i\boldsymbol{k}'\cdot\boldsymbol{r}} V(\boldsymbol{r}) e^{i\boldsymbol{k}\cdot\boldsymbol{r}} dv &= \Omega^{-1} \sum_{\boldsymbol{R}} \int_{\Omega_{\boldsymbol{R}}} e^{i(\boldsymbol{k}-\boldsymbol{k}')\cdot\boldsymbol{r}} V(\boldsymbol{r}) dv \\ &= \Omega^{-1} \sum_{\boldsymbol{R}} e^{i(\boldsymbol{k}-\boldsymbol{k}')\cdot\boldsymbol{R}} \int_{\Omega_0} e^{i(\boldsymbol{k}-\boldsymbol{k}')\cdot\boldsymbol{r}} V(\boldsymbol{r}) dv. \end{aligned}$$

Here, $\Omega_{\boldsymbol{R}}$ denotes that the region of integration is given by the unit cell that is centered at $\boldsymbol{R}$. In the last transformation, we used the substitution $\boldsymbol{r} \rightarrow \boldsymbol{r} + \boldsymbol{R}$, and used the periodicity of $V(\boldsymbol{r})$. By eq. (4.22), the above equation reduces to

$$= \omega_0{}^{-1} \sum_{\boldsymbol{K}} \delta_{\boldsymbol{k}-\boldsymbol{k}',\boldsymbol{K}} \int_{\Omega_0} e^{i\boldsymbol{K}\cdot\boldsymbol{r}} V(\boldsymbol{r}) dv$$
$$= \sum_{\boldsymbol{K}} \delta_{\boldsymbol{k}-\boldsymbol{k}',\boldsymbol{K}} V_{-\boldsymbol{K}} = \sum_{\boldsymbol{K}} \delta_{\boldsymbol{k}',\boldsymbol{k}+\boldsymbol{K}} V_{\boldsymbol{K}}. \tag{4.23}$$

In other words, the periodic potential can only link the plane waves whose $\boldsymbol{k}$-vectors differ by $\boldsymbol{K}$. Here, we used eq. (4.11). $V_{\boldsymbol{K}}$ are the expansion coefficients of $V(\boldsymbol{r})$ when it is expanded in the form of eq. (4.10). From this result, by ordinary perturbation theory, the wavefunction and energy are obtained as

$$\left. \begin{array}{l} e^{i\boldsymbol{k}\cdot\boldsymbol{r}} + \displaystyle\sum_{\boldsymbol{K}\neq 0} \frac{V_{\boldsymbol{K}}}{E_{\boldsymbol{k}} - E_{\boldsymbol{k}+\boldsymbol{K}}} e^{i(\boldsymbol{k}+\boldsymbol{K})\cdot\boldsymbol{r}} \\ E_{\boldsymbol{k}} + V_0 + \displaystyle\sum_{\boldsymbol{K}\neq 0} \frac{|V_{\boldsymbol{K}}|^2}{E_{\boldsymbol{k}} - E_{\boldsymbol{k}+\boldsymbol{K}}} \end{array} \right\}. \tag{4.24}$$

When the $V_{\boldsymbol{K}}$ are small, the perturbation is small in general, but when the denominator is zero, the perturbative expansion breaks down regardless of the size of the $V_{\boldsymbol{K}}$. This occurs when $E_{\boldsymbol{k}} = E_{\boldsymbol{k}+\boldsymbol{K}}$. Since $\boldsymbol{k}^2 = (\boldsymbol{k}+\boldsymbol{K})^2$, this condition may be expressed as

$$\boldsymbol{k} \cdot \boldsymbol{K} = -\frac{1}{2} \boldsymbol{K} \cdot \boldsymbol{K}. \tag{4.25}$$

The values of $\boldsymbol{k}$ that satisfy this, as shown in Fig. 4.3, are points on the plane that perpendicularly bisects $-\boldsymbol{K}$. Here, when $\boldsymbol{K}$ is a reciprocal vector, so is $-\boldsymbol{K}$.

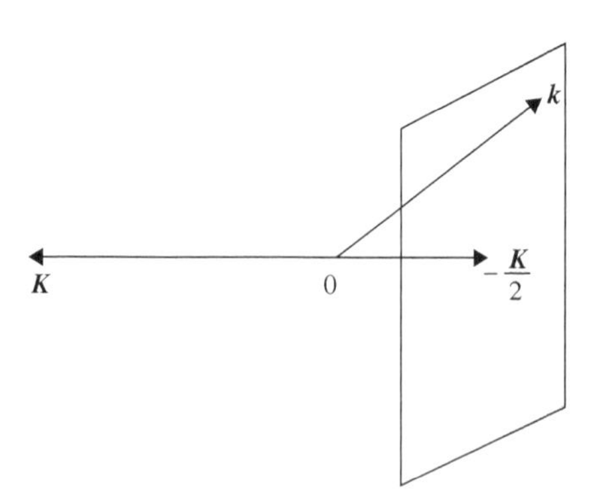

Figure 4.3 The values of $\boldsymbol{k}$ that satisfy eq. (4.25) lie on the plane that perpendicularly bisects $-\boldsymbol{K}$.

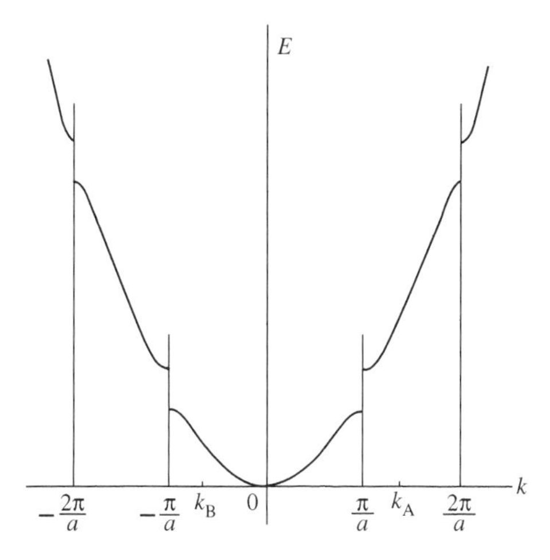

Figure 4.4 The influence of the periodic potential on the spectrum of the one-dimensional free electron. Gaps occur at values of k that are the multiples of π/a.

When we draw all such perpendicularly bisecting planes, eq. (4.24) diverges on the planes and in their vicinity.

Perturbation theory cannot be used in such regions. Let us discuss here a method, for the one-dimensional case, which does not use perturbation theory. In one dimension, the reciprocal vectors are $K = 2\pi m/a$ (m is an integer), and the points $\pi m/a$ play the role of the perpendicularly bisecting planes. As shown in Fig. 4.4, if we take k to be in the vicinity of π/a, the energies of ψ_k and $\psi_{k-2\pi/a}$ are almost equal, and so we need to take both of them into account in the same way. We thus write

$$\psi = c_1\psi_k + c_2\psi_{k-2\pi/a}, \tag{4.26}$$

and substitute this into eq. (4.12). We then multiply the result by $\psi_k{}^*$ and $\psi_{k-2\pi/a}{}^*$ from the left, where ψ_k denotes a plane wave, and integrate them. This leads to

$$c_1(E_k - E) + c_2 V_{2\pi/a} = 0, \tag{4.27}$$

$$c_1 V_{2\pi/a}{}^* + c_2(E_{k-2\pi/a} - E) = 0. \tag{4.28}$$

We adopted $V_0 = 0$. The solution, as is well known, is then given by

$$E = \frac{1}{2}\left\{E_k + E_{k-2\pi/a} \pm \sqrt{(E_k - E_{k-2\pi/a})^2 + 4|V_{2\pi/a}|^2}\right\}. \tag{4.29}$$

The behavior of this solution is shown in Fig. 4.4, and it exhibits gaps at $k = \pm \pi/a$. The influence from the waves other than ψ_k and $\psi_{k-2\pi/a}$ is small as long as V_K are small, so that it can be taken account of perturbationally. This leads to only small corrections in Fig. 4.4. We have also shown the behavior in the vicinity of the other values of $K/2$ in Fig. 4.4.

The first Brillouin zone in one dimension is the region $-\pi/a \leqq k \leqq \pi/a$. For example, k_{A} in Fig. 4.4 is translated to k_{B} when shifted by $-2\pi/a$. Since $e^{ik_{\mathrm{A}}R} = e^{ik_{\mathrm{B}}R}$ $(R = na)$, they are equivalent as a vector that appears in eq. (4.19). Hence, if we move all points outside the first Brillouin zone inside by using the reduced zone scheme, we obtain Fig. 4.5.

We see that there is more than one energy eigenvalue for each value of k inside the first Brillouin zone. As we vary k continuously, each eigenvalue also varies continuously. As we do so, there are values of energy that can be adopted as eigen-energies and others that cannot be. The energy range that belongs to the latter category is called the energy gap. The energy range that belongs to the former category is called the energy band. In the case of three dimensions also, there appear, in general, gaps in the energy range, that is, energy gaps, in the

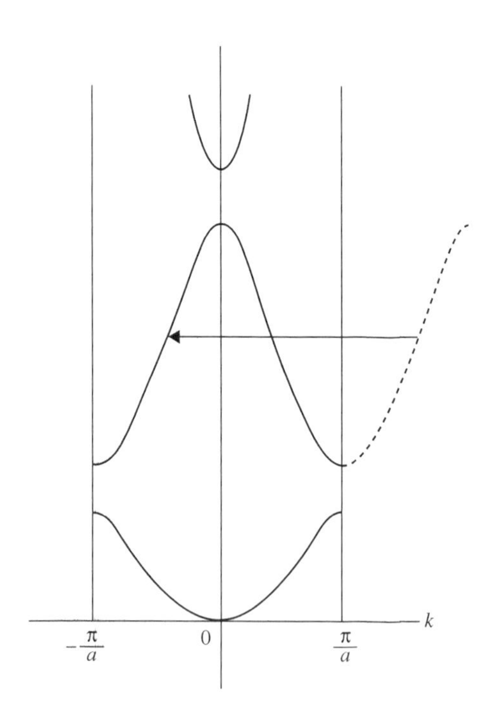

Figure 4.5 The energy spectrum in one dimension in the reduced zone scheme.

Figure 4.6 Equienergy curves in the two-dimensional $\boldsymbol{k}$-space.

vicinity of the planes that satisfy eq. (4.25). In Fig. 4.6, we present an example of equienergy curves in the case of two dimensions. Note the behavior at the boundaries of the Brillouin zones where the gaps arise.

4.4 The Bloch orbital as a linear combination of atomic orbitals

Let us now discuss the case where there is a strong periodic potential. This situation is applicable particularly to the electrons in the deep levels, 3d electrons in the transition metals, and so on. Since, in the neighborhood of an atom, we may suppose that the periodic potential is similar to the potential of the atom itself, the wavefunction is expected to resemble the atomic wavefunction. We may hence approximate the Bloch function by a linear combination of atomic orbitals. This corresponds to the LCAO of §2.1.

Let us formulate this problem using the second quantization method. Consider an s orbital. When an electron is in this orbital, it has energy ε_{s}. Writing the creation and annihilation operators as $a_{\boldsymbol{R}\sigma}{}^{\dagger}$, $a_{\boldsymbol{R}\sigma}$ ($\sigma = \uparrow$ or $\downarrow$) for an s orbital located at $\boldsymbol{R}$ defined by eq. (4.1), the energy when the electron is in this orbital corresponds to the term

$$\varepsilon_{\mathrm{s}} \sum_{\boldsymbol{R}\sigma} a_{\boldsymbol{R}\sigma}{}^{\dagger} a_{\boldsymbol{R}\sigma} \tag{4.30}$$

in the Hamiltonian. In addition, as in the case of the molecule, there is a term that transfers an electron in between the neighboring s orbitals. This is written as

$$t\sum_{\boldsymbol{R}\rho\sigma} a_{\boldsymbol{R}+\rho\sigma}{}^{\dagger} a_{\boldsymbol{R}\sigma}. \tag{4.31}$$

Here, $\boldsymbol{\rho}$ is one of the vectors that link the origin to the nearest neighbor lattice points, of which there are 6 in the simple cubic lattice, 8 in the body-centered cubic lattice and 12 in the face-centered cubic lattice. The atomic orbitals are defined to be mutually orthogonal. The sum of eqs. (4.30) and (4.31) is adopted as our total Hamiltonian. This is diagonalized by

$$a_{\boldsymbol{k}\sigma} = N^{-1/2}\sum_{\boldsymbol{R}} e^{-i\boldsymbol{k}\cdot\boldsymbol{R}} a_{\boldsymbol{R}\sigma}, \tag{4.32}$$

where $\boldsymbol{k}$ is a $\boldsymbol{k}$-vector in the first Brillouin zone, and $N = N_1N_2N_3$.

To demonstrate this result, we perform the inverse transformation of eq. (4.32) using eq. (4.21). This yields

$$a_{\boldsymbol{R}\sigma} = N^{-1/2}\sum_{\boldsymbol{k}} e^{i\boldsymbol{k}\cdot\boldsymbol{R}} a_{\boldsymbol{k}\sigma}.$$

Substituting this into eqs. (4.30) and (4.31), and using eq. (4.22), we obtain as the total Hamiltonian

$$\sum_{\boldsymbol{k}\sigma}\left(\varepsilon_{\mathrm{s}} + t\sum_{\rho} e^{-i\boldsymbol{k}\cdot\boldsymbol{\rho}}\right) a_{\boldsymbol{k}\sigma}{}^{\dagger} a_{\boldsymbol{k}\sigma}. \tag{4.33}$$

This implies that the orbital that is represented by eq. (4.32) is an eigenstate with energy

$$\varepsilon_{\boldsymbol{k}} = \varepsilon_{\mathrm{s}} + t\sum_{\rho} e^{-i\boldsymbol{k}\cdot\boldsymbol{\rho}}. \tag{4.34}$$

Taking $\psi_{\boldsymbol{R}}$ as the s-orbital at the point $\boldsymbol{R}$, we find, from eq. (4.32), that

$$\psi_{\boldsymbol{k}} = N^{-1/2}\sum_{\boldsymbol{R}} e^{i\boldsymbol{k}\cdot\boldsymbol{R}} \psi_{\boldsymbol{R}} \tag{4.35}$$

is the orbital that corresponds to $a_{\boldsymbol{k}\sigma}$. This can be seen by comparing eq. (2.51) with eq. (2.53). The Bloch orbital has thus been expressed as a linear combination of the atomic orbitals. We can easily prove that eq. (4.35) satisfies the Bloch theorem, eq. (4.19), by using $\psi_{\boldsymbol{R}'}(\boldsymbol{r}+\boldsymbol{R}) = \psi_{\boldsymbol{R}'-\boldsymbol{R}}(\boldsymbol{r})$.

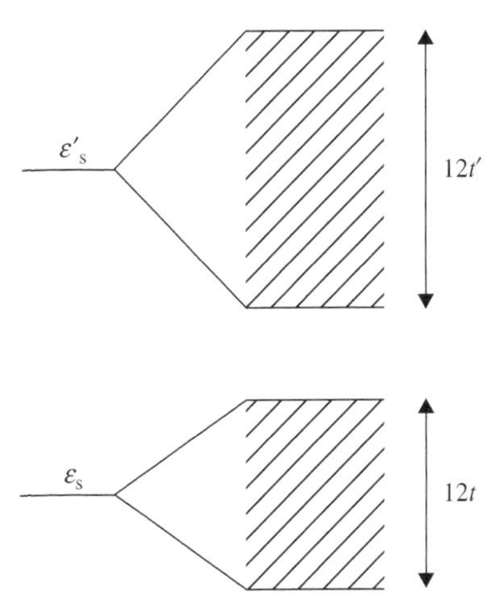

Figure 4.7 s-bands given by the LCAO method.

In the case of the simple cubic lattice, the energy spectrum of eq. (4.34) becomes

$$\varepsilon_{\boldsymbol{k}} = \varepsilon_s + 2t(\cos k_x a + \cos k_y a + \cos k_z a). \tag{4.36}$$

The fundamental vectors of the reciprocal lattice of this lattice have length $2\pi/a$ and the same directions as $\boldsymbol{a}_i$. From eq. (4.18), we therefore have $k_x = (2\pi/a)l_1/N_1$, $k_y = (2\pi/a)l_2/N_2$ and $k_z = (2\pi/a)l_3/N_3$, and the first Brillouin zone is given by $-\pi/a \leqq k_x,\ k_y,\ k_z \leqq \pi/a$. Hence, from eq. (4.36), we obtain an energy band which is centered at ε_s and whose width is given by $|12t|$. Corresponding to the other atomic orbitals, we obtain the other bands. So long as these bands do not overlap with each other, there are gaps in between them (see Fig. 4.7).

4.5 Metals and insulators

Taking metallic Na as an example, let us consider its electronic structure. In the band that corresponds to the 1s orbital of the atom, the Bloch orbitals, eq. (4.35), are filled for all values of $\boldsymbol{k}$ (in the first Brillouin zone). The wavefunction is thus written in the second quantization method as

$$\prod_{\boldsymbol{k}\sigma} a_{\boldsymbol{k}\sigma}{}^{\dagger}|0\rangle, \tag{4.37}$$

where we consider all N points for $\boldsymbol{k}$ in the first Brillouin zone. Now, if we express eq. (4.37) in terms of $a_{\boldsymbol{R}\sigma}{}^{\dagger}$ using eq. (4.32), all terms in which $a_{\boldsymbol{R}\sigma}{}^{\dagger}$

appears twice or more vanish, and only those terms in which all $\boldsymbol{R}$s are different survive. Since the number of the factors is N per spin, only the wavefunction in which all atomic orbitals are occupied survives. The above equation is thus proved to be equivalent to

$$\prod_{\boldsymbol{R}\sigma} a_{\boldsymbol{R}\sigma}{}^{\dagger}|0\rangle, \tag{4.38}$$

where $\boldsymbol{R}$ is for all atoms in the crystal. That is, eq. (4.37) represents a state in which the 1s orbital of each atom is filled with two electrons. This situation is applicable up to $(2p)^6$. Generally, when a band that corresponds to a closed-shell structure is filled by electrons, the wavefunction is identical to that in which the closed-shell atoms are placed in regular order, if we employ the approximation in which the Bloch function is expressed as a linear combination of atomic orbitals.

The situation differs in the case of the band that corresponds to the 3s orbital of the atom. It is not a good approximation to express this band in terms of linear combinations of atomic orbitals, but we use this approach for qualitative discussions. This band can accommodate $2N$ electrons taking account of spin, while there are only N electrons in metallic Na. As a result, the electrons fill the lower half of the band, as shown in Fig. 4.8. For this reason, the energy is lower in the metal than in the state consisting of isolated atoms, where all electrons carry the same energy ε_{3s}. This can be explained by the fact that the electrons can move around through the whole crystal (see §3.1.4).

If we describe this state in terms of $a_{\boldsymbol{R}\sigma}{}^{\dagger}$ using eq. (4.32), all combinations occur in which $N/2$ electrons are allocated over the N atomic orbitals per spin. For each atomic orbital, we have all of the following cases: where it is occupied

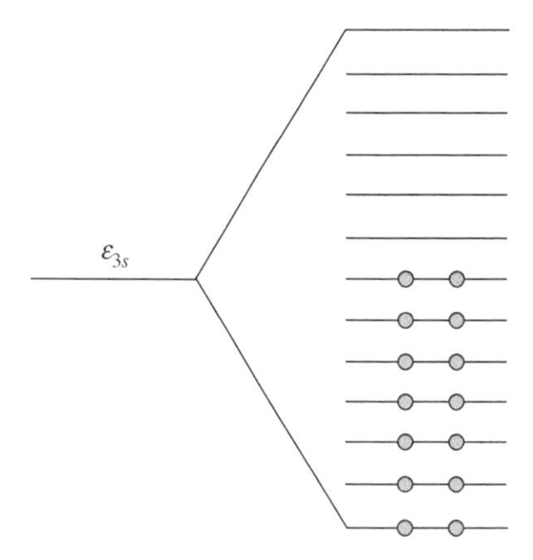

Figure 4.8 The conduction band of Na.

by two electrons; where it is occupied by a single electron with a spin $\uparrow$ or $\downarrow$; and where it is empty. In general, such states of various valences appear when the band is not completely filled. This is a characteristic feature of metals. Such a band is called the conduction band. On the other hand, in a material in which all bands are completely filled, its electronic structure is identical to that in which closed-shell atoms are arranged into regular order, and the material is then an insulator.

4.6 The Wigner–Seitz theory

We have discussed two methods to obtain the Bloch orbital: first, an approximation that starts from plane waves and, second, a method based on the linear combination of atomic orbitals. However, these are two limiting cases, and are not sufficient for discussing real materials. A method which is closer to reality was proposed by Slater, and Wigner and Seitz. This became the basis for the band calculations that are presently carried out for many materials. Further, Wigner and Seitz clarified the essential nature of the metallic bond based on the results of this calculation.

Let us consider the alkaline metals, Na for example. We specify the mean field that a conduction electron is subject to in the crystal as follows. First, the whole crystal is divided into atomic polyhedra. When an electron is in an atomic polyhedron, it is subject to the atomic potential of the atom that is contained in the polyhedron. When the electron moves to a neighboring atomic polyhedron, it is then subject to the atomic potential of the atom that is contained therein. As a result, we obtain the potential that is illustrated in Fig. 4.9.

This is a rather ingenious idea. In reality, there are contributions in the form of the atomic potentials of the atoms in other polyhedra than that in which the electron is located, and also the potential due to the other electrons. However, since, on average, there is one electron in each polyhedron, each atomic polyhedron

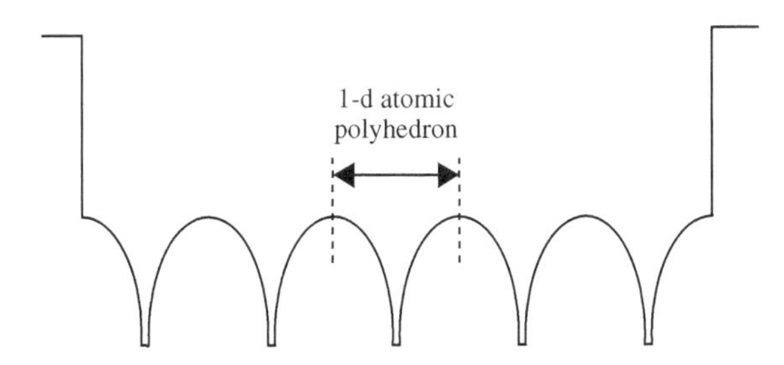

Figure 4.9 The Wigner–Seitz potential (in one dimension).

is charge neutral, and therefore exerts only a weak potential. Hence we may neglect them altogether. When an electron moves to a neighboring atomic polyhedron, the electron that was lying there moves away to somewhere, so that all of the atomic polyhedra other than that in which the electron that is presently under attention has entered can again be assumed to be charge neutral. This potential is thus able to account for much of the electronic correlation.

We now denote the potential in an atomic polyhedron by $V_a(\boldsymbol{r})$, and consider

$$\left[-\frac{\hbar^2}{2m}\Delta + V_a(\boldsymbol{r})\right]\psi = E\psi \tag{4.39}$$

in this polyhedron. If we fix E to be at a certain value, we can numerically evaluate the value of ψ by starting from $\boldsymbol{r}=0$. When the value of ψ inside the polyhedron has been determined in this way, inside a polyhedron that is at a displacement of $\boldsymbol{R}$ from it, the value must be given by $e^{i\boldsymbol{k}\cdot\boldsymbol{R}}\psi$ according to Bloch's theorem. The value of ψ is determined over the whole crystal by this procedure. Next, we impose the boundary condition. In the present case, we require that the values of the wavefunction and its derivatives match at the boundary of the polyhedra. This can be satisfied for certain values of E that are chosen at the beginning. These values are then the eigenvalues, and these depend on $\boldsymbol{k}$.

Here, it is helpful to make a comparison with the case of the isolated atom. In that case, the eigenvalue was determined by first giving a value to E and then integrating the equation. The values of E for which $\psi \to 0$ as $r \to \infty$ are the eigenvalues for the isolated atom. Thus the difference in the boundary conditions is responsible for the effect that is due to the electron not being bound to just one atom.

In order to solve eq. (4.39), assuming $V_a(\boldsymbol{r})$ is spherically symmetric, we adopt the following expansion:

$$\psi = \sum_{lm} c_{lm} R_l(r) Y_{lm}(\theta, \varphi).$$

For R_l, we obtain eq. (1.7), with ε_{nl} replaced by E. When we assign a value to E, $R_l(r)$ is obtained by numerical integration. For l and m we can, in general, choose only a finite number of combinations. As a result, the boundary condition can only be satisfied at a finite number of points on the boundary.

Now, with the 3s band of Na in mind, let us consider the case $l=0$. We first discuss the point $\boldsymbol{k}=0$. According to Bloch's theorem, ψ in the neighboring polyhedron is given simply by translating ψ by a fundamental vector. Therefore, as is clear from Fig. 4.10, $dR_0(r)/dr$ must vanish at the boundary. There then

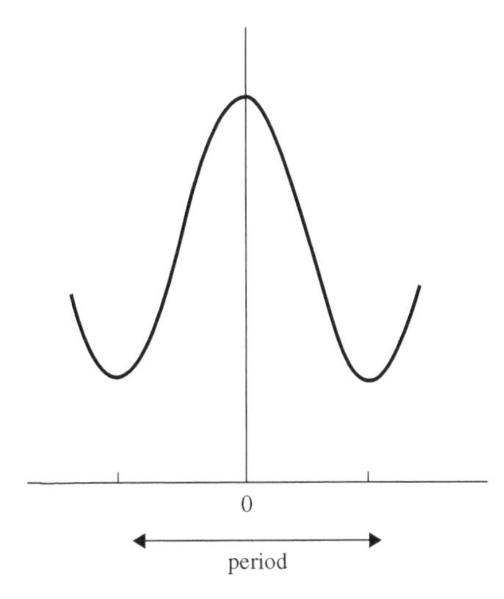

Figure 4.10 If the derivative of an even periodic function is continuous at the boundary, it must be zero.

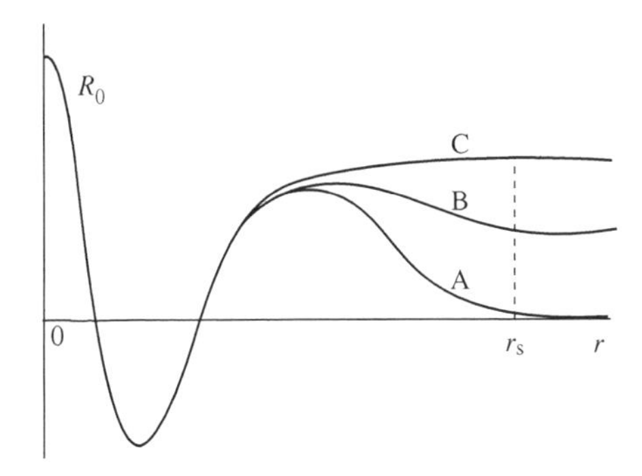

Figure 4.11 A schematic diagram illustrating the 3s orbital of Na. The differential equation, eq. (1.7), is integrated by giving some specific values to ε_{nl}. Curve A is for $\varepsilon_{nl} = \varepsilon_{3s}$. Curves B and C are for deeper (more negative) energies. The derivative of C vanishes at $r = r_s$.

arises the problem of at which positions on the polyhedron this condition should be satisfied. Na has a body-centered cubic lattice and, as seen in Fig. 4.2a, its atomic polyhedron is comparatively close to a sphere. We may thus choose a sphere whose volume is equal to that of the atomic polyhedron, and impose the condition $dR_0(r)/dr = 0$ at $r = r_s$, where r_s is the radius of this sphere.

In Fig. 4.11, we first show the values of R_0 when E is set equal to ε_{3s} of the isolated atom as curve A. This satisfies $R_0 \to 0$ as $r \to \infty$. When the value of E

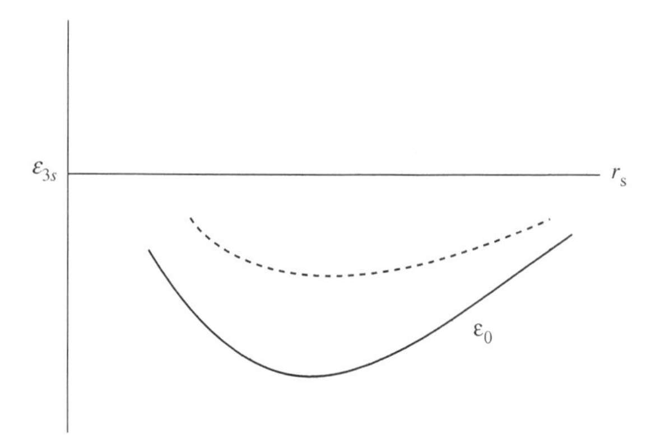

Figure 4.12 A schematic plot of the level ε_0 of the bottom of the conduction band as a function of r_s. The dashed curve corresponds to the values given by eq. (4.43). The difference between this and the 3s level ε_{3s} of the isolated atom corresponds to the cohesive energy.

is decreased slightly, we obtain curve B. With further decrease of E, at a certain value, curve C, that is, the solution that satisfies $R_0{}' = 0$ at $r = r_s$, is obtained. This value corresponds to the energy of the 3s band at $\boldsymbol{k} = 0$, that is, it represents the position of the bottom of the conduction band. Let us call this energy ε_0. As is seen from the figure, R undergoes a large oscillation in the neighborhood of $r = 0$, and this is influenced by the strong atomic potential. On the other hand, it is almost constant in the vicinity of r_s, so that, in this sense, it behaves like a plane wave with $\boldsymbol{k} = 0$.

We may imagine modifying r_s. The resulting eigen-energy can be calculated, and this gives rise to ε_0 that is as shown in Fig. 4.12. The result that it is lower than ε_{3s} is due to the electron not being bound to a single atom.

Let us now move on to the case of finite $\boldsymbol{k}$. In general, as the Bloch orbital satisfies eq. (4.19), we may consider a quantity, denoted as $u(\boldsymbol{r})$, which is defined by

$$\psi = e^{i\boldsymbol{k}\cdot\boldsymbol{r}} u(\boldsymbol{r}). \tag{4.40}$$

$u(\boldsymbol{r})$ is then invariant under $\boldsymbol{r} \to \boldsymbol{r} + \boldsymbol{R}$. That is, it behaves in the same way as the Bloch orbital does at $\boldsymbol{k} = 0$. From eq. (4.39), we can derive an equation for u, which reads

$$\left[-\frac{\hbar^2}{2m}\Delta + V_a(\boldsymbol{r}) - i\frac{\hbar^2}{m}\boldsymbol{k}\cdot\nabla \right] u = \left(E - \frac{\hbar^2 k^2}{2m} \right) u. \tag{4.41}$$

The third term in the Hamiltonian on the left-hand side is small, and so we may treat it using perturbation theory. Since the boundary condition for u is

$u(\boldsymbol{r}) = u(\boldsymbol{r} + \boldsymbol{R})$, if we omit the third term, the solution for u is equal to the solution that we obtained earlier for $\boldsymbol{k} = 0$.

In order to treat the third term as a perturbation, we need the solutions with l equal to 1 or more, in addition to the $l = 0$ solution discussed above. Denoting their energies as ε_l and the functions as u_{lm}, we obtain, as the energy perturbation,

$$E = \varepsilon_0 + \frac{\hbar^2 k^2}{2m} - \sum_{lm} \frac{\left| \int u_{lm}{}^* \frac{\hbar^2}{m} i\boldsymbol{k} \cdot \nabla u_0 dv \right|^2}{\varepsilon_l - \varepsilon_0}. \tag{4.42}$$

Here ε_0 is the energy shown in Fig. 4.12. For practical purposes, taking $l = 1$ is sufficient. As the third term is proportional to k^2, the above equation can be written as $\varepsilon_0 + \hbar^2 k^2/2m^*$. This is valid when k is not very large. Hence, with ε_0 as the bottom of the band, there is a nearly continuous distribution of levels starting from it. The ground state is obtained by filling these levels, starting from the lowest level upwards with two in each level, and with all the conduction electrons.

As we have already carried this out in §3.2, we may quote the result, but replacing m by m^*. The energy per electron is given from eqs. (3.7) and (3.8a) by

$$\varepsilon_0 + \frac{3}{5} \frac{\hbar^2}{2m^*} \left(3\pi^2 \frac{N}{\Omega} \right)^{2/3} \tag{4.43}$$

(the dashed curve in Fig. 4.12). The difference between this and ε_{3s} gives the cohesive energy of Na. Wigner and Seitz calculated ε_0, m^* and N/Ω as functions of r_s and, further, obtained the value of r_s that minimizes eq. (4.43) and the cohesive energy that corresponds to this value. In the case of Na and Li, these were found to be in extremely good agreement with the measured values.

The origin of the metallic bond may hence be attributed to the fact that the electron in the outermost shell of the metallic atom is not bound to a single atom and can move around over the whole of the crystal. However, the above calculation is still less than satisfactory.

In the above calculation, the wavefunction is taken as a Slater determinant that consists of $\psi(\boldsymbol{k})$ in eq. (4.40), and its mean energy is expressed in terms of a large number of integrals, as seen in eq. (1.22). Many of these integrals can be combined together and reduced to the form of eq. (4.43), but there are still residual terms. As we discussed in §4.5, if we consider a single atomic polyhedron, the wavefunction is a mixture of the cases with zero, one, two and even three or more electrons in the polyhedron. In the case where there are more than two electrons, especially in the case of two electrons with opposite spin, the Coulomb energy between them becomes very large. This cannot be avoided so long as we are making use of the Slater determinant. Because of this, the method gives a much worse value (higher value) than the value in eq. (4.43), which was obtained by assuming that only one electron is present in a polyhedron, and hence leads to a mismatch with the measured values.

This can, in general, be rectified by reducing the probability of two or more electrons residing in a polyhedron by using the configuration interaction as in the case of the H_2 molecule. However, it is quite unreasonable to suppose that such a procedure may be carried out for systems with, say, 10^{22} electrons. This is an important problem which has been discussed for some time as the problem of electronic correlation. The oldest approach was that of Wigner. A number of other methods followed, such as that by Bohm and Pines based on the extraction of collective modes due to plasma oscillations, diagrammatic perturbation expansions, and so on. These results corroborate our discussion in that analyzing the cohesive energy of alkaline metals using eq. (4.43) is correct at least in the outline.

5
Magnetic impurities in metals

In this chapter, we discuss the electronic states of a single impurity atom in metals. In particular, when the impurity atom is a 3d transition metal, its 3d orbital tends to assume the character of an isolated atom, and has a non-zero spin due to a similar mechanism to Hund's rule in atoms. We first describe the Friedel–Anderson theory regarding the emergence of this localized spin.

In alloys with a small amount of 3d transition metals, we have a long-standing problem which is known as the resistance minimum phenomenon. This is the phenomenon that the electrical resistance starts to rise as the temperature falls to around the boiling point of helium. We explain that this phenomenon is due to the exchange interaction between a localized spin and conduction electrons. This result suggests that we need to refine our discussion of the emergence of localized spin further, and many theoretical studies have been done. These theoretical works will be discussed later in Chapters 6 to 8. In this chapter, we compare the speed of the fluctuation of localized spin against the timescale of observation. We emphasize that when the latter is greater than the former, localized spin appears to vanish.

5.1 Local charge neutrality

In the discussion of the electronic states of an impurity atom in metals, the overall electric charge neutrality becomes an important issue. The potential due to a single impurity needs to fall to zero sufficiently fast as the distance r from the impurity becomes large. The integral of the net charge in a sphere of radius R_0 centered on the impurity should go to zero when R_0 is much larger than the microscopic scale. If not, the potential would affect the conduction electrons that are far away, and the electronic distribution would be altered, finally leading to local charge neutrality.

We thus need to treat the problem of electronic charge neutrality through a self-consistent method. That is, firstly, we introduce an electronic distribution and determine the potential according to the Poisson equation. Secondly, we solve the Schrödinger equation for the electrons in this potential, and fill the energy levels starting from the lowest level. We then calculate the electronic distribution, and this needs to match the distribution that was introduced at first.

Thus the treatment of the Coulomb interaction, which plays an essential role in our discussion, needs to proceed in a way that respects self-consistency if we are to work on the basis of mean-field approximation. However, it is not easy to obtain a strictly self-consistent result. Even when we fail to do so, the final result is required to satisfy the local neutrality condition, which is the most basic requirement in this problem.

As an example, let us consider adding a Zn atom to Cu metal. If we add only a Zn nucleus, the metal will be charged by $30e$, and this charge will be distributed on the surface. The electronic distribution near the nucleus will not be different from the case when a Zn atom is introduced. We then try to obtain a self-consistent potential around the Zn nucleus, which should be almost unchanged from the potential due to an isolated Zn atom. Far away from the nucleus, the potential quickly falls to 'zero'. The 'zero' level should be the bottom of the conduction band in the Cu metal. We may assume that the potential is spherically symmetric.

Then, by evaluating the electronic wavefunctions, we obtain 1s, 2s, 2p, etc., in ascending order. These functions are expected to be almost the same as those for the isolated atom. The situation is almost unchanged up to the 3d level. As a result, 28 electrons in total fill the levels that are lower than the bottom of the Cu conduction band.

On the other hand, the level that corresponds to 4s emerges inside the conduction band, and this becomes a problem that is characteristically associated with the impurity level in metals. If we adopt the free electron approximation for the Cu conduction electron, the effect of the Zn atom can be replaced with that of a $Z = 1$ positive charge and an extra electron located in the free electron gas. In the same way, if the impurity consists of a Ga atom, its effect will be equivalent to that of a $Z = 2$ positive charge and two extra electrons, because of the difference between the Ga and Cu atomic numbers. We can proceed in the same way up to As.

When the problem has been simplified in this manner, let us consider how to impose self-consistency as discussed above. The first step, i.e., evaluating the potential on the basis of the electronic distribution, can be carried out easily using electromagnetism. However, the second step of solving the Schrödinger equation in this potential is not necessarily easy.

Here, let us make use of the Thomas–Fermi approximation. This approximation was used to obtain the electronic distribution of an isolated atom based on the assumption that the spatial variation of the potential is small. We now apply this approximation to our problem.

Let us say that the density of electrons is n_0 before the addition of the impurity. That is, the electronic density is n_0 after the electrons have filled the levels up to the Fermi level. If the potential energy at $\boldsymbol{r}$ due to the impurity is $V(\boldsymbol{r})$ (< 0), the levels from $V(\boldsymbol{r})$ to the Fermi level will be filled.

Because the Fermi level is constant, the energy range that is filled by the electrons increases by $-V(\boldsymbol{r})$. The density of electrons thus also increases by $\Delta n(\boldsymbol{r})$:

$$\Delta n(\boldsymbol{r}) = -V(\boldsymbol{r})N(\varepsilon_F).$$

Here, $N(\varepsilon_F)$ is the density of states per unit volume at the Fermi level, and $N(\varepsilon_F) = 2\rho(\varepsilon_F)/\Omega$, using the notation of §3.2. Within this approximation, the density of electrons is determined only by the value of the potential at $\boldsymbol{r}$. We then make use of the Poisson equation. When Z is the difference between the atomic numbers of Cu and the impurity, $V(\boldsymbol{r})$ comprises the contributions of charges $+Ze$ at the origin and $-e\Delta n(\boldsymbol{r})$, that is,

$$-\nabla^2 V(\boldsymbol{r}) = 4\pi e^2 \Delta n(\boldsymbol{r}) - 4\pi e^2 Z\delta(\boldsymbol{r}).$$

From these two equations, we obtain

$$\nabla^2 V(\boldsymbol{r}) = \lambda^2 V(\boldsymbol{r}) + 4\pi e^2 Z\delta(\boldsymbol{r}), \tag{5.1}$$

where

$$\lambda^2 = 4\pi e^2 N(\varepsilon_F). \tag{5.2}$$

The solution to this equation is:

$$V(\boldsymbol{r}) = -\frac{Ze^2}{r} e^{-\lambda r}. \tag{5.3}$$

This is the so-called screened Coulomb potential, and λ is called the screening constant. We can easily verify that $\Delta n(\boldsymbol{r})$ satisfies the local neutrality condition,

$$\int_{r<R_0} \Delta n(\boldsymbol{r}) dv = Z.$$

In eq. (5.3), the characteristics of the host metal are encoded in λ, and those of the impurity are reflected in Z. The scattering cross-sections due to the screened Coulomb potential, calculated within the Born approximation, are proportional

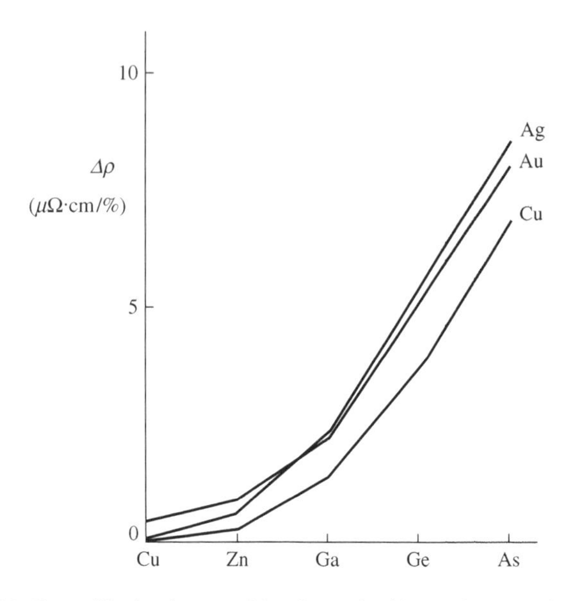

Figure 5.1 The residual resistance of the alloys of noble metals contaminated with 1 atomic percent of Cu, Zn, Ga, Ge, As. From the data of J. O. Linde, *Ann. Physik* **10** (1931) 52; **15** (1932) 219.

to Z^2. This explains the fact that the residual resistance at low temperature of the copper alloy with 1 atomic percent impurities is proportional to Z^2 when the impurity is any one of Zn, Ga, Ge or As (see Fig. 5.1). However, the resistance as expected from the calculated scattering cross-sections is too large compared with the experimental results. It has been found that the Thomas–Fermi approximation is not sufficient to explain many of the other experimental results. In order to improve our approximation, we have to find a consistent method of solving the Schrödinger equation. This becomes essential in the case where the impurity is any one of the atoms with atomic number less than that of Cu.

5.2 The spherical representation

This section and the two sections that follow are mainly based on Friedel (1952, 1954, 1958).

In order to solve the Schrödinger equation, let us define the potential as follows. We take a length r_0, which is about the radius of the impurity atom. We take the potential to be identical to that of the isolated impurity atom for $r < r_0$, and to take the value of the bottom of the conduction band for $r > r_0$.

As indicated by eq. (5.3), the potential is not constant outside r_0, but let us disregard the r-dependence for now. We can then solve the equation analytically for $r > r_0$. For $r < r_0$, on the other hand, we can integrate, numerically, the potential of the isolated atom which is also given numerically. The two solutions need to match at $r = r_0$.

An important quantity here is the logarithmic derivative of the $r < r_0$ solution at $r = r_0$, which determines the character of the solution outside. This can be replaced by the so-called 'phase shift'.

Let us consider a metallic sphere of radius R, containing an impurity atom which is located at the center. The potential being spherically symmetric, we write our wavefunction in the form $\psi(r)Y_{lm}(\theta, \varphi)$, where Y_{lm} are the lth-order spherical harmonics. ψ satisfies the following equation:

$$\left[-\frac{d^2}{dr^2} - \frac{2}{r}\frac{d}{dr} + U(r) + \frac{l(l+1)}{r^2} \right] \psi = k^2 \psi, \tag{5.4}$$

where

$$U(r) = \frac{2m}{\hbar^2} V(r), \tag{5.5}$$

and

$$\varepsilon = \frac{\hbar^2 k^2}{2m}. \tag{5.6}$$

Here, $V(r)$ is the potential energy as shown in Fig. 5.2, and ε is the energy eigenvalue. The energy is defined to be zero at the bottom of the conduction band.

In the case of an Fe impurity atom inside Cu, the energy eigenvalues for the 1s to 3p levels are negative, and the wavefunctions are almost the same as those of the Fe atom. The wavefunctions can be thought to be localized inside $r = r_0$. However, we expect that the 3d level is positive. Although the 3d level is certainly negative for Cu, the atomic potential of Fe is higher than Cu because the atomic number of Fe is smaller, and as a result the 3d level is correspondingly higher.

For Mn, Cr, etc., whose atomic numbers are less than that of Fe, the 3d level is even higher. In contrast, the 3d level is lower for Co and Ni, whose atomic numbers are greater than that of Fe. When we replace Fe by Zn, the 3d level falls beneath the bottom of the Cu band, so that only the 4s level remains positive, as discussed at the beginning of this chapter.

Hereafter, let us focus on the levels with positive energy eigenvalues. We integrate eq. (5.4) numerically from $r = 0$ to r_0 for a given value of k. Since $V(r) \propto r^{-1}$ near $r = 0$, we adopt $\psi \propto r^l$ as the initial boundary condition,

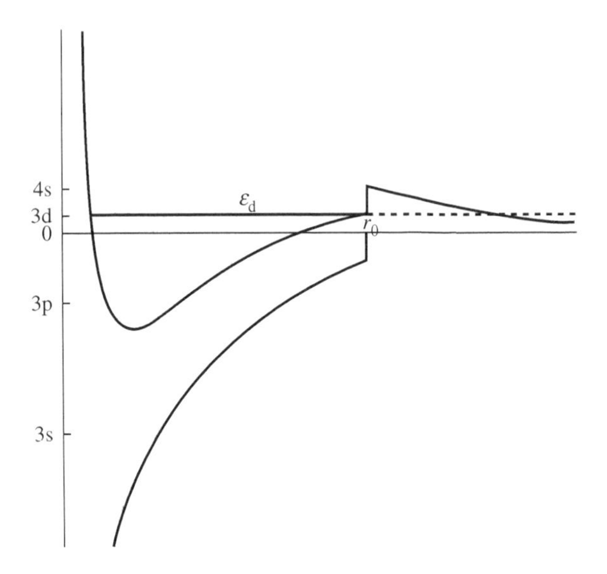

Figure 5.2 A schematic diagram of the impurity potential due to an Fe atom in Cu (the bold line). Zero corresponds to the bottom of the conduction band on the vertical scale. The 3d and 4s levels of an isolated Fe atom have positive energies, and the levels including and below 3p have negative energies. The 3d orbital is affected by a large centrifugal potential. The thin line indicates the potential for the 3d orbital when the centrifugal potential is added. The 3d orbital is confined in the neighborhood of the origin by the barrier due to the centrifugal potential. However, unlike in atoms, it can leak out of the barrier.

and integrate from $r = 0$ to $r = r_0$. The ψ thus obtained is denoted as $u_k(r)$. Its logarithmic derivative at r_0 is denoted as $L(k)$:

$$L(k) \equiv \frac{u_k{}'(r_0)}{u_k(r_0)}. \tag{5.7}$$

Since $U = 0$ outside $r = r_0$, the solutions to eq. (5.4) are $j_l(kr)$ and $n_l(kr)$, which are the lth-order spherical Bessel and Neumann functions, respectively. We then consider the linear combination of these, in a way such that its logarithmic derivative at r_0 is equal to $L(k)$. Furthermore, we set the boundary condition that it vanishes at $r = R$. These two conditions determine the coefficients of the linear combination and the eigenvalues of k.

When we write the linear combination as

$$v_k(r) = A_k[j_l(kr) - c(k)n_l(kr)], \tag{5.8}$$

the latter condition is satisfied when

$$c(k) \equiv \frac{j_l(kR)}{n_l(kR)}. \tag{5.9}$$

On the other hand, the former condition implies

$$c(k) = \frac{j_l{}'(kr_0) - j_l(kr_0)L(k)}{n_l{}'(kr_0) - n_l(kr_0)L(k)}. \tag{5.10}$$

Here, $j_l{}'(kr) = (d/dr)j_l(kr)$ etc. The eigenvalues of k are determined as the solutions to eq. (5.10).

The right-hand side of eq. (5.10) is a smoothly varying function of k, but the left-hand side goes from $+\infty$ to $-\infty$ each time k increases by π/R. We can see this from the asymptotic forms of j_l and n_l as x approaches infinity, i.e.:

$$j_l(x) \longrightarrow \frac{1}{x}\sin\left(x - \frac{l\pi}{2}\right),$$

$$n_l(x) \longrightarrow -\frac{1}{x}\cos\left(x - \frac{l\pi}{2}\right). \tag{5.11}$$

Hence eq. (5.10) has roots that are separated by approximately π/R.

Let us consider the special case of there being no impurities, that is, $U(r) = 0$, in eq. (5.4). The solution then becomes $j_l(kr)$ for all r, since $n_l(kr)$, being singular at $r = 0$, is excluded. The boundary condition, $j_l(kR) = 0$, determines the eigenvalues κ_n of k, and these satisfy:

$$\kappa_n R - \frac{l\pi}{2} = n\pi. \tag{5.12}$$

Here, we made use of eq. (5.11) assuming that R is sufficiently large. However, eq. (5.11) will no longer be appropriate when l becomes large. Let us denote the eigenfunctions as ϕ_n, which are normalized in $0 \leqq r \leqq R$. Eigenfunctions ϕ_n are given by

$$\phi_n(r) = \left(\frac{2\kappa_n{}^2}{R}\right)^{1/2} j_l(\kappa_n r). \tag{5.13}$$

These form a complete and orthogonal system, that is,

$$\sum_n r^2 \phi_n(r)\phi_n(r') = \delta(r - r'). \tag{5.14}$$

This relation holds for any choice of l.

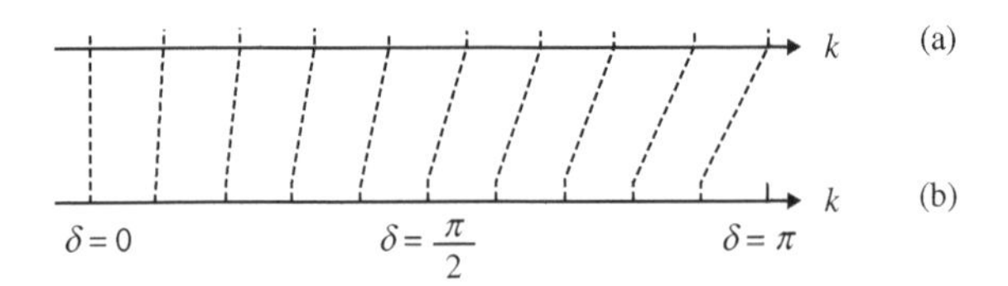

Figure 5.3 The eigenvalues of k. (a) represents the case without the scattering potential. The solutions, as given by eq. (5.12), are spaced equally. (b) represents the case with the potential. In the case of attractive interaction, the levels are more dragged down for larger k, as shown. The shift in k is represented by the phase shift δ; $\delta = \pi$ corresponds to the shift by one full step.

We see that, even in the unperturbed system, there is one solution for k for each π/R. Let us therefore introduce the phase shift δ_l, which is a measure of the deviation of the solutions for $U \neq 0$ compared with the unperturbed ones. This is defined as

$$k_n R - \frac{l\pi}{2} + \delta_l = n\pi. \tag{5.15}$$

$\delta_l = 0$ corresponds to there being no shift, and $\delta_l = \pi$ corresponds to a shift by one full step, which increases the number of states below that energy by 1. This is shown in Fig. 5.3.

Now, let us solve these equations for δ_l instead of k. Substituting k_n as given by eq. (5.15) into eq. (5.10), and noting that $c(k_n) = \tan \delta_l$, we obtain

$$\tan \delta_l = \frac{j_l{}'(kr_0) - j_l(kr_0)L(k)}{n_l{}'(kr_0) - n_l(kr_0)L(k)}. \tag{5.16}$$

$L(k)$ is given by numerical integration as a function of k. The right-hand side of eq. (5.16) does not vary very fast with k. This determines δ_l as a function of k. After substituting $c(k) = \tan \delta_l$ into eq. (5.8), we obtain the asymptotic form of $v_k(r)$ as $r \to \infty$, as

$$v_k(r) \longrightarrow \frac{A_k}{kr \cos \delta_l} \sin\left(kr + \delta_l - \frac{1}{2} l\pi\right), \tag{5.17}$$

and hence δ_l is called the phase shift.

Let us determine the value of A_k in eq. (5.8) using the normalization condition

$$\int_0^R r^2 \psi_k{}^2 dr = 1.$$

To do so, we introduce $\psi_{k'}$, which satisfies an equation that is analogous to eq. (5.4),

$$\left[-\frac{d^2}{dr^2} - \frac{2}{r}\frac{d}{dr} + U(r) + \frac{l(l+1)}{r^2}\right]\psi_{k'} = k'^2\psi_{k'}.$$

We multiply this by ψ_k, and subtract eq. (5.4) multiplied by $\psi_{k'}$ from it. This leads to

$$\frac{d}{dr}\left[r^2\left(\frac{d\psi_k}{dr}\psi_{k'} - \frac{d\psi_{k'}}{dr}\psi_k\right)\right] + (k^2 - k'^2)r^2\psi_k\psi_{k'} = 0. \tag{5.18}$$

The $k' \to k$ limit of this equation is

$$2kr^2{\psi_k}^2 = \frac{\partial}{\partial r}\left[r^2\left(\frac{\partial\psi_k}{\partial r}\frac{\partial\psi_k}{\partial k} - \psi_k\frac{\partial^2\psi_k}{\partial k\partial r}\right)\right]. \tag{5.19}$$

We then integrate this equation with respect to r from 0 to R, apply the normalization condition to the left-hand side, and make use of eq. (5.17) on the right-hand side. Noting that $\psi_k(R) = 0$, we obtain

$${A_k}^2 = \frac{2k^2}{R}\cos^2\delta_l\left[1 + \frac{1}{R}\frac{d\delta_l}{dk}\right]^{-1}. \tag{5.20}$$

This procedure should be carried out for each value of l. When $U = 0$, the solution is ordinarily written in terms of the plane waves, and the state in which all levels below k_F are filled with electrons is the ground state. All the levels with $k_n \leqq k_F$ in eq. (5.15) need to be filled also in the spherical representation, which we are adopting now.

When l becomes greater than about $k_F R$, even the minimum k_n given by eq. (5.15) exceeds k_F, so that this value of l becomes irrelevant. On the other hand, if we only consider the effect of the impurity, that is, if we are only interested in those values of l whose corresponding δ_l are non-zero, we find that only those values of l up to about 10 are relevant. Large l implies that the wavefunction is concentrated far away from the origin, and thus the wavefunction is hardly affected by the impurity potential, which is local. In other words, in terms of equations, when l is large, $j_l(kr_0) \sim j_l'(kr_0) \sim 0$ in eq. (5.16).

It often happens, however, that $\delta_l \sim 0$ even for small l. This occurs when $j_l'(kr_0)/j_l(kr_0) \sim L(k)$. The meaning of this relation is that, even though the value of $U(r)$ is quite large near the origin, and, correspondingly, $u_k(r)$ oscillates vigorously, the situation resembles that of the unperturbed system at $r = r_0$. In this case, it appears as if there is no effect of the impurity potential outside

$r = r_0$. The cases $l = 0$ (4s) and $l = 1$ (4p) in Cu doped with any impurity from Zn to As corresponds roughly to this case.

If the impurity potential is treated in the Born approximation in these cases, a strong scattering appears due to the large value of the potential near the origin. This is one of the reasons why the estimate for the residual resistance was found to be too large. The scattering is actually not so marked since $j_l'/j_l \sim L$. On the other hand, when Cu is doped with Fe or Cr, which are on the left-hand side of the periodic table, quite dissimilar circumstances occur for $l = 2$ (3d). Here, $n_l'(kr_0)/n_l(kr_0) \sim L(k)$ is satisfied for k near the Fermi surface and, therefore, $\delta_l \sim \pi/2$. Before discussing this point in detail, let us explain how we can estimate $\Delta n(r)$ etc. when δ_l has been obtained as a function of k using eq. (5.16).

5.3 Charge distribution and the density of states

Let us first consider the density of states (DOS). The number of roots of eq. (5.15) in between k and $k + dk$ is given by $dn = (R/\pi)(1 + R^{-1} d\delta_l/dk)dk$. Since there are $2l + 1$ degenerate states corresponding to one particular value of l, the total number of states, per spin, between k and $k + dk$, is given by $\rho(k)dk$, where

$$\rho(k) = \sum_l (2l+1)\frac{R}{\pi}\left(1 + \frac{1}{R}\frac{d\delta_l}{dk}\right). \tag{5.21}$$

The change of DOS, $\Delta\rho(k)$, due to the impurity is given by

$$\Delta\rho(k) = \frac{1}{\pi}\sum_l (2l+1)\frac{d\delta_l}{dk}. \tag{5.22}$$

We then evaluate the contribution, due to all wavefunctions with k_n in between k and $k + dk$, to the electronic density at r. We write this, including the contribution of both spin orientations, as $n(r,k)dk$. We make use of

$$\sum_{m=-l}^{l} |Y_{lm}|^2 = \frac{2l+1}{4\pi},$$

to obtain

$$n(r,k) = \sum_l \frac{2(2l+1)}{4\pi}\,\psi_k{}^2\frac{R}{\pi}\left(1 + \frac{1}{R}\frac{d\delta_l}{dk}\right). \tag{5.23}$$

For $r > r_0$, we substitute eqs. (5.8) and (5.20) to obtain

$$n(r,k) = \frac{k^2}{\pi^2} \sum_l (2l+1) \left[j_l(kr) \cos\delta_l - n_l(kr) \sin\delta_l \right]^2 .$$

When $U=0$, we should set $\delta_l=0$. Hence, making use of eq. (5.11) for sufficiently large r, the change $\Delta n(r,k)$ of $n(r,k)$ due to the impurity is given by

$$\Delta n(r,k) = \frac{1}{\pi^2 r^2} \sum_l (2l+1) \sin\delta_l \sin(2kr - l\pi + \delta_l). \tag{5.24}$$

The change of DOS, $\Delta\rho(k)$, is simply $\Delta n(r,k)$ integrated over the whole space, but, as seen in the above equation, $\Delta n(r,k)$ is localized near the origin.

Integrating $\Delta n(r,k)$ over k, on the other hand, from 0 to k_F yields $\Delta n(r)$, the change of the electronic density at r. Let us carry out this integration. We use eq. (5.24) when r is sufficiently large, and note that $\sin\delta_l$ is a slowly varying function of k. When r is sufficiently large, $\sin 2kr$ oscillates vigorously with k, and the resulting cancellations suppress the integral. However, there is a cut-off at the upper limit k_F, and a contribution arises from there.

In general, when $G(k)$ is a smoothly varying function of k (and $G(0)=0$), we may integrate by parts repeatedly, that is,

$$\int_0^{k_F} G(k) \sin 2kr dk = \frac{-1}{2r} G(k_F) \cos 2k_F r + \frac{1}{2r} \int_0^{k_F} \frac{dG}{dk} \cos 2kr dr,$$

to obtain an expansion in terms of r^{-1}. Taking the first term, we have

$$\Delta n(r) = -\frac{1}{2\pi^2 r^3} \sum_l (2l+1) \sin\delta_l(k_F) \cos(2k_F r - l\pi + \delta_l(k_F)). \tag{5.25}$$

We thus see that $\Delta n(r)$ exhibits a damped oscillation which is proportional to r^{-3}. The integral of $\Delta n(r)$ over the whole space is equal to the integral of $2\Delta\rho(k)$ from 0 to k_F. Therefore, using eq. (5.22), we obtain

$$\Delta N \equiv \int \Delta n(r) dv = 2 \int_0^{k_F} \Delta\rho(k) dk = \frac{2}{\pi} \sum_l (2l+1) \delta_l(k_F), \tag{5.26}$$

when $\delta_l(0)=0$.

5.4 Virtual bound states

Coming back to the problem of the Fe or Cr impurities in Cu, we would like to focus particularly on the 3d ($l = 2$) level. As shown in eq. (5.4), the centrifugal potential, $l(l+1)/r^2$, becomes very large in this case. The total potential energy which includes this potential is shown in Fig. 5.2, alongside the 3d level, ε_{d}, of an isolated atom.

As discussed earlier, since the atomic number of Fe is smaller than that of Cu, the Fe 3d level is raised above the bottom of the Cu conduction band, but remains below the barrier due to the centrifugal potential. The 3d wavefunction is therefore strongly localized inside the barrier, and is similar to that of the atom. The wavefunction penetrates the barrier by the tunneling effect, and mixes with the external spherical waves. This state is usually called the virtual bound state.

Now the wavefunction that is obtained by integrating eq. (5.4) with ε set near ε_{d} in eq. (5.6) exhibits behaviors as shown in Fig. 5.4(a) depending on the value of ε. That is, if we set $\varepsilon = \varepsilon_{\mathrm{d}}$, the wavefunction is expected to be sufficiently small and flat at r_0 that it is possible to find ε_1 and ε_2 near ε_{d}, where $u'(r_0) = 0$ and $u(r_0) = 0$, respectively. The logarithmic derivative L of u then behaves in the way shown in Fig. 5.4(b).

On the other hand, n_2'/n_2 also behaves in the same way, and so they cross each other at ε_0, which is found between ε_1 and ε_2. According to eq. (5.16), we have $\delta_2 = \pi/2$ at this point. Furthermore, as L varies fast near ε_0, so does δ_2, and we obtain the behavior shown in Fig. 5.5(a). According to eq. (5.22),

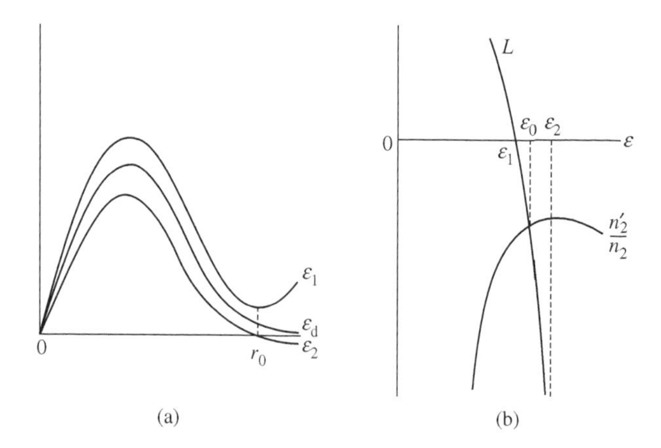

Figure 5.4 (a) 3d orbitals for various values of ε. (b) Dependence of L on ε. If the orbital is strongly localized as in (a), L varies from 0 to $-\infty$ in a narrow range of ε.

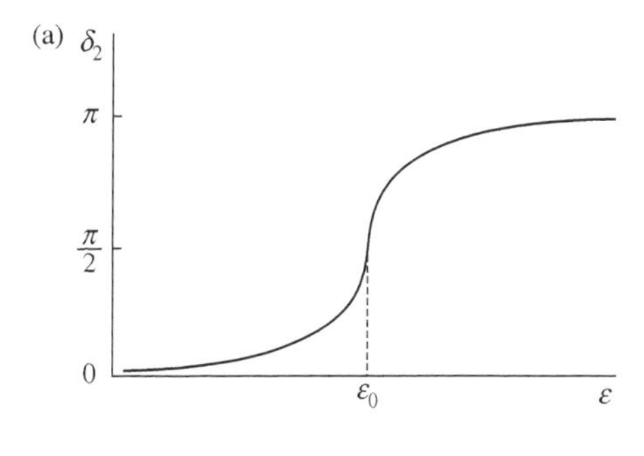

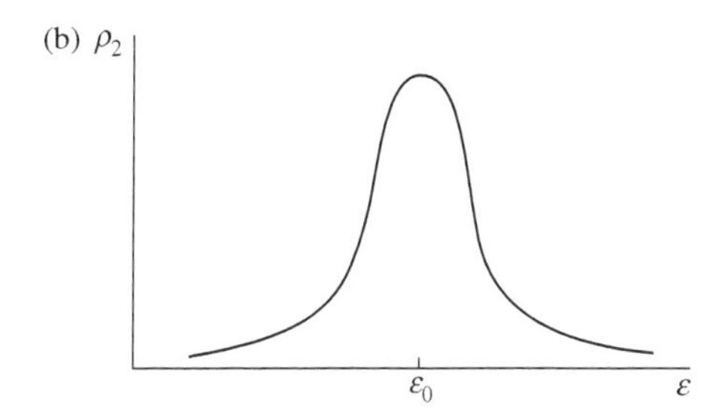

Figure 5.5 (a) The ε dependence of the phase shift δ_2 of 3d orbitals. (b) The density of states for the 3d orbital.

the change of DOS, $\Delta\rho(k)$, in this case has the behavior shown in Fig. 5.5(b), that is, localized with a strong peak near ε_0. This behavior can be interpreted as being brought about by the 3d level of the isolated atom at ε_{d} which penetrates out and mixes with the states in the Cu conduction band, giving rise to the broadening of the peak.

When the DOS exhibits a narrow peak at ε_0, as in the above case, change is induced in physical quantities depending on the positional relationship between the Fermi level ε_F and ε_0. While the position of ε_0 varies according to the atomic number of the impurity atom, ε_F is almost completely determined by the host metal. As mentioned in §3.2, since the specific heat of electrons is proportional to the DOS at ε_F, the specific heat is maximized when ε_0 approaches ε_F. Furthermore, as shown in eq. (5.25), $\Delta n(r)$ is proportional to $\sin\delta_l$, and is maximized when $\delta_l = \pi/2$. Thus, the influence on the electronic density distribution is also maximized when ε_0 approaches ε_F.

It is well known that the phase shift was originally introduced in order to treat the problem of the scattering of electrons, and so we can discuss the scattering of electrons due to an impurity when δ_l is determined in our problem. It is the electrons near the Fermi level that are involved in the electrical resistance of metals, and so we only need to consider the phase shift near the Fermi level in the following.

When electrons approach an impurity from far away, the differential scattering cross-section as a function of the scattering angle θ is given by

$$\sigma(\theta) = k^{-2} \left| \sum (2l+1) e^{i\delta_l} \sin \delta_l P_l(\cos\theta) \right|^2,$$

where P_l is the Legendre function. We have defined it so that the integral of $\sigma(\theta)$ with respect to the solid angle yields the total scattering cross-section, which is proportional to the probability of the scattering of electrons. However, as far as calculating the electrical resistance is concerned, what we are interested in is by what amount the velocity of an electron changes due to the scattering.

Scattering by angle θ changes the component of velocity in the direction of travel by $v(1-\cos\theta)$, and so the cross-section which contributes to electrical resistance is expressed as

$$\sigma_{\text{tr}} = \int \sigma(\theta)(1-\cos\theta)\, d\Omega = \frac{4\pi}{k^2} \sum (l+1) \sin^2(\delta_l - \delta_{l+1}).$$

When there are n_i impurities per 1 cm^3 with this cross-section, the probability, τ^{-1}, of the collision of electrons with the target impurities per second is given by $v_F n_i \sigma_{\text{tr}}$. Denoting the density of electrons as n (cm^3), the electrical resistance due to the impurities is given by

$$R = \frac{m}{ne^2\tau} = \frac{4\pi \hbar n_i}{ne^2 k_F} \sum (l+1) \sin^2(\delta_l - \delta_{l+1}), \tag{5.27}$$

where δ_l refers to the value at the Fermi level. When only δ_2 is significant and the other δ_l are small, as in the above case, we have $R \propto \sin^2 \delta_2$, and the electrical resistance is maximized when $\delta_2 = \pi/2$. As mentioned before, this situation occurs when the 3d level of the impurity is found near the Fermi level.

As an example, let us consider the residual resistance of Al when 1% 3d transition metal is added. (If 3d transition metal is added to Cu, the situation becomes more complex because of electron–electron repulsion, which will be discussed later.)

As shown in Fig. 5.6, the residual resistance is relatively small when Ti is added. One possible explanation for this is that the atomic number of Ti is small and so the 3d level of the isolated atom is above the Fermi level of Al. As the atomic number increases, the 3d level goes down, and at Cr, it crosses the Fermi level. As the atomic number increases further, the 3d level goes further away

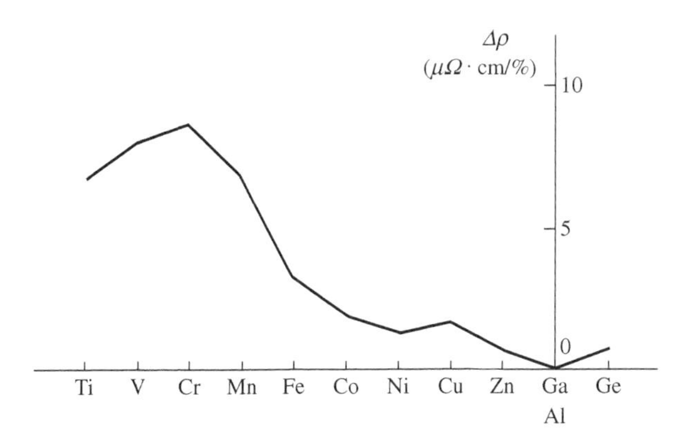

Figure 5.6 Residual resistance of Al with 1 atomic percent of iron group elements. From J. Friedel, *Nuovo Cim., Suppl.* **2** (1958) 287, reproduced with kind permission of Società Italiana di Fisica.

from the Fermi level. This situation is indicated by the peak of the residual resistance at Cr.

Returning to eq. (5.26), while Al supplies three electrons to the conduction band, an impurity, Cr for example, supplies six. In general, when the difference between the atomic valence of the host and the impurity is Z, Z electrons are expected to be localized near the impurity by the condition of local charge neutrality. Since the number of electrons localized near the impurity is given by ΔN in eq. (5.26), we have

$$Z = \frac{2}{\pi} \sum_l (2l+1)\delta_l(k_F). \tag{5.28}$$

This is called the Friedel sum rule, and is important as it gives a self-consistent constraint for the impurity potential. We may also deduce the value of δ_l to some extent using it.

When the wavefunction for a given l becomes close to zero at r_0 as shown in Fig. 5.4(a), that is, in a strongly localized case, we can prove (see Appendix D) that

$$\int_{r \geqq r_0} n(r,k)dv = \int n_0(r,k)dv. \tag{5.29}$$

n and n_0 are for each value of l under consideration; n_0 refers to n in the case of there being no impurities.

This equation implies that, for the particular value of l under consideration, the same number of electrons as existed inside r_0 in the case without impurity, have been pushed out of r_0. As a result, when there are impurities, the number

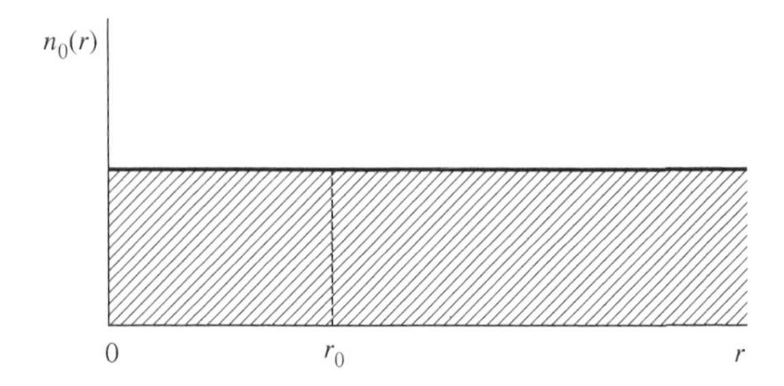

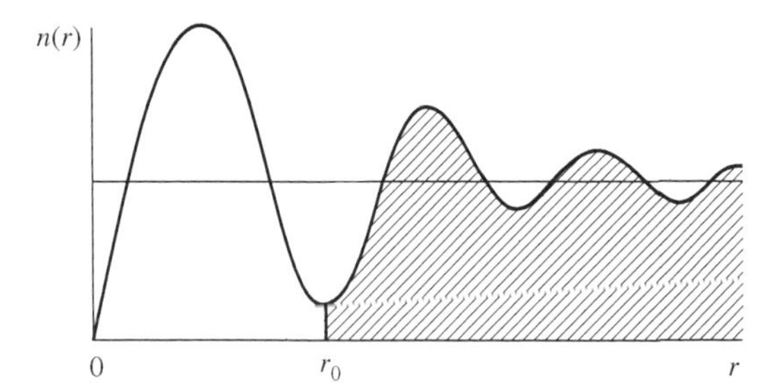

Figure 5.7 The r-dependence of the electronic density. The electronic density is uniform in the case without impurity (upper figure). When orbitals in the impurity are strongly localized, the shaded area in the lower figure should be the same as that in the upper figure. The blank area in the lower figure hence determines the increment ΔN due to the impurity.

of electrons inside r_0 is equal to the increment in the number of electrons due to the impurity, that is:

$$\int_0^{r_0} n(r,k) 4\pi r^2 dr = 2\Delta\rho(k). \tag{5.30}$$

That is, for each value of l under consideration, the number of electrons inside r_0 whose orbitals have the character of the isolated atom is given by ΔN in eq. (5.26). The electrons that were located inside r_0 when there were no impurities, and which were pushed out of r_0 by the impurity, contribute to the oscillating term in eq. (5.25) (see Fig. 5.7).

5.5 The Anderson model I

In this section we derive the results of Kanamori *et al.* (1969) in a more direct way.

Equation (5.4) can be solved for the particular choice of $V(r)$ introduced above, as we have discussed. When the impurity is a transition metal with $l=2$, or when it is a rare-earth metal with $l=3$, the orbitals are confined to inside r_0 by the centrifugal potential. They then maintain the characteristics of the orbitals of isolated atoms, with a small amount of mixing with the external spherical waves through the tunneling effect.

In these cases, the problem becomes more transparent when we adopt as the bases the atomic orbitals and the spherical waves of eq. (5.13), and expand the wavefunction ψ in terms of them. The coefficients are determined by solving the secular equation, as will be discussed below. Since the spherical waves of eq. (5.13) comprise a complete set, we also need to justify the additional inclusion of the atomic orbitals. In order to achieve these objectives, we start by rewriting the results of the previous section in another way.

In §5.2, we first integrated eq. (5.4) from 0 to r_0 to obtain $u_k(r)$. We then calculated the logarithmic derivative, that is, eq. (5.7), at $r=r_0$. Using this, we determined $c(k)$ on the basis of eq. (5.10), to construct $v_k(r)$ as given by eq. (5.8). This yields eigenfunctions which are connected smoothly with u_k at r_0 and vanish at $r=R$.

Let us now define $v_k(r)$ as

$$v_k(r) = \begin{cases} A_k\left[j_l(kr) - c(k)n_l(kr)\right] & r > r_0, \\ 0 & r < r_0. \end{cases} \tag{5.31}$$

We also define $u_k(r)$ in the opposite manner, to be identically 0 for $r > r_0$. The solution to eq. (5.4) for $0 < r < R$ can then be written as

$$\psi_k = u_k + v_k. \tag{5.32}$$

If k satisfies eq. (5.10), we have

$$\left.\begin{aligned} u_k(r_0) &= v_k(r_0) \\ {u_k}'(r_0) &= {v_k}'(r_0) \end{aligned}\right\}. \tag{5.33}$$

To be more precise, if k satisfies eq. (5.10), we may multiply u or v by an appropriate constant factor to make them satisfy eq. (5.33).

For orbitals with strong localization, such as the d levels of a transition metal or the f levels of a rare-earth metal, u_k contains components with large wavenumbers, while v_k is expected to have relatively small wavenumbers. Let us therefore consider the expansion of v_k in eq. (5.31), rather than expanding ψ_k itself in terms of ϕ_n in eq. (5.13).

The expansion in terms of ϕ_n in eq. (5.13) is called the Fourier–Bessel expansion, and the expansion coefficients are determined according to the formula given for it. However, we may carry out the expansion more easily by defining the following Green's function:

$$G_k(r,r') = \sum_n \frac{\phi_n(r)\phi_n(r')}{k^2 - \kappa_n{}^2}. \tag{5.34}$$

Here, κ_n and ϕ_n are defined by eqs. (5.12) and (5.13), respectively. In terms of the standard formula for Fourier–Bessel expansion, this is expressed (see Appendix E) as follows:

$$G_k(r,r') = kj_l(kr_<)\left[n_l(kr_>) - c(k)^{-1}j_l(kr_>)\right]. \tag{5.35}$$

$r_<$ is the lesser of r and r', and $r_>$ is the greater of the two; $c(k)$ is defined by eq. (5.9). Using this expression, we can show that v_k in eq. (5.31) satisfies

$$v_k(r) = r_0{}^2 v_k{}'(r_0)G_k(r_0,r) - r_0{}^2 v_k(r_0)\frac{\partial}{\partial r_0}G_k(r_0,r), \tag{5.36}$$

for both $r > r_0$ and $r < r_0$. We can derive this easily by using eq. (5.35) for G_k, making use of the fact that $v_k(r_0)$ is a linear combination of $j_l(kr_0)$ and $n_l(kr_0)$, and further noting that

$$j_l(x)n_l{}'(x) - j_l{}'(x)n_l(x) = x^{-2}. \tag{5.37}$$

We then employ eqs. (5.33) and (5.34) on the right-hand side of eq. (5.36) to obtain

$$v_k(r) = r_0{}^2 \sum_n \frac{u_k{}'(r_0)\phi_n(r_0) - u_k(r_0)\phi_n{}'(r_0)}{k^2 - \kappa_n{}^2}\phi_n(r). \tag{5.38}$$

This is precisely the expansion that we need. In particular, we note that the right-hand side of eq. (5.38) is equal to zero for $r < r_0$. When v_k in eq. (5.32) is replaced by eq. (5.38), ψ_k will have been expanded in terms of u_k and ϕ_n. Since the ϕ_n form a complete set, the additional inclusion of u_k as the bases of the expansion entails redundancy. Nevertheless, we have been able to show that such an expansion is possible.

Next, eq. (5.10), which determines k, is rewritten as:

$$\sum_n \frac{\left[L(k)\phi_n(r_0) - \phi_n{}'(r_0)\right]^2}{k^2 - \kappa_n{}^2} = 0, \tag{5.39}$$

as shown in Appendix E. $j_l(kr_0)$ and so on in eq. (5.10) have been replaced by $\phi_n(r_0)$ and its derivative. The results of §5.2 are thus represented, in terms of the localized orbital u_k and ϕ_n, by eqs. (5.32), (5.38) and (5.39).

Let us expand this result around some point k_0. We substitute $k' = k_0$ in eq. (5.18) and integrate this equation with respect to r from 0 to r_0, to obtain

$$r_0{}^2\left[u_{k_0}(r_0)u_k{}'(r_0) - u_k(r_0)u_{k_0}{}'(r_0)\right] + (k^2 - k_0{}^2)\int_0^{r_0} u_{k_0}(r)u_k(r)r^2 dr = 0.$$

When k is close to k_0, we make use of the normalization condition for u_{k_0} in the form

$$\int_0^{r_0} u_{k_0}(r)^2 r^2 dr = 1. \tag{5.40}$$

We then divide both terms of the equation by $u_{k_0}(r_0)u_k(r_0)$ to obtain the expansion,

$$L(k) - L(k_0) = -\frac{k^2 - k_0{}^2}{r_0{}^2 u_{k_0}(r_0)^2} \equiv \Delta L, \tag{5.41}$$

up to the first order in $k - k_0$.

By substituting $L(k) = L(k_0) + \Delta L$ into eq. (5.39), retaining terms to the first order in ΔL, and replacing ΔL by eq. (5.41), we obtain (see Appendix E)

$$k^2 - k_0{}^2 = r_0{}^4 \sum_n \frac{\left[u_{k_0}{}'(r_0)\phi_n(r_0) - u_{k_0}(r_0)\phi_n{}'(r_0)\right]^2}{k^2 - \kappa_n{}^2}.$$

We then substitute $\varepsilon = \hbar^2 k^2/2m$, $\varepsilon_n \equiv \hbar^2 \kappa_n{}^2/2m$ and $\varepsilon_0 \equiv \hbar^2 k_0{}^2/2m$, to obtain

$$\varepsilon - \varepsilon_0 = \sum_n \frac{V_n{}^2}{\varepsilon - \varepsilon_n}, \tag{5.42}$$

where

$$V_n = \frac{\hbar^2 r_0{}^2}{2m}\left[u_{k_0}{}'(r_0)\phi_n(r_0) - u_{k_0}(r_0)\phi_n{}'(r_0)\right]. \tag{5.43}$$

Here, it should be noted that u_{k_0} satisfies eq. (5.40). Furthermore, by using V_n as defined by this equation, we obtain

$$\psi_k(r) = u_k(r) + \sum_n \frac{V_n}{\varepsilon - \varepsilon_n}\phi_n(r), \tag{5.44}$$

from eqs. (5.32) and (5.38). We have disregarded the difference between k and k_0 in V_n; u_k may be replaced by u_{k_0} to the same level of accuracy.

We have now shown that the results in §5.2 can be rewritten so that the eigenvalues are determined by the secular equation of eq. (5.42) and the wave-functions are given by eq. (5.44). By examining these two equations, we find that they are actually equivalent to the following problem. Let us now adopt a complete orthonormal system of u_{k_0} and ϕ_n, expand ψ as

$$\psi = u_{k_0} + \sum c_n \phi_n, \tag{5.45}$$

and substitute this into the Schrödinger equation,

$$H\psi = \varepsilon\psi. \tag{5.46}$$

When the matrix elements of the Hamiltonian H are given by

$$\left.\begin{aligned} &\langle u_{k_0} | H | u_{k_0} \rangle = \varepsilon_0 \\ &\langle \phi_n | H | \phi_n \rangle = \varepsilon_n \\ &\langle \phi_n | H | u_{k_0} \rangle = \langle u_{k_0} | H | \phi_n \rangle = V_n \\ &(\text{other elements}) = 0 \end{aligned}\right\}, \tag{5.47}$$

we obtain

$$V_n + c_n \varepsilon_n = c_n \varepsilon,$$

$$\varepsilon_0 + \sum_n c_n V_n = \varepsilon,$$

so that eqs. (5.44) and (5.42) follow. Equation (5.47) has a clear interpretation that a localized orbital with energy ε_0 and a spherical wave ϕ_n with energy ε_n mix with each other through the matrix element V_n. What we have been able to show is that the corresponding secular equation and its eigen-wavefunction are precisely the same as eqs. (5.42) and (5.44), which are obtained by modifying the results of §5.2.

As $u_{k_0}(r)$ is constrained by eq. (5.40), $u_{k_0}(r_0)$ and $u_{k_0}{}'(r_0)$ are small when they are strongly localized. In this case, by eq. (5.43), V_n is small. We shall mainly discuss this case from here on.

We may ordinarily adopt a k_0 that satisfies the following condition:

$$L(k_0) = \frac{n_l{}'(k_0 r_0)}{n_l(k_0 r_0)}. \tag{5.48}$$

By eq. (5.16), we then obtain $\delta_l = \pi/2$ at k_0.

When $k_0 r_0$ satisfies this condition and its value is sufficiently greater than unity, V_n is found to be independent of κ_n near $\kappa_n = k_0$. To see this, we make

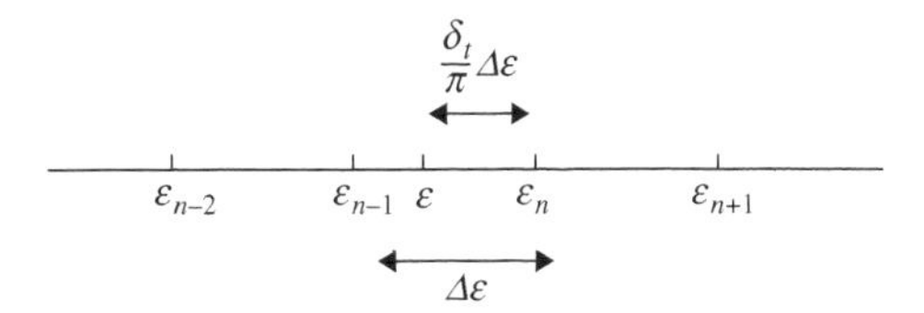

Figure 5.8 When the unperturbed energy ε_n is lined up with constant interval $\Delta\varepsilon$, the perturbed solution ε is smaller than ε_n by $(\delta_l/\pi)\Delta\varepsilon$, and δ_l is obtained using eq. (5.49).

use, in eq. (5.43), of the fact that $j_l{}'$ asymptotically tends to $-n_l$, and make use of eq. (5.48). When V_n is independent of n, we can obtain the solutions to eq. (5.42) as follows. Near the ε that is to be obtained, let us say that ε_n are lined up with constant interval $\Delta\varepsilon$, and that ε is smaller than ε_n by $\Delta\varepsilon \cdot \delta_l/\pi$, as shown in Fig. 5.8. This is equivalent to the prior definition of δ_l. Writing $\varepsilon_n = \Delta\varepsilon \cdot n$, eq. (5.42) reduces to

$$\varepsilon - \varepsilon_0 = V_n{}^2 \sum_{n'} \frac{1}{\Delta\varepsilon\,(n - (\delta_l/\pi) - n')} = \frac{-\pi V_n{}^2 \cot\delta_l}{\Delta\varepsilon}.$$

We then transform this into

$$\delta_l = \frac{\pi}{2} + \tan^{-1}\frac{\varepsilon - \varepsilon_0}{\Delta_l}, \tag{5.49}$$

with

$$\Delta_l = \pi V_n{}^2 \rho_l. \tag{5.50}$$

Here, $\rho_l = 1/\Delta\varepsilon$, and this is the unperturbed density of states for the symmetry state l, ignoring its $(2l + 1)$-fold degeneracy. By eq. (5.22), the change of the density of states due to the impurity has the form of a Lorentzian function with center ε_0 and width Δ_l. Equation (5.22) gives us the change of the number of states between k and $k + dk$. When we denote the change of the number of states per spin between ε and $\varepsilon + d\varepsilon$ as $\Delta\rho(\varepsilon)d\varepsilon$, we obtain

$$\begin{aligned}\Delta\rho(\varepsilon) &= \frac{1}{\pi}(2l+1)\frac{d\delta_l}{d\varepsilon} \\ &= (2l+1)\frac{\Delta_l}{\pi}\left[(\varepsilon - \varepsilon_0)^2 + \Delta_l{}^2\right]^{-1}.\end{aligned} \tag{5.51}$$

When the state is strongly localized, V_n is small, and therefore Δ_l is small by eq. (5.50). In this case, eq. (5.51) takes the form of a function with a sharp peak centered around ε_0. Furthermore, as mentioned previously, u_{k_0} is responsible for all of this change in the density of states. The model that is described by eqs. (5.45) (5.46), and (5.47) is called the Anderson model (without Coulomb repulsion).

5.6 The Anderson model II

Up to now, we have employed the spherical-wave representation. However, the plane-wave representation is often more useful. Let us consider using the plane-wave representation for the Anderson model (Anderson, 1961). Although the plane waves compose a complete set, we add $u_{k_0}(r)Y_{lm}(\theta,\varphi)$ $(m = -l,\ldots,l)$ as extra basis functions.

Here, we should choose only those values of l that correspond to strongly localized states. It is inappropriate to introduce localized orbitals for those values of l that correspond to poorly localized states. For these states, the form of the impurity potential should be retained as it is, but we omit such terms for present. We treat the plane waves and localized orbitals as if they compose a complete set. The annihilation and creation operators for these localized functions are designated by a_{lm} and $a_{lm}{}^{\dagger}$, respectively.

Let us consider the matrix element between this function and the plane wave $e^{i\boldsymbol{k}\cdot\boldsymbol{r}}$,

$$\left\langle \frac{1}{\sqrt{\Omega}} e^{i\boldsymbol{k}\cdot\boldsymbol{r}} \, |H| \, u_{k_0} Y_{lm} \right\rangle. \tag{5.52}$$

In order to express this in terms of V_n, we make use of

$$e^{i\boldsymbol{k}\cdot\boldsymbol{r}} = \sum_{lm} 4\pi\, i^l j_l(kr) Y_{lm}(\hat{\boldsymbol{r}}) Y_{lm}{}^*(\hat{\boldsymbol{k}}). \tag{5.53}$$

$Y_{lm}(\hat{\boldsymbol{k}})$ refers to $Y_{lm}(\theta_k, \varphi_k)$, and similarly for $Y_{lm}(\hat{\boldsymbol{r}})$. When V_n in eq. (5.47) is independent of n, we can write $\langle j_l\, |H|\, u_{k_0}\rangle = (R/2k^2)^{1/2} V_n$. After integrating over the solid angle, we obtain

$$(5.52) = (4\pi)^{1/2} V_l Y_{lm}(\hat{\boldsymbol{k}}),$$

where

$$V_l \equiv \left(\frac{4\pi R}{2k^2\Omega}\right)^{1/2} (-i)^l\, V_n.$$

The second-quantized Hamiltonian is then written as

$$H = \sum_k \varepsilon_k a_k{}^\dagger a_k + \sum_{lm} \varepsilon_l a_{lm}{}^\dagger a_{lm}$$
$$+(4\pi)^{1/2} \sum_{lmk} \left[V_l Y_{lm}(\hat{\boldsymbol{k}}) a_k{}^\dagger a_{lm} + V_l{}^* Y_{lm}{}^*(\hat{\boldsymbol{k}}) a_{lm}{}^\dagger a_k \right]. \quad (5.54)$$

Here, ε_l refers to ε_0, which appeared in the previous section. The change of notation is due to the fact that ε_0 is a function of l in general. This is called the Anderson Hamiltonian (without Coulomb repulsion).

If V_l does not depend on k, we may make use of

$$c_{lm} = (4\pi)^{1/2} \sum_k Y_{lm}{}^*(\hat{\boldsymbol{k}}) a_k ,$$

to write the last term of eq. (5.54) as $\sum_{lm}(V_l c_{lm}{}^\dagger a_{lm} + V_l^* a_{lm}{}^\dagger c_{lm})$, where c_{lm} is the counterpart of the mixing of the localized orbital a_{lm}.

The summation in the above equation is not over all $\boldsymbol{k}$. Just after eq. (5.48), we said that V_n is independent of k. However, this is true only when k is near k_0. When k becomes sufficiently large, $j_l(kr_0)$ becomes small, and hence V_n must also become small. We therefore need to limit the summation over $\boldsymbol{k}$ in c_{lm} to the region up to $k \lesssim r_0{}^{-1}$. This will be implied in all summations over $\boldsymbol{k}$ that will appear hereafter.

According to eq. (2.51), the real-space wavefunction that corresponds to c_{lm} is proportional to

$$\sum_k Y_{lm}(\hat{\boldsymbol{k}}) e^{i\boldsymbol{k}\cdot\boldsymbol{r}},$$

and is localized in the region $r \lesssim r_0$ near the origin.

We would like to find the linear combination of the form:

$$a_n = \sum_k a_k \langle \boldsymbol{k}|n\rangle + \sum_{lm} a_{lm} \langle lm|n\rangle \quad (5.55)$$

that diagonalizes eq. (5.54). Such a result was obtained previously in eq. (5.44), but it is useful to reproduce the same result from eq. (5.54). We denote $\boldsymbol{k}$ and lm together by α or β. Since eq. (5.55) represents a unitary transformation, the coefficients $\langle\alpha|n\rangle$ satisfy:

$$\sum_n \langle\alpha|n\rangle\langle n|\beta\rangle = \delta_{\alpha\beta}, \quad (5.56)$$

$$\sum_\alpha \langle n|\alpha\rangle\langle\alpha|n'\rangle = \delta_{nn'}. \quad (5.57)$$

Here, $\langle n|\alpha\rangle = \langle\alpha|n\rangle^*$. By inverting eq. (5.55), we obtain

$$a_\alpha = \sum_n a_n \langle n|\alpha\rangle \quad (\alpha = \boldsymbol{k} \text{ or } lm). \tag{5.58}$$

Substituting this into eq. (5.54), we obtain

$$\begin{aligned} H = \sum_{nn'} \Bigg[& \sum_{\boldsymbol{k}} \langle n'|\boldsymbol{k}\rangle \varepsilon_{\boldsymbol{k}} \langle \boldsymbol{k}|n\rangle + \sum_{lm} \langle n'|lm\rangle \varepsilon_l \langle lm|n\rangle \\ & + (4\pi)^{1/2} \sum_{lm\boldsymbol{k}} V_l \langle n'|lm\rangle Y_{lm}(\hat{\boldsymbol{k}}) \langle \boldsymbol{k}|n\rangle \\ & + (4\pi)^{1/2} \sum_{lm\boldsymbol{k}} V_l^* \langle n'|\boldsymbol{k}\rangle Y_{lm}^*(\hat{\boldsymbol{k}}) \langle lm|n\rangle \Bigg] a_n{}^\dagger a_{n'}. \end{aligned} \tag{5.59}$$

Hence, defining ε_n as the eigen-energy corresponding to a_n, we obtain the following secular equations:

$$\left.\begin{aligned} (4\pi)^{1/2} V_l \sum_{\boldsymbol{k}} Y_{lm}(\hat{\boldsymbol{k}}) \langle \boldsymbol{k}|n\rangle + (\varepsilon_l - \varepsilon_n)\langle lm|n\rangle = 0 \\ (4\pi)^{1/2} \sum_{lm} V_l^* Y_{lm}^*(\hat{\boldsymbol{k}}) \langle lm|n\rangle + (\varepsilon_{\boldsymbol{k}} - \varepsilon_n)\langle \boldsymbol{k}|n\rangle = 0 \end{aligned}\right\}. \tag{5.60}$$

If these equations are satisfied, we can make use of eq. (5.57) to transform eq. (5.59) into

$$H = \sum \varepsilon_n a_n{}^\dagger a_n. \tag{5.61}$$

Now, instead of solving eq. (5.60) directly, let us first define the Green's functions as follows:

$$G_{\alpha\beta}(\omega) \equiv \sum_n \langle\alpha|n\rangle (\omega - \varepsilon_n)^{-1} \langle n|\beta\rangle. \tag{5.62}$$

Here, α and β denote either $\boldsymbol{k}$ or lm, and ω may be complex in general.

In eq. (5.60), we write $\varepsilon_l - \varepsilon_n = (\varepsilon_l - \omega) - (\varepsilon_n - \omega)$ and so on, and divide the whole equation by $\varepsilon_n - \omega$. We then employ eq. (5.56) to obtain

$$(\omega - \varepsilon_l) G_{lml'm'} = \delta_{ll'}\delta_{mm'} + (4\pi)^{1/2} V_l \sum_{\boldsymbol{k}} Y_{lm}(\hat{\boldsymbol{k}}) G_{\boldsymbol{k}l'm'}, \tag{5.63}$$

$$(\omega - \varepsilon_{\boldsymbol{k}}) G_{\boldsymbol{k}l'm'} = (4\pi)^{1/2} \sum_{lm} V_l^* Y_{lm}^*(\hat{\boldsymbol{k}}) G_{lml'm'}, \tag{5.64}$$

$$(\omega - \varepsilon_k)G_{kk'} = \delta_{kk'} + (4\pi)^{1/2}\sum_{lm} V_l{}^* Y_{lm}{}^*(\hat{k})G_{lmk'}, \tag{5.65}$$

$$(\omega - \varepsilon_l)G_{lmk'} = (4\pi)^{1/2}V_l\sum_k Y_{lm}(\hat{k})G_{kk'}. \tag{5.66}$$

It is easy to solve these equations. We first obtain G_{klm} from eq. (5.64), and substitute this into eq. (5.63). We then use the orthogonality of Y_{lm}. This yields:

$$\left[\omega - \varepsilon_l - |V_l|^2\sum_k(\omega - \varepsilon_k)^{-1}\right]G_{lml'm'} = \delta_{ll'}\delta_{mm'}. \tag{5.67}$$

In the same way, we obtain $G_{kk'}$ from eq. (5.65) and substitute this into eq. (5.66). This yields:

$$\left[\omega - \varepsilon_l - |V_l|^2\sum_k(\omega - \varepsilon_k)^{-1}\right]G_{lmk} = \frac{(4\pi)^{1/2}V_l Y_{lm}(\hat{k})}{\omega - \varepsilon_k}. \tag{5.68}$$

It is more convenient to find the Green's functions, $G_{\alpha\beta}(\omega)$, than to solve for the coefficients $\langle\alpha|n\rangle$ directly, for the purpose of calculating various physical quantities. As an example, we define the density of states that has the component α as

$$\rho_\alpha(\varepsilon) = \sum_n \langle\alpha|n\rangle\delta(\varepsilon - \varepsilon_n)\langle n|\alpha\rangle. \tag{5.69}$$

This can also be expressed using $G_{\alpha\alpha}(\varepsilon + i\delta)$, where δ is positive infinitesimal. To do so, we make use of the relation:

$$\delta(\varepsilon - \varepsilon_n) = \frac{1}{\pi}\frac{\delta}{(\varepsilon - \varepsilon_n)^2 + \delta^2} = -\frac{1}{\pi}\mathrm{Im}\frac{1}{\varepsilon - \varepsilon_n + i\delta}$$

(Im means the operation of taking the imaginary part of the quantity to the right), to write

$$\rho_\alpha(\varepsilon) = -\frac{1}{\pi}\mathrm{Im}\, G_{\alpha\alpha}(\varepsilon + i\delta). \tag{5.70}$$

Now, in order to obtain $\rho_{lm}(\varepsilon)$ from eq. (5.67), we evaluate the sum appearing within it as follows:

$$\begin{aligned}\sum_k \frac{1}{\varepsilon + i\delta - \varepsilon_k} &= \int \rho(\varepsilon_k)\frac{\varepsilon - \varepsilon_k - i\delta}{(\varepsilon - \varepsilon_k)^2 + \delta^2}d\varepsilon_k \\ &= P\int \frac{\rho(\varepsilon_k)}{\varepsilon - \varepsilon_k}d\varepsilon_k - i\pi\rho(\varepsilon).\end{aligned}$$

Here, P stands for the operation of taking the principal value of the integral:

$$P\int \frac{\rho(\varepsilon')}{\varepsilon-\varepsilon'}d\varepsilon' = \lim_{\delta\to 0}\left[\int^{\varepsilon-\delta}\frac{\rho(\varepsilon')}{\varepsilon-\varepsilon'}d\varepsilon' + \int_{\varepsilon+\delta}\frac{\rho(\varepsilon')}{\varepsilon-\varepsilon'}d\varepsilon'\right].$$

This term is not very important. It merely shifts ε_l when its weak dependence on ε is ignored. The more important term is the imaginary part. When we omit the former, we obtain

$$G_{lmlm}(\varepsilon+i\delta) = (\varepsilon-\varepsilon_l+i\Delta_l)^{-1}, \tag{5.71}$$

where

$$\Delta_l = \pi|V_l|^2\rho. \tag{5.72}$$

We hence obtain

$$\rho_{lm}(\varepsilon) = \frac{1}{\pi}\frac{\Delta_l}{(\varepsilon-\varepsilon_l)^2+\Delta_l{}^2}. \tag{5.73}$$

We can easily confirm that eq. (5.72) coincides with eq. (5.50). Equation (5.73) represents the part of the density of states that has the character of atomic orbitals.

Next, we calculate the part of the density of states that has the character of plane waves $\rho_{\boldsymbol{k}}(\varepsilon)$. When we substitute eq. (5.68) into eq. (5.65) and use the relation

$$\sum_m |Y_{lm}(\theta,\varphi)|^2 = \frac{2l+1}{4\pi},$$

we obtain

$$G_{\boldsymbol{kk}}(\omega) = \frac{1}{\omega-\varepsilon_{\boldsymbol{k}}} + \frac{1}{(\omega-\varepsilon_{\boldsymbol{k}})^2}\sum_l \frac{(2l+1)V_l{}^2}{\omega-\varepsilon_l+i\Delta_l}. \tag{5.74}$$

The first term on the right-hand side (RHS) of eq. (5.74), of course, gives the unperturbed density of states:

$$\begin{aligned}\sum_{\boldsymbol{k}}\rho_{\boldsymbol{k}}(\varepsilon) &= -\frac{1}{\pi}\sum_{\boldsymbol{k}}\mathrm{Im}\frac{1}{\varepsilon-\varepsilon_{\boldsymbol{k}}+i\delta}\\ &= -\frac{1}{\pi}\mathrm{Im}\int\frac{\rho(\varepsilon_{\boldsymbol{k}})}{\varepsilon-\varepsilon_{\boldsymbol{k}}+i\delta}d\varepsilon_{\boldsymbol{k}}\\ &= \rho(\varepsilon).\end{aligned}$$

The second term in the RHS of eq. (5.74) has a second-order pole, and so its contribution is small. In fact, when we evaluate the integral

$$\int\frac{\rho(\varepsilon_{\boldsymbol{k}})}{(\varepsilon-\varepsilon_{\boldsymbol{k}}+i\delta)^2}d\varepsilon_{\boldsymbol{k}}$$

with, say, $\rho(\varepsilon) = (D/\pi)/[(\varepsilon - \varepsilon_F)^2 + D^2]$, we obtain $(\varepsilon - \varepsilon_F + iD)^{-2}$. Here, the quantity D is comparable to ε_F. When we assign this to the second term in the RHS of eq. (5.74), and take its imaginary part, we find that it is suppressed by about Δ_l/D compared with eq. (5.73). Omitting this contribution, the part of the DOS that has the character of plane waves is invariant even with the impurity. Hence, the total change of DOS is given by

$$\left.\begin{aligned} \Delta\rho(\varepsilon) &= \sum_{lm} \rho_{lm}(\varepsilon) = \sum_{l} (2l+1)\rho_l(\varepsilon) \\ \rho_l(\varepsilon) &\equiv \frac{1}{\pi} \frac{\Delta_l}{(\varepsilon - \varepsilon_l)^2 + {\Delta_l}^2} \end{aligned}\right\}. \tag{5.75}$$

This result is identical to eq. (5.51). As we mentioned near eq. (5.51), and also near eq. (5.30), the change of DOS caused by an impurity is mainly attributable to localized orbitals when there is strong localization (when $d\delta_l/d\varepsilon$ is large, or Δ_l/D is less than 1). Now eq. (5.75) also shows that the change of DOS is due only to lm and not to $\boldsymbol{k}$. The relation between the area of $\rho_l(\varepsilon)$ below ε and the value of $\rho_l(\varepsilon)$ at ε, which will be employed later, is described as follows. Defining $n_l(\varepsilon)$ as:

$$n_l(\varepsilon) = \int_{-\infty}^{\varepsilon} \rho_l(\varepsilon) d\varepsilon = \frac{1}{\pi}\left(\frac{\pi}{2} + \tan^{-1}\frac{\varepsilon - \varepsilon_l}{\Delta_l}\right),$$

we can easily show that

$$\pi\Delta_l\rho_l(\varepsilon) = \sin^2 \pi n_l(\varepsilon). \tag{5.75a}$$

The wavefunction that corresponds to a_n in eq. (5.55) is written as

$$\psi_n = \sum_{\boldsymbol{k}} \frac{1}{\sqrt{\Omega}} e^{i\mathbf{k}\cdot\mathbf{r}} \langle n|\boldsymbol{k}\rangle + \sum_{lm} u_{k_0}(r) Y_{lm}(\theta,\varphi)\langle n|lm\rangle,$$

making use of the coefficients employed there (see eqs. (2.51) and (2.53)). When we decompose this into spherical waves, the lm element is precisely eq. (5.45). When $r \to \infty$, as shown in eq. (5.17), we describe this using the phase shift δ_l, given by eq. (5.49):

$$\delta_l = \frac{\pi}{2} + \tan^{-1}\frac{\varepsilon - \varepsilon_l}{\Delta_l}.$$

This gives the phase shift of ψ_n. We have stated before that the electrical resistance caused by an impurity is given by the phase shift at the Fermi level, as shown in eq. (5.27). In particular, when a certain δ_l is large and the other $\delta_{l'}$ are negligible, we obtain

$$R = \frac{4\pi\hbar n_i}{ne^2 k_F}(2l+1)\sin^2\delta_l. \tag{5.75b}$$

Then, by using eq. (5.75a) and $\delta_l = \pi n_l$, we obtain

$$R = \frac{4\pi\hbar n_i}{ne^2 k_F}(2l+1)\pi\Delta_l \cdot \rho_l(\varepsilon_F), \tag{5.76}$$

where n_i is the concentration of the impurity. As we have discussed in §5.4 taking the case of Al alloys as an example, the electrical resistance is maximized when ε_F is equal to ε_l.

Since ψ_n has energy eigenvalue ε_n, the probability of its occupation is $f(\varepsilon_n - \mu)$. Here, f is defined by eq. (3.18), and μ represents the chemical potential. On the other hand, since $\alpha(=lm$ or $\boldsymbol{k})$ is not an eigenstate, the probability of the occupation of the state α is given by

$$\langle a_\alpha{}^\dagger a_\alpha \rangle = \sum_n \langle \alpha|n\rangle\langle n|\alpha\rangle f(\varepsilon_n - \mu) = \int \rho_\alpha(\varepsilon) f(\varepsilon - \mu) d\varepsilon, \tag{5.77}$$

where we employed eq. (5.69). When $\alpha = lm$, and at absolute zero temperature, we obtain (for $\varepsilon_l \gg \Delta_l$)

$$\langle a_{lm}{}^\dagger a_{lm} \rangle = \frac{\Delta_l}{\pi}\int_0^{\varepsilon_F} \frac{d\varepsilon}{(\varepsilon - \varepsilon_l)^2 + \Delta_l{}^2} = \frac{1}{\pi}\left(\frac{\pi}{2} + \tan^{-1}\frac{\varepsilon_F - \varepsilon_l}{\Delta_l}\right). \tag{5.78}$$

This is almost equal to unity when ε_l is sufficiently lower than ε_F (i.e., $\Delta_l \ll |\varepsilon_l - \varepsilon_F|$), and almost equal to zero when ε_l is sufficiently higher than ε_F (i.e., $\Delta_l \ll \varepsilon_l - \varepsilon_F$). It is 1/2 when $\varepsilon_l = \varepsilon_F$.

Let us now consider the case of $\varepsilon_l = \varepsilon_F$. When the DOS is given in the form of ρ_l in eq. (5.75), the increments of specific heat and susceptibility are given by

$$\Delta c = \frac{1}{k_B T^2}\int \varepsilon^2 2\rho_l(\varepsilon) f(\varepsilon)[1 - f(\varepsilon)]\, d\varepsilon, \tag{5.79}$$

and

$$\Delta\chi = \frac{\mu_B{}^2}{k_B T}\int 2\rho_l(\varepsilon) f(\varepsilon)[1 - f(\varepsilon)]\, d\varepsilon, \tag{5.80}$$

respectively, according to eqs. (3.27) and (3.31). Hereafter, in order to simplify our discussions, we adopt $l = 0$. If $k_B T$ is sufficiently smaller than Δ_0, we can consider ρ_l to be constant and take it out of these integrals. The result is then equivalent to that for the degenerate electronic system, in which $\Delta\chi$ is constant and Δc is proportional to T.

When k_BT increases, $\Delta\chi$ decreases; Δc also decreases after it reaches its maximum at a certain value of T. When k_BT is sufficiently greater than Δ_0, $\Delta\chi \sim {\mu_B}^2/2k_BT$ and $\Delta c \sim 0$. These are exactly the same as the results that would be obtained by putting $\Delta_0 = 0$, that is, $\rho_l(\varepsilon) = \delta(\varepsilon - \varepsilon_F)$, in eqs. (5.79) and (5.80). When $\Delta_0 = 0$, the atomic orbital and the plane waves are completely independent. As the energy level of the isolated atomic orbital is precisely at the Fermi level ($\varepsilon_l = \varepsilon_F$) here, the spin $\uparrow$ and $\downarrow$ electrons each occupy the level for exactly half of the time. The Curie susceptibility hence arises.

Δ_0 is a measure of how fast the atomic orbital and the plane waves mix with one another. When the timescale is much shorter than that indicated by Δ_0, we are able to observe the isolated spin $\uparrow$ or $\downarrow$. Hence Curie susceptibility is obtained when $\Delta_0 \ll k_BT$. In contrast, when $\Delta_0 \gg k_BT$, we observe the result of their mixing, and so it appears as if the spin of the atomic orbital has been lost. Hence weak Pauli-like susceptibility is obtained.

5.7 The Coulomb interaction: UHF

Up to the previous section, we have discussed the case in which iron group transition metal elements are present as an impurity in a noble metal like Cu. We first saw that the 3d orbital of the transition metal is almost the same as that of an isolated atom, as the orbital is localized due to centrifugal potential; we then saw that a small portion of the orbital tunnels out of the potential and the orbital mixes with the plane waves, leading to the spectral width Δ.

In the case of transition metal elements other than the iron group, Δ becomes greater, and thus this picture becomes a poor approximation. However, in the case of the rare-earth metal elements, the 4f orbital is strongly localized due to its angular momentum $l = 3$, and Δ is correspondingly small. We shall hereafter call the orbital of an isolated atom the 'localized orbital', and call the plane waves the 'conduction electrons'.

All discussions so far have been limited to finding one-electron wavefunctions. Interelectronic Coulomb interaction has been considered only in the context of the determination of the self-consistent potential. This is sufficient for fixing the value of ε_l, for instance. However, when the effect of the Coulomb interaction becomes greater, there arise a number of significant consequences, which cannot be treated using this one-body approximation.

For a given set of bases composed of the localized orbital and the conduction-band wavefunctions, let us now consider the many-body problem that includes the Coulomb interaction. Regarding the Coulomb interaction, the most important effects are the effect inside the localized orbital and that between the localized orbital and the conduction band. Out of these, the former is particularly significant, and so let us only consider that here.

For the sake of simplicity, let us adopt $l = 0$. That is, we have a single localized orbital, and the energy is increased by U when both spin ↑ and ↓ electrons occupy the orbital. In eq. (5.54), we take the case $l = 0$ and bring spin into consideration, and add the Coulomb repulsion energy U. This yields

$$H = \sum_{k\sigma} \varepsilon_k a_{k\sigma}{}^\dagger a_{k\sigma} + \varepsilon_0 \sum_\sigma a_{0\sigma}{}^\dagger a_{0\sigma} + U a_{0\uparrow}{}^\dagger a_{0\uparrow} a_{0\downarrow}{}^\dagger a_{0\downarrow}$$
$$+ V_0 \sum_{k\sigma} \left(a_{k\sigma}{}^\dagger a_{0\sigma} + a_{0\sigma}{}^\dagger a_{k\sigma} \right), \tag{5.81}$$

where σ is ↑ or ↓.

Hereafter, we take the Fermi level to be zero on the energy scale. Let us now say that ε_0 is sufficiently lower than the Fermi level and $|\varepsilon_0| \gg \Delta_0$ (see Fig. 5.9). Then, for instance, a spin ↑ electron fills the localized orbital, and $\langle a_{0\uparrow}{}^\dagger a_{0\uparrow} \rangle$ is approximately equal to 1. If we now set $a_{0\uparrow}{}^\dagger a_{0\uparrow} = 1$ in the U term in eq. (5.81), the role assumed by ε_0 so far will be taken over by $\varepsilon_0 + U$ for the spin ↓ electron. When U is sufficiently large and this level is sufficiently higher than the Fermi level ($U + \varepsilon_0 \gg \Delta_0$), $\langle a_{0\downarrow}{}^\dagger a_{0\downarrow} \rangle$ almost vanishes.

The same argument holds when ↑ and ↓ are swapped. Hence, in the case shown in Fig. 5.9, the localized orbital is occupied by an electron with either

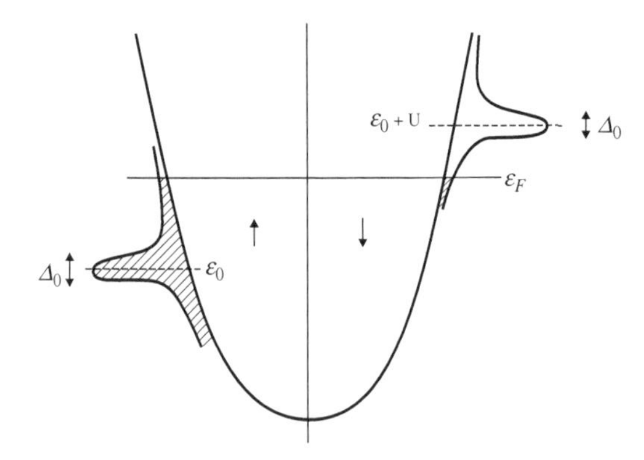

Figure 5.9 Emergence of localized spin. When ε_0 is sufficiently lower than ε_F, and the level is filled by a spin ↑ electron, the energy level of the spin ↓ electron is raised by U. If this raised energy level is sufficiently above ε_F, the number of spin ↓ electrons will be almost nil. Hence localized spin emerges.

spin $\uparrow$ or $\downarrow$, and we may say that a localized spin has been generated. When both ε_0 and $\varepsilon_0 + U$ are below the Fermi level, both spin $\uparrow$ and $\downarrow$ electrons occupy the localized orbital. When both ε_0 and $\varepsilon_0 + U$ are above the Fermi level, the localized orbital is unoccupied by either of these electrons. In these two cases, localized spin does not emerge.

Strictly speaking, however, whether the localized spin is present or not depends on the timescale of observation. As discussed at the end of the last section, even if $\varepsilon_0 = 0$ and $U = 0$, Curie susceptibility is observed when the temperature is sufficiently high. Here, Δ_0 is a measure of the timescale of observation. Even in the case shown in Fig. 5.9, the spin of the electron in the localized orbital is not always $\uparrow$ but is fluctuating. For instance, a spin $\downarrow$ electron may leap into the localized orbital, which has been filled by a spin $\uparrow$ electron, from the conduction band. This may be followed by a spin $\uparrow$ electron going back to the conduction band, resulting in an overall spin exchange. Such spin fluctuation occurs much more slowly than the speed corresponding to Δ_0. However, if the observation timescale is sufficiently greater than that of spin fluctuation (i.e., at sufficiently low temperatures), the localized spin appears to have vanished. This is a central issue concerning the localized spin in metals, but here let us pursue in more detail the picture obtained with a faster observation timescale, where the spin fluctuation appears to have stopped.

For this purpose, let us solve eq. (5.81) using the unrestricted Hartree–Fock (UHF) scheme. This is a generalized mean-field approximation, in which we assume that the mean field for the spin $\uparrow$ electron and that for the spin $\downarrow$ electron are different in general. Denoting the number of localized electrons with spin σ as n_σ, the U term can be regarded as $n_\sigma U a_{0-\sigma}{}^\dagger a_{0-\sigma}$ for the electron with spin $-\sigma$. For treating spin $-\sigma$ electrons, we may hence apply the results of the last section directly by replacing ε_0 with $\varepsilon_0 + n_\sigma U$. Adopting $\varepsilon_F = 0$, eq. (5.78) is reduced to

$$n_{-\sigma} = \langle a_{0-\sigma}{}^\dagger a_{0-\sigma} \rangle = \frac{1}{\pi}\left(\frac{\pi}{2} - \tan^{-1}\frac{\varepsilon_0 + n_\sigma U}{\Delta_0}\right),$$

for $T = 0$.

From the two equations for $\sigma = \uparrow$ and $\downarrow$, we obtain $n_\uparrow$ and $n_\downarrow$ respectively. $n_\uparrow = n_\downarrow$ is always a solution, and in addition, for some values of the parameters, we may also have $n_\uparrow \neq n_\downarrow$. The corresponding parameter region, for $\varepsilon_0 < 0$ and $\varepsilon_0 + U > 0$, is indicated by the hatched area in Fig. 5.10.

The same result is also obtained through the following consideration. Let us assume that the $n_\uparrow = n_\downarrow$ solution is determined by the self-consistent condition:

$$n = \frac{1}{\pi}\left(\frac{\pi}{2} - \tan^{-1}\frac{\varepsilon_0 + nU}{\Delta_0}\right),$$

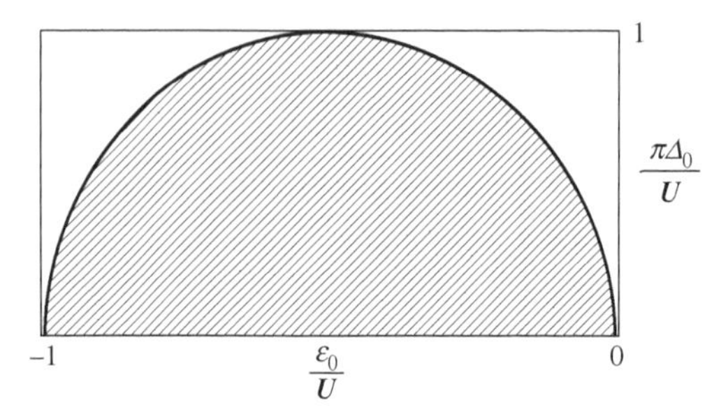

Figure 5.10 The parameter region (hatched area) in which localized spin emerges.

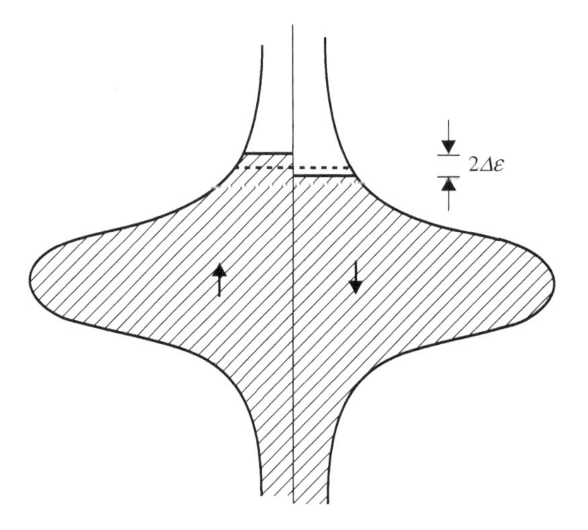

Figure 5.11 Moving Δn spin $\downarrow$ electrons at the Fermi level to the spin $\uparrow$ state causes energy to rise by $\Delta n \cdot \Delta\varepsilon$. Here, we have the relation $\Delta n = \rho_0 \cdot \Delta\varepsilon$.

and that the density of states for both spin orientations is determined as

$$\rho_0 = \frac{\Delta_0}{\pi} \cdot \frac{1}{(\varepsilon - \varepsilon_0 - nU)^2 + {\Delta_0}^2}.$$

(see Fig. 5.11). Now we increase $n_\uparrow$ by a small amount, $n_\uparrow = n + \Delta n$, and reduce $n_\downarrow$ by $n_\downarrow = n - \Delta n$. In this case, the Coulomb energy goes down to $U\left(n^2 - (\Delta n)^2\right)$, but, as shown in Fig. 5.11, Δn electrons need to be raised to higher energy by amount $\Delta\varepsilon$. Therefore there is $\Delta n \cdot \Delta\varepsilon$ increment in energy. Since $\Delta n = \rho_0(\varepsilon_F) \cdot \Delta\varepsilon$, we have $\Delta n \cdot \Delta\varepsilon = (\Delta n)^2/\rho_0(\varepsilon_F)$. We thus see that the energy is lower for $n_\uparrow \neq n_\downarrow$ than for $n_\uparrow = n_\downarrow$ when $U > 1/\rho_0(\varepsilon_F)$. The threshold for this is given by

$$\frac{\Delta_0}{\pi}\frac{U}{(\varepsilon_0+nU)^2+{\Delta_0}^2}=1,$$

where we made use of $\varepsilon_F=0$. By substituting $x=-\varepsilon_0/U$ and $y=\pi\Delta_0/U$, we then obtain

$$y=y^2+(\pi n-\pi x)^2,$$

$$y\cot\pi n=\pi n-\pi x.$$

When we eliminate n from these equations, we obtain the boundary of the hatched area in Fig. 5.10. In particular, the case of $-\varepsilon_0=U/2$ is called the symmetric state, where ε_0 and ε_0+U are respectively above and below the Fermi level by the same amount. In this case, according to Fig. 5.10, an $n_\uparrow\neq n_\downarrow$ solution emerges when U is greater than $\pi\Delta_0$. In particular, when $-\varepsilon_0\gg\pi\Delta_0$ and $\varepsilon_0+U\gg\pi\Delta_0$, we obtain $n_\uparrow\sim1$ and $n_\downarrow\sim0$, or vice versa (see Fig. 5.9).

After n_σ is determined in this manner, the increment of the spin σ density of states is given by:

$$\rho_\sigma(\varepsilon)=\frac{\Delta_0}{\pi}\cdot\frac{1}{(\varepsilon-\varepsilon_\sigma)^2+{\Delta_0}^2},$$

where

$$\varepsilon_\sigma=\varepsilon_0+n_{-\sigma}U.$$

Let us now consider the electrical resistance due to the impurity. A spin $\uparrow$ conduction electron can be scattered by the impurity atom with either spin $\uparrow$ or $\downarrow$. Therefore, according to eq. (5.76), we have

$$R\propto\rho_\uparrow(\varepsilon_F)+\rho_\downarrow(\varepsilon_F).$$

Thus the electrical resistance is maximal when either $\varepsilon_\uparrow$ or $\varepsilon_\downarrow$ is equal to ε_F.

In §5.4, we discussed the residual resistance of Al alloys in which a 3d transition metal is introduced as an impurity. Al is trivalent and, furthermore, its $\rho(\varepsilon_F)$ is large. Hence Δ_2 is large, and we may consider $\pi\Delta_2/U$ to be greater than 1. As a result, in view of Fig. 5.10, localized spin does not emerge. Therefore $\varepsilon_\uparrow=\varepsilon_\downarrow$. We discussed in §5.4 that these levels are above ε_F for Ti, approximately equal to ε_F for Cr, and much lower than ε_F for Ni and Cu.

What, then, would be the situation with Cu alloys? In this case, Δ_2 is not so large, and we may consider $\pi\Delta_2/U$ to be less than 1. In the symmetric state, which is the simplest, the situation is as described by Fig. 5.9, and this seems

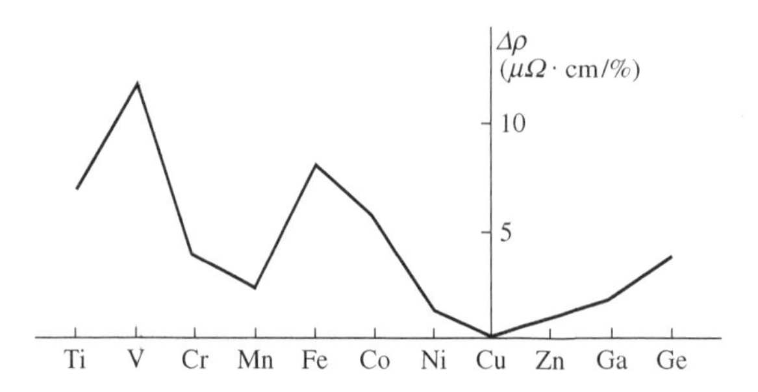

Figure 5.12 Residual resistance of Cu to which 1 atomic percent of iron group elements has been added. From J. Friedel, *Nuovo Cim., Suppl.* **2** (1958) 287, reproduced with kind permission of Società Italiana di Fisica.

to be realized when the impurity is Mn. As both $\rho_\uparrow(\varepsilon_F)$ and $\rho_\downarrow(\varepsilon_F)$ are small in this particular case, the residual resistance is also small.

For Cr, the density of states depicted in Fig. 5.9 is raised as a whole with reference to ε_F, whereas for Fe, the density of states is lowered as a whole. In these cases, either of the $\rho_\sigma(\varepsilon_F)$ is increased, and so is the residual resistance. Proceeding further, all the energy levels move above ε_F for Ti, and below ε_F for Ni. The localized spin then vanishes and the residual resistance is small, as shown in Fig 5.12.

The above treatment gives us a simple picture to explain the spin acquired by a localized orbital, but is incomplete in the sense that it gives different answers for spin ↑ and ↓. Let us now adopt the symmetric state for the sake of simplicity, and consider increasing U starting from 0 while maintaining the symmetry.

We start from $U = 0$ and $\varepsilon_0 = 0$. Since Δ_0 is usually greater than 0.1 eV, we have $k_B T \ll \Delta_0$ for an ordinary range of temperatures. Hence there is no localized spin, and the Pauli paramagnetic susceptibility, ${\mu_B}^2 \rho_l(0)$, appears. As U is increased, the Pauli paramagnetism is enhanced. This is because when $n_\uparrow$ is raised and $n_\downarrow$ lowered by the external magnetic field, the spin ↓ one-electron level, $\varepsilon_\downarrow = \varepsilon_0 + n_\uparrow U$, rises. As a result, $n_\downarrow$ goes down further, and the spin ↑ one-electron level, $\varepsilon_\uparrow = \varepsilon_0 + n_\downarrow U$, falls further. Hence $n_\uparrow$ goes up.

A measure of the rise of Pauli paramagnetism is $U/\pi\Delta_0$. When this is less than 1, the Pauli paramagnetic susceptibility increases little and there is not much temperature dependence. When this exceeds 1, the Pauli paramagnetic susceptibility increases by more than 100%. When this value is much larger than 1, Curie paramagnetism should be observed due to the emergence of localized spin. Since Curie paramagnetic susceptibility is given by ${\mu_B}^2/k_B T$, the susceptibility is expected to approach this value as U is increased. This situation is shown in Fig. 5.13.

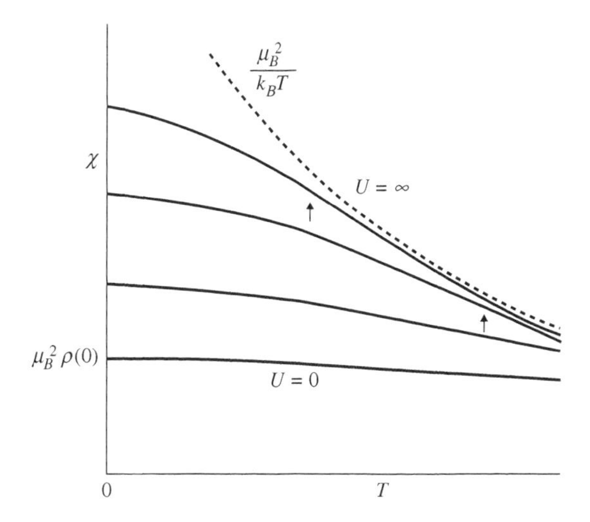

Figure 5.13 Temperature dependence of susceptibility for various values of U. For $U = 0$, χ tends to a half of the Curie susceptibility $\mu_B{}^2/2k_BT$ when $k_BT > \Delta_0$, but this occurs far away from the temperature range of this figure. χ goes up with increasing U, and it tends to the Curie susceptibility $\mu_B{}^2/k_BT$ as U tends to infinity. For finite U, χ departs from $\mu_B{}^2/k_BT$ below a certain temperature (indicated by arrows). This temperature gives a measure of the frequency of spin fluctuation.

Let us repeat the above discussion for the case of slightly smaller values of k_BT. In this case, the susceptibility rises with the increase of U. On the other hand, for fixed U and decreasing temperature, the susceptibility goes up by the Curie law at first, and departs from it at a certain point and reaches saturation. k_BT at this point gives a measure of the frequency of spin fluctuation.

Let us adapt the discussion to the context of electrical resistance (see Fig. 5.14). Extending eq. (5.76) to finite temperatures, we obtain

$$\frac{1}{R} = \frac{ne^2k_F}{4\pi\hbar n_i}\int\frac{1}{\pi\Delta_0\cdot\rho_0(\varepsilon)}\left(-\frac{df}{d\varepsilon}\right)d\varepsilon. \tag{5.82}$$

As before, we consider the symmetric state where $U = 0$ and $\varepsilon_0 = 0$. When $k_BT \ll \Delta_0$, we can set $\rho_0(\varepsilon)$ equal to $\rho_0(0)$, so that the integral yields 1 by eq. (5.75). This is of course the same as the result of taking $\delta_0 = \pi/2$ in eq. (5.27) when $l = 0$. This is the largest possible value of the electrical resistance when $l = 0$.

As the temperature increases, the electrical resistance starts to fall due to the form of $\rho_0(\varepsilon)$ near $k_BT \sim \Delta_0$. We may then ask what will happen if U goes up. Since the temperature around which the electrical resistivity starts to fall

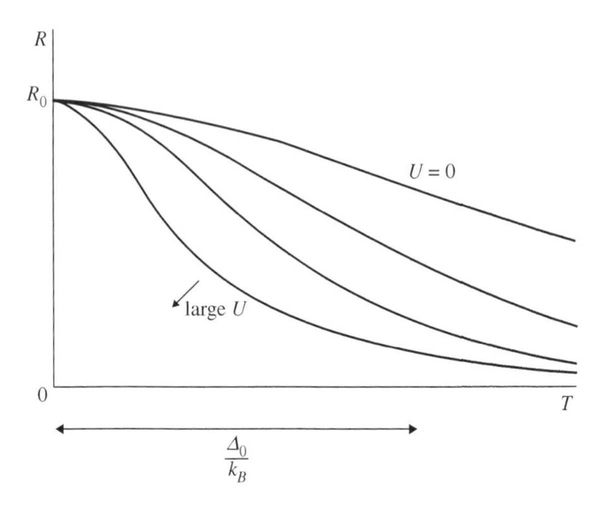

Figure 5.14 Temperature dependence of electrical resistivity for various values of U. When $U = 0$, a measure of the decrease of electrical resistivity is given by the temperature Δ_0/k_B. As U increases, this characteristic temperature decreases. The electrical resistivity remains small down to low temperatures, starts to rise below this characteristic temperature, and approaches R_0 at $T = 0$.

decreases, the electrical resistivity itself decreases for fixed T and increasing U. When U is increased such that the condition shown in Fig. 5.9 is realized, both of the $\rho_\sigma(\varepsilon_F)$ go down, and so does the electrical resistivity. However, even in this case, when T is lowered further, we cannot consider the spin fluctuation to be absent as is the case in Fig. 5.9, and the electrical resistivity goes up. We thus expect the behavior shown in Fig. 5.14. An important objective for us is to obtain a reliable theory about such behavior of electrical resistivity, susceptibility and the specific heat.

5.8 Expansion in powers of U

This section is mainly based on Yamada (1975) and Yosida and Yamada (1975).

When we compare eq. (5.81) against eq. (5.54), we find that the former includes the two-electron interaction U, which is what makes the problem essentially more difficult.

In order to investigate the behavior with respect to increasing U, let us consider the perturbative expansion in powers of U. Putting $U = 0$ in eq. (5.81) is equivalent to including spin and adopting $l = 0$ in eq. (5.54), and so the solution is as described by eqs. (5.67), (5.68), and so on. We may then treat this case as the unperturbed system, and consider the U term as a perturbation, by which we may make a power expansion.

In general, even in systems with interelectronic interaction, the Green's function is defined in a different way than in eq. (5.62). This Green's function can be derived as a perturbative power expansion in U, and the first term is, of course, the same as eq. (5.62). Let us omit the details of this derivation, and present the results.

For the symmetric state in which $\varepsilon_0 = -U/2$, we can obtain the Green's function and calculate the density of states of the localized orbital from it using eq. (5.70). This yields

$$\rho_0(\varepsilon) = \frac{1}{\pi\Delta_0}\left\{1 - \alpha_I\left[\left(\frac{\varepsilon}{\Delta_0}\right)^2 + \left(\frac{\pi k_B T}{\Delta_0}\right)^2\right] - (\alpha_R + 1)^2\left(\frac{\varepsilon}{\Delta_0}\right)^2\right\}, \tag{5.83}$$

when both ε and $k_B T$ are less than ε_F. Here, α_I and α_R are given by:

$$\alpha_R = \left(3 - \frac{\pi^2}{4}\right)\left(\frac{U}{\pi\Delta_0}\right)^2 + 0.0553\left(\frac{U}{\pi\Delta_0}\right)^4 + \cdots,$$

$$\alpha_I = 2\left\{\left(\frac{U}{\pi\Delta_0}\right)^2 + 6\left(5 - \frac{\pi^2}{2}\right)\left(\frac{U}{\pi\Delta_0}\right)^4 + \cdots\right\}.$$

As for the susceptibility at $T = 0$ due to the impurity, and γ, where γT represents the specific heat, these are given by:

$$\chi = \frac{2\mu_B{}^2}{\pi\Delta_0}(\tilde{\chi}_{\text{even}} + \tilde{\chi}_{\text{odd}}),$$

$$\gamma = \frac{2\pi^2}{3}\frac{\tilde{\gamma} k_B{}^2}{\pi\Delta_0},$$

where

$$\tilde{\chi}_{\text{even}} = \tilde{\gamma} = 1 + \alpha_R, \tag{5.84}$$

$$\tilde{\chi}_{\text{odd}} = \frac{U}{\pi\Delta_0} + 3\left(5 - \frac{\pi^2}{2}\right)\left(\frac{U}{\pi\Delta_0}\right)^3 + \cdots. \tag{5.85}$$

Although these are expansions in powers of $U/\pi\Delta_0$, they are valid up to large values of $U/\pi\Delta_0$, because the higher-order coefficients are small. Thus we see that χ and γ increase rapidly with increasing U and, at the same time, the width of ρ_0 falls rapidly. Substituting eq. (5.83) into eq. (5.82), we obtain:

$$R = R_0 \left\{ 1 - \frac{\pi^2}{3} \left(\frac{k_B T}{\Delta_0} \right)^2 [4\alpha_I + (\alpha_R + 1)^2] \right\}, \tag{5.86}$$

where R_0 is given by eq. (5.82), with the integral normalized to unity. Hence R falls rapidly with temperature. These results show that the frequency of spin fluctuation, which was Δ_0 to start with, falls rapidly with increasing U. As discussed earlier, localized spin emerges if $k_B T$ is greater than this frequency.

Yamada (1975) and Yosida and Yamada (1975) investigated the general terms of the perturbative expansion in powers of U and found that eq. (5.84) holds at any order of perturbation. We thus see the following. Since perturbative expansion holds regardless of the sign of U, it also extends to the region of large and negative U. Then the localized level is either doubly occupied or unoccupied (note that $\varepsilon_0 = -U/2$), and so the susceptibility must be zero. Hence,

$$\lim_{U \to \infty} \chi(-U) \propto \lim_{U \to \infty} [\tilde{\chi}_{\text{even}}(U) - \tilde{\chi}_{\text{odd}}(U)] = 0.$$

This shows that when U is positive and large, $\tilde{\chi}_{\text{even}}(U)$ and $\tilde{\chi}_{\text{odd}}(U)$ become equal. Defining r, the Wilson ratio, as

$$r = \frac{\chi T}{c} \frac{\pi^2}{3} \frac{k_B{}^2}{\mu_B{}^2}, \tag{5.87}$$

we see that $r = (\tilde{\chi}_{\text{even}} + \tilde{\chi}_{\text{odd}})/\tilde{\gamma}$. This is equal to 1 when $U = 0$, and the limiting value for large U is

$$r = 2. \tag{5.88}$$

r is an important parameter related to localized spin.

5.9 s–d interaction

This section is mainly based on Kondo (1969).

Let us now ask, how should χ and γ be expressed for larger values of U? And how should $\rho_0(\varepsilon)$ behave for large values of ε? The answer will not simply be a Lorentzian function with a narrower width, as localized spin causes singular behavior. In order to answer these questions, we adopt a different approach from the perturbative expansion in powers of U.

In eq. (5.81), let us consider the case in which $-\varepsilon_0 \gg \pi\Delta_0$ and $\varepsilon_0 + U \gg \pi\Delta_0$. We consider the term that includes U as the unperturbed system, and the term that includes V_0 is considered as a perturbation.

In the ground state of the unperturbed system, the Fermi sphere is occupied by the conduction electrons and the localized level is occupied by an electron of spin $\uparrow$ or $\downarrow$, and so the ground state is doubly degenerate. The low-lying excitations are due to the excitations of the Fermi sphere, and the number of electrons occupying the localized level remains 1. Changing this number to 0 or 2 corresponds to high-energy excitations. Operating twice by the V_0 term on a low-energy excited state causes transition to another low-energy excited state. We may thus obtain the effective Hamiltonian, which is only effective among the low-energy excited states, in the following way.

The matrix element for the transition from initial state i to final state f via the intermediate state m is given, in general, by

$$\sum_m \frac{H'_{fm}H'_{mi}}{E_i - E_m}, \tag{5.89}$$

according to second-order perturbation theory. Here, H'_{fm} and H'_{mi} are the matrix elements of the perturbative Hamiltonian, and $E_m - E_i$ is equal to the excitation energy from i to m.

Let us say that we have a localized spin $\uparrow$, and that a $\boldsymbol{k}\downarrow$ conduction electron is then transferred to the localized orbital by the V_0 term. The energy of this excitation is $\varepsilon_0 + U - \varepsilon_{\boldsymbol{k}}$, and its matrix element is V_0. Next, the spin $\downarrow$ electron that has been transferred to the localized orbital is subsequently transferred back into a $\boldsymbol{k}'\downarrow$ conduction-electron state by the V_0 term. The matrix element of the overall, second-order, process is then given by $V_0{}^2/(\varepsilon_{\boldsymbol{k}} - \varepsilon_0 - U)$. This is nothing other than the scattering of an electron from $\boldsymbol{k}\downarrow$ to $\boldsymbol{k}'\downarrow$. Since all values of momenta are allowed for both $\boldsymbol{k}$ and $\boldsymbol{k}'$, the effective Hamiltonian reads

$$\sum_{\boldsymbol{k}\boldsymbol{k}'} \left(\frac{V_0{}^2}{\varepsilon_{\boldsymbol{k}} - \varepsilon_0 - U} \right) a_{\boldsymbol{k}'\downarrow}{}^{\dagger} a_{\boldsymbol{k}\downarrow} \left(\frac{1}{2} + S_z \right).$$

Here, S_z is the z-component of the operator that represents localized spin, and the factor at the end is due to the fact that we have a localized spin $\uparrow$. When we have a localized spin $\downarrow$, the conduction electron is scattered from $\boldsymbol{k}\uparrow$ to $\boldsymbol{k}'\uparrow$. Correspondingly, $\downarrow$ should be replaced by $\uparrow$, and the factor at the end should be replaced by $\frac{1}{2} - S_z$,

It is possible, furthermore, that the localized spin is flipped before and after the scattering. For instance, in the process

$$(\boldsymbol{k}\uparrow, \boldsymbol{0}\downarrow) \longrightarrow (\boldsymbol{0}\uparrow, \boldsymbol{0}\downarrow) \longrightarrow (\boldsymbol{0}\uparrow, \boldsymbol{k}'\downarrow),$$

the conduction electron $\boldsymbol{k}\uparrow$ is scattered into $\boldsymbol{k}'\downarrow$ and, at the same time, the localized spin flips from $\downarrow$ to $\uparrow$. Corresponding to this process, the effective Hamiltonian term

$$-\sum_{kk'}\left(\frac{V_0{}^2}{\varepsilon_k-\varepsilon_0-U}\right)a_{k'\downarrow}{}^{\dagger}a_{k\uparrow}S_+$$

arises. In the other process

$$(\boldsymbol{k}\uparrow,\boldsymbol{0}\downarrow)\longrightarrow(\boldsymbol{k}\uparrow,\boldsymbol{k}'\downarrow)\longrightarrow(\boldsymbol{0}\uparrow,\boldsymbol{k}'\downarrow),$$

both the initial and final states are identical to those in the last process, but the intermediate state is different. In other words, there are two intermediate states contributing to the scattering in which the spin orientation is flipped. When we consider all such possible (low-energy excited state $\longrightarrow$ low-energy excited state) scattering processes, we see that they are represented by the following effective Hamiltonian:

$$\begin{aligned}H_{\text{sd}}=&-J\sum_{kk'}\left[S_z(a_{k'\uparrow}{}^{\dagger}a_{k\uparrow}-a_{k'\downarrow}{}^{\dagger}a_{k\downarrow})+S_+a_{k'\downarrow}{}^{\dagger}a_{k\uparrow}+S_-a_{k'\uparrow}{}^{\dagger}a_{k\downarrow}\right]\\&+V\sum_{kk'\sigma}a_{k'\sigma}{}^{\dagger}a_{k\sigma},\end{aligned}\tag{5.90}$$

where

$$\left.\begin{aligned}J&=V_0{}^2\left(\frac{1}{\varepsilon_0}-\frac{1}{\varepsilon_0+U}\right)\\V&=-\frac{V_0{}^2}{2}\left(\frac{1}{\varepsilon_0}+\frac{1}{\varepsilon_0+U}\right)\end{aligned}\right\}.\tag{5.91}$$

Here, we have put ε_k equal to zero, since $\boldsymbol{k}$ is close to the Fermi surface.

The term that is proportional to J takes the form of the exchange interaction between a localized spin and conduction electrons, and is called the s–d interaction. J in eq. (5.91) is negative. Although $S=1/2$ in this derivation, the generalized s–d interaction, where S is greater than $1/2$ in eq. (5.90), is also an object of study.

Out of the two-body interaction terms that have been omitted previously and not included in eq. (5.81), there is a term that takes the form of the J term in eq. (5.90). This term is called the direct exchange interaction between a localized orbital and the conduction band. If we include this, the total J becomes the sum of eq. (5.91) and the positive contribution due to this term. The direct exchange interaction is not large in transition metals, but is thought to be significant in rare-earth metals. The V term in eq. (5.90) represents the part of the scattering that is independent of spin, and is zero in the symmetric state.

Let us now consider how the specific heat, susceptibility and electrical resistance are modified by the s–d interaction of eq. (5.90). The total Hamiltonian is given by

$$H = H_0 + H_{sd}, \tag{5.92}$$

where

$$H_0 = \sum_{k\sigma} \varepsilon_k a_{k\sigma}{}^{\dagger} a_{k\sigma}. \tag{5.93}$$

We then consider H_{sd} as the perturbation, in terms of which we carry out perturbative expansion. We omit the V term, and consider the expansion in J.

We first calculate the scattering probability of electrons. The $\boldsymbol{k}\uparrow$ electron can be scattered in one out of two ways, namely, the case in which it is transferred to $\boldsymbol{k}'\uparrow$, and the case in which it is transferred to $\boldsymbol{k}'\downarrow$.

The matrix element for the former case is

$$H(\boldsymbol{k}\uparrow \longrightarrow \boldsymbol{k}'\uparrow) = -JM,$$

according to eq. (5.90). Here, we have set $S_z = M$. The probability of the transition from $\boldsymbol{k}\uparrow$ to arbitrary $\boldsymbol{k}'\uparrow$ in unit time is given by

$$\begin{aligned} &\frac{2\pi}{\hbar}\sum_{k'} \left|H(\boldsymbol{k}\uparrow \longrightarrow \boldsymbol{k}'\uparrow)\right|^2 \delta(\varepsilon_k - \varepsilon_{k'}) \\ &\quad = \frac{2\pi}{\hbar} J^2 M^2 \sum_{k'} \delta(\varepsilon_k - \varepsilon_{k'}) = \frac{2\pi}{\hbar} J^2 M^2 \rho(\varepsilon_k). \end{aligned}$$

Here, ρ is the density of states per spin in the conduction band.

Next, the matrix element for the latter case is

$$-J\sqrt{S(S+1) - M(M+1)},$$

and the transition probability in unit time is given by

$$(2\pi/\hbar) J^2 \left[S(S+1) - M(M+1)\right] \rho.$$

When we sum the two contributions and take the average over them in the region $M = S$ to $M = -S$, we obtain

$$W(\varepsilon) = \frac{2\pi}{\hbar} J^2 S(S+1) \rho(\varepsilon). \tag{5.94}$$

Here, we have put $\varepsilon_{\boldsymbol{k}} = \varepsilon$. We obtain the same result for the $\boldsymbol{k} \downarrow$ electron.

We then calculate the next-order term in the J expansion. For the transition $\boldsymbol{k} \uparrow \longrightarrow \boldsymbol{k}' \uparrow$, we can consider the transition via intermediate states, in the form $\boldsymbol{k} \uparrow \longrightarrow \boldsymbol{k}'' \uparrow \longrightarrow \boldsymbol{k}' \uparrow$. In order to obtain the matrix element, we make use of eq. (5.89) again. During this transition, according to the Pauli principle, the $\boldsymbol{k}''$ state should not be occupied by electrons. The probability for $\boldsymbol{k}''$ to be unoccupied is $1 - f(\varepsilon_{\boldsymbol{k}''})$. Hereafter we shall use the abbreviations $f(\varepsilon_{\boldsymbol{k}''}) = f''$ etc.

Using eq. (5.89), we therefore obtain

$$(-J)^2 M^2 \sum_{\boldsymbol{k}''} \frac{1 - f''}{\varepsilon_{\boldsymbol{k}} - \varepsilon_{\boldsymbol{k}''}}$$

as the matrix element for the transition $\boldsymbol{k} \uparrow \longrightarrow \boldsymbol{k}' \uparrow$. Another possibility for the intermediate state is the following. A $\boldsymbol{k}'' \uparrow$ electron is first scattered into $\boldsymbol{k}' \uparrow$, and then $\boldsymbol{k} \uparrow$ is scattered into $\boldsymbol{k}'' \uparrow$. This also contributes to $\boldsymbol{k} \uparrow \longrightarrow \boldsymbol{k}' \uparrow$. However, in this case, writing the initial state as $|\boldsymbol{k} \uparrow \boldsymbol{k}'' \uparrow\rangle$, the final state is represented by $|\boldsymbol{k}'' \uparrow \boldsymbol{k}' \uparrow\rangle$, which is equal to $-|\boldsymbol{k}' \uparrow \boldsymbol{k}'' \uparrow\rangle$ by the antisymmetry of the wavefunction. Taking into account this fact and the condition that $\boldsymbol{k}''$ needs to be occupied by electrons, we obtain

$$-(-J)^2 M^2 \sum_{\boldsymbol{k}''} \frac{f''}{\varepsilon_{\boldsymbol{k}''} - \varepsilon_{\boldsymbol{k}'}}$$

as the matrix element. Given that energy is conserved in the scattering, we have $\varepsilon_{\boldsymbol{k}} = \varepsilon_{\boldsymbol{k}'}$. Hence, adding the two matrix elements, we obtain

$$(-J)^2 M^2 \sum_{\boldsymbol{k}''} \frac{1}{\varepsilon_{\boldsymbol{k}} - \varepsilon_{\boldsymbol{k}''}}.$$

Thus the dependence on f'' has vanished, and the result is identical with that which is obtained without considering the Pauli principle in the intermediate states.

In cases such as the scattering due to the V term in eq. (5.90), we are dealing with a one-body problem, and we do not need to take into consideration the contributions of the other electrons. The result obtained in our present case is analogous to that case.

The situation changes, however, when we consider spin-flipped intermediate states. The transition $\boldsymbol{k} \uparrow \longrightarrow \boldsymbol{k}' \uparrow$ may take place through the following process:

$$\boldsymbol{k} \uparrow M \longrightarrow \boldsymbol{k}'' \downarrow M + 1 \longrightarrow \boldsymbol{k}' \uparrow M.$$

The corresponding matrix element, according to eqs. (5.89) and (5.90), is given by

$$(-J)^2\,[S(S+1)-M(M+1)]\sum_{k''}\frac{1-f''}{\varepsilon_k-\varepsilon_{k''}}.$$

In addition, the following process also contributes to the transition $\boldsymbol{k}\uparrow\longrightarrow\boldsymbol{k}'\uparrow$:

$$\boldsymbol{k}''\downarrow M\longrightarrow\boldsymbol{k}'\uparrow M-1;\quad \boldsymbol{k}\uparrow M-1\longrightarrow\boldsymbol{k}''\downarrow M.$$

Going over the same considerations as those involved in the cases without the spin-flip, we obtain

$$-(-J)^2\,[S(S+1)-M(M-1)]\sum_{k''}\frac{f''}{\varepsilon_{k''}-\varepsilon_{k'}}$$

as the matrix element. Adding togther all these matrix elements and including the first-order term $-JM$, we obtain

$$H(\boldsymbol{k}\uparrow\longrightarrow\boldsymbol{k}'\uparrow)=-JM\left(1+J\sum_{k''}\frac{1-2f''}{\varepsilon_k-\varepsilon_{k''}}\right)+(-J)^2S(S+1)\sum_{k''}\frac{1}{\varepsilon_k-\varepsilon_{k''}}. \tag{5.95}$$

This time, the dependence on f'' does not cancel. The matrix element of the process in which the z-component of the localized spin is increased and then decreased by 1 differs from that of the process in which it is decreased and then increased by 1. In other words, $S_+S_- - S_-S_+ \neq 0$, and so f'' survives. $S_+S_- - S_-S_+ \neq 0$ means that the spin is a system with an internal degree of freedom. When an electron is scattered by such a system, we cannot consider it as a one-body problem, and the scattering is affected by the presence of the other electrons. The non-cancellation of f'' is a reflection of this fact.

The term that includes f'' makes a significant contribution, and this will be discussed subsequently. On the other hand, the second term of eq. (5.95) will be omitted hereafter, as it is not important when J is small.

In the same way, we can evaluate the higher-order contributions to the scattering $\boldsymbol{k}\uparrow\longrightarrow\boldsymbol{k}'\downarrow$. The result reads

$$-J\sqrt{S(S+1)-M(M+1)}\,[1+2Jg(\varepsilon_k)], \tag{5.96}$$

where

$$g(\varepsilon)=\sum_{k''}\frac{f''-1/2}{\varepsilon_{k''}-\varepsilon}. \tag{5.97}$$

In the same way as in the derivation of eq. (5.94), we then obtain

$$W(\varepsilon) = \frac{2\pi}{\hbar} J^2 S(S+1)\rho(\varepsilon)[1 + 4Jg(\varepsilon)]. \tag{5.98}$$

This is a third-order expansion in J.

Let us now calculate g. In eq. (5.97), there does not seem to be an upper limit in the summation with respect to $\boldsymbol{k}''$. However, V_0, and hence J, depend on $\boldsymbol{k}$ to start with. Furthermore, as mentioned before, they become small when $\boldsymbol{k}$ becomes sufficiently large. The sum with respect to $\boldsymbol{k}''$ should thus be cut off at a certain point.

Using the density of states ρ, we can transform the sum over $\boldsymbol{k}''$ into the following integral:

$$g(\varepsilon) = \int_{-\varepsilon_F}^{D} \left(f(\varepsilon') - \frac{1}{2} \right) \frac{\rho(\varepsilon')}{\varepsilon' - \varepsilon} d\varepsilon'. \tag{5.99}$$

Here, the Fermi level is taken to be zero. D is a characteristic energy for cutting off the integral, and it is of the same order as ε_F. We shall adopt $D = \varepsilon_F$ hereafter.

g is a function of ε and T. Let us consider the case where both $|\varepsilon|$ and $k_B T$ are much less than ε_F. In this case, since the region where ε' is near the origin makes a crucial contribution to the integral, we set $\rho(\varepsilon')$ equal to $\rho(0)$ and take it out of the integral. For $T = 0$, we then obtain

$$g(\varepsilon) = \rho \log \frac{|\varepsilon|}{D}. \tag{5.100}$$

When $\varepsilon = 0$, we obtain

$$g(0) = \rho \left[\log \frac{k_B T}{D} - C - \log \frac{\pi}{2} \right] \qquad (C = 0.577215). \tag{5.101}$$

Here, $\rho = \rho(0)$. As indicated by this result, g is a peculiar function that behaves like the logarithm of whichever is greater between ε and $k_B T$.

$W(\varepsilon)$, as obtained by eq. (5.98), is the relaxation rate of the conduction electrons per unit time. In fact, this function has the following relationship with the density of states, which was discussed in §5.7, of the localized orbital $\rho_0(\varepsilon)$:

$$\rho_0(\varepsilon) = \frac{\hbar W(\varepsilon)}{2\pi {V_0}^2}. \tag{5.102}$$

Although we shall not go through this in detail, for the derivation of this equation, we need to make use of the following three points: first, ρ_0 is the imaginary part of the Green's function of the localized electron; second, $\hbar W$ is the imaginary part of the t matrix of the Green's function of the conduction electrons; third, the t matrix of the Green's function of the conduction electrons is the Green's function of the localized electron times $V_0{}^2$.

Using the result for $T = 0$, that is, eq. (5.100), we obtain

$$\pi \Delta_0 \cdot \rho_0(\varepsilon) = \pi^2 \rho^2 J^2 S(S+1) \left[1 + 4J\rho \log \frac{|\varepsilon|}{D} \right]. \tag{5.103}$$

Here, we adopt $S = 1/2$. According to eq. (5.103), ρ_0 increases as $|\varepsilon|$ decreases because $J < 0$ in the present situation. Although $|J\rho|$ is ordinarily expected to be much smaller than 1, if $|\varepsilon|$ is very small, the second term in square brackets in eq. (5.103) may become greater than 1. In this case, we cannot terminate the expansion in $J\rho$ at the first order, and we need to consider the higher-order terms.

At the same time, according to eq. (5.83), $\rho_0 = 1/\pi\Delta_0$ when $\varepsilon = 0$, and ρ_0 decreases rapidly when ε increases. Although eq. (5.83) is valid when U is small, this equation remains valid even for large U if we choose appropriate values of α_R and α_I. Equations (5.83) and (5.103) should therefore be considered to be connected with each other.

As we mentioned earlier, the frequency of spin fluctuation decreases when U is large, and localized spin appears to emerge when the observation timescale becomes shorter. The region where $|\varepsilon|$ is greater than the frequency of spin fluctuation is the region in which eq. (5.103) is applicable. Here, the spin is seen clearly, and its internal degree of freedom brings the logarithmic singularity into the scattering. On the other hand, if $|\varepsilon|$ is smaller than the frequency of spin fluctuation, eq. (5.83) is applicable, with α_R and α_I taken as the corresponding quantities for large U.

Next, let us consider the electrical resistance. We substitute eq. (5.103) into eq. (5.82) (note that deriving the relaxation time from eq. (5.98) also leads to the same result, that is, eq. (5.104)). Here we do not adopt $T = 0$ and leave $g(\varepsilon)$ as it is, using eq. (5.99). We consider $\rho(\varepsilon)$ to be constant, take it out of the integral, and obtain:

$$g(\varepsilon) = \rho(0) \left[-\frac{1}{2} \log(D^2 - \varepsilon^2) - \int \log|\varepsilon' - \varepsilon| \frac{df(\varepsilon')}{d\varepsilon'} d\varepsilon' \right]$$

by partial integration.

When $1/\rho_0(\varepsilon)$ in eq. (5.82) is expanded in powers of J, the following integral emerges:

$$\int g(\varepsilon) \left(-\frac{df}{d\varepsilon} \right) d\varepsilon.$$

Making use of the above equation, this is transformed into

$$\rho(0)\left[-\log D+\int \log|\varepsilon'-\varepsilon|\frac{df}{d\varepsilon}\frac{df'}{d\varepsilon'}d\varepsilon d\varepsilon'\right]$$

$$=\rho(0)\Bigg[-\log D+\log k_BT$$

$$+\int_{-\infty}^{\infty}\log|x'-x|\frac{1}{(e^x+1)(e^{-x}+1)(e^{x'}+1)(e^{-x'}+1)}dx\,dx'\Bigg].$$

The last integral yields a constant which is equal to 0.26066, and is hence omitted. In this way, we obtain

$$R=R_0\pi^2\rho^2J^2S(S+1)\left[1+4J\rho\log\frac{k_BT}{D}\right]. \tag{5.104}$$

This increases with decreasing temperature, and it crosses over to eq. (5.86) with the disappearance of localized spin. We show an example of the $\log T$ dependence of electrical resistivity in Fig. 5.15.

Our problem thus possesses a certain characteristic energy, that is, the frequency of spin fluctuation, and when k_BT is less than this energy, there is essentially no difference from the case with $U=0$ and $\varepsilon_0=0$, which was discussed in the final part of §5.6. When k_BT exceeds this energy, localized spin

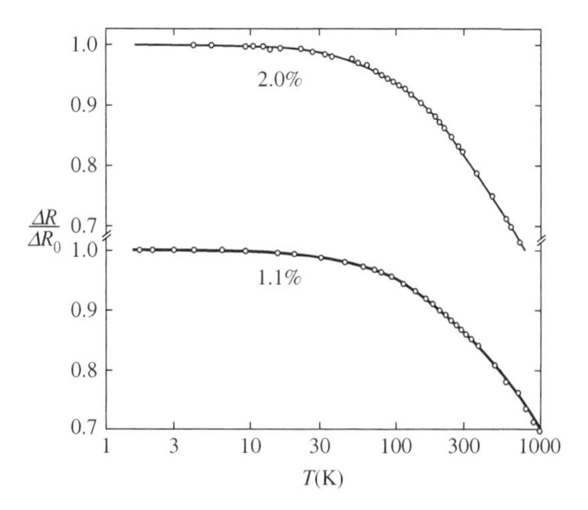

Figure 5.15 The temperature dependence of the electrical resistivity of gold alloys with small quantities of vanadium. The phonon contribution has been subtracted. The $\propto \log T$ behavior above room temperature and the saturation at low temperature are seen. From K. Kume, *J. Phys. Soc. Japan* **23** (1967) 1226.

begins to be visible, and the logarithmic singularity emerges. The case $U = 0$ and $\varepsilon_0 = 0$ also has a characteristic energy Δ_0. When $k_B T$ exceeds Δ_0, localized spin begins to be visible, but this does not bring about a logarithmic singularity. Thus, in this problem, we may say that U is large, the problem is essentially a many-body one, and this brings about the logarithmic singularity.

As discussed near Fig. 5.13, the susceptibility is expected to follow Curie's law at high temperature and to depart from it with decreasing temperature. To see this, we expand the susceptibility in terms of J by a standard method (see Appendix F). This yields

$$\chi = \frac{{\mu_B}^2}{k_B T}\left[1 + 2J\rho + 4J^2\rho^2 \log \frac{k_B T}{0.6001 D}\right]. \tag{5.105}$$

We used $S = 1/2$. For the sake of later reference, we have included all terms up to order J^2 with and without $\log T$. The last term in square brackets represents the departure from Curie's law. In the same way, we can obtain the specific heat as a perturbative expansion in J:

$$c = 16\pi^2 S(S+1) J^4 \rho^4 k_B \left[1 + 8J\rho \log \frac{k_B T}{D}\right]. \tag{5.106}$$

These results show that when we expand for physical observables in powers of J, we obtain terms that are proportional to $\log T$. In general, raising the order by 1 leads to one extra power of $J\rho \log(k_B T/D)$. Therefore, at low temperatures, where the absolute value of $J\rho \log(k_B T/D)$ exceeds ~ 1, the expansion in powers of J becomes a poor approximation.

We have discussed the qualitative behavior at low temperatures earlier. Various attempts at extrapolating the high-temperature behavior to low temperatures by summing as many higher-order terms in the perturbative expansion as possible have appeared, but none reached a satisfactory conclusion. In fact, a completely different method was employed to formulate a theory that provides a satisfactory description of the high-temperature-to-low-temperature crossover. We shall elaborate further on this in Chapters 7 and 8.

5.10 Case with orbital degeneracy

In eq. (5.81) we adopted $l = 0$, but when $l \geqq 1$ we have $2l + 1$ localized orbitals and need to consider the Coulomb interaction among these different orbitals. Moreover, if the potential considered in Fig. 5.2 is not spherically symmetric, orbitals with the same l and different m will, in general, have different energies.

Because of these reasons, it may so happen that the occupation number of electrons depends on m. However, when Δ_l is greater than the relevant energies, in general, either these situations do not arise or even if they do, they give little effect.

Let us assign the Coulomb energies for the interaction between orbitals of different m as follows: $U - J$ between electrons with parallel spin; and U between electrons with antiparallel spin. J is called the exchange integral and is positive. Although this is a rather rough model, it is sufficient for us to see the outline of the discussion. At first, let us say that the number of electrons in each orbital is independent of both m and spin, and that it is determined by the self-consistency condition:

$$n = \frac{1}{\pi}\left(\frac{\pi}{2} - \tan^{-1}\frac{\varepsilon_l{}'}{\Delta_l}\right), \tag{5.107}$$

where

$$\varepsilon_l{}' = \varepsilon_l + [(2l+1)U + 2l(U-J)]\,n. \tag{5.108}$$

An orbital with a certain value of m sees $2l+1$ orbitals with spin opposite to it, and $2l$ orbitals with the same spin to it. Hence $\varepsilon_l{}'$ gives the self-consistent localized energy level.

Let us now consider exciting Δn electrons with spin $\downarrow$ in all orbitals to the spin $\uparrow$ state. The energy required for this is $(\Delta n)^2/\rho_l(\varepsilon_F)$ per orbital, which is as given in §5.7. The decrease of the Coulomb energy can be calculated easily, and is given by

$$[-(2l+1)U + 2l(U-J)](\Delta n)^2 = -(U+2lJ)(\Delta n)^2$$

per orbital. Hence spin emerges when $(U+2lJ)\cdot\rho_l(\varepsilon_F) > 1$. Now, in this situation, where spin has appeared, let us consider moving Δn spin σ electrons from orbital m to orbital m'. The decrease of the Coulomb energy in this case is given by $(U-J)(\Delta n)^2$. Therefore the occupation number per orbital depends on m when $(U-J)\cdot\rho_\sigma(\varepsilon_F) > 1$, or in other words, orbital angular momentum emerges. Here, $\rho_\sigma(\varepsilon_F)$ denotes the density of states of orbital m and spin σ at ε_F.

In this way, we have the following three cases:

1. The occupation number of electrons is independent of both spin and m.
2. The occupation number depends on spin but is independent of m.
3. The occupation number depends on both spin and m.

Generally speaking, impurities in the transition metals fall into the the first two categories, whereas the case of the rare-earth elements corresponds to the third category. In the case of the impurities in rare-earth elements, the deviation from spherical symmetry of the potential, that is, the crystal field, can be significant. In the case of the impurities in transition metals, the crystal field is generally insignificant.

Now, on the basis of this discussion about orbital degeneracy, let us discuss in some detail the residual resistance due to iron group transition metal elements added to Cu (Shiba, 1973). In this system, d-wave scattering is the most important. At high temperature we may safely base our discussions on Fig. 5.9, with the caveat that the localized orbitals are fivefold degenerate (in the case of the second category in the above). Denoting the area of density of states below $\varepsilon_F(=0)$ as n_σ, we obtain $n_\sigma = \delta_\sigma/\pi$, where δ_σ denotes the phase shift of the spin-σ 3d orbital at ε_F. The total number of 3d electrons with spin σ is therefore equal to $5\delta_\sigma/\pi$.

Denoting the total number of 3d electrons as Z and the magnitude of 3d spin as S, we obtain

$$\left.\begin{aligned} Z &= \frac{5}{\pi}(\delta_\uparrow + \delta_\downarrow) \\ 2S &= \frac{5}{\pi}(\delta_\uparrow - \delta_\downarrow) \end{aligned}\right\} . \tag{5.109}$$

Let us now analyze the electrical resistance at high temperature on the basis of eq. (5.75b). Since localized spin can be either spin $\uparrow$ or $\downarrow$, a conduction electron experiences the left-hand side and the right-hand side of Fig. 5.9 with equal frequency. Hence, at high temperature, we obtain

$$\begin{aligned} R_1 &= \frac{4\pi\hbar n_i}{ne^2 k_F} \cdot 5 \cdot \frac{1}{2}\left(\sin^2\delta_\uparrow + \sin^2\delta_\downarrow\right) \\ &= \frac{4\pi\hbar n_i}{ne^2 k_F} \cdot 5 \cdot \frac{1}{2}\left[\sin^2\frac{\pi}{10}(Z+2S) + \sin^2\frac{\pi}{10}(Z-2S)\right]. \end{aligned} \tag{5.110}$$

The electrical resistance at absolute zero should obey the discussion in §5.8. There, all quantities, including the density of states, were independent of spin. Keeping this point in mind, we may proceed with our discussion as follows. Even for $l \neq 0$, the scattering probability of conduction electrons can be shown to be proportional to the density of states of the localized orbital, similarly to eq. (5.102). Thus, the electrical resistance is also expressed in terms of the density of states of the localized orbital. Equation (5.76) is precisely this expression. Here, $\rho_l(\varepsilon)$ means the density of states in the case with Coulomb interaction. On the other hand, the density of states and the area n below ε_F are related via

Table 5.1 *For Cu doped with iron group atoms: the electrical resistance R_1 at room temperature, and R_0 at absolute zero; the number of 3d electrons Z, and the magnitude of 3d spin S, calculated using R_1 and R_0. From H. Shiba, Prog. Theor. Phys.* **50** *(1973) 1797.*

	Cr	Mn	Fe	Co	Ni
$R_0(\mu\Omega\,\mathrm{cm/at\%})$	19.5	(18.5)	14.5	5.9	1.1
$R_1(\mu\Omega\,\mathrm{cm/at\%})$	4.1	3.0	8.0	(5.9)	(1.1)
Z	5.0	5.73	6.69	8.15	9.24
S	1.74	1.95	1.55	0.0	0.0

eq. (5.75a), and so we obtain

$$R_0 = \frac{4\pi\hbar n_i}{ne^2k_F}\cdot 5\cdot \sin^2\pi n. \tag{5.111}$$

As the degree of degeneracy of orbitals and spin is $2(2l+1)=10$, $10n$ must be equal to Z, the number of 3d electrons. We therefore obtain

$$R_0 = \frac{4\pi\hbar n_i}{ne^2k_F}\cdot 5\cdot \sin^2\frac{\pi}{10}Z. \tag{5.112}$$

As shown above, the electrical resistance at both high temperature and absolute zero are expressed in terms of Z, the number of 3d electrons, and S, the magnitude of the 3d spin at high temperature. Turning our attention to the experimental results in Fig. 5.15, the value of R_0 may be thought of as the saturated value at low temperature. Taking 300 K as the value of R_1, the values of Z and S can be evaluated by inverting the above relationships. The value of S thus obtained represents the magnitude of spin at 300 K.

The first factor that appears in both eqs. (5.110) and (5.112) takes the value $19\,\mu\Omega\,\mathrm{cm}$ for the case of the Cu host material when $n_i = 1$ atomic %. Using this, we obtain the numbers shown in Table 5.1. For example, as the difference between the atomic valence of Cr and that of Cu is 5, the result shown in Table 5.1 means that the electrons in the doped Cr are localized for the most part, and this seems plausible. Similar statements may be made concerning the other doped elements. Moreover, the magnitude of the spin of doped Cr is consistent with that estimated from the size of the Curie susceptibility measured close to room temperature. This holds also for the other doped elements.

References and further reading

Anderson, P. W. (1961) *Phys. Rev.* **124**: 41.
Friedel, J. (1952) *Phil. Mag.* **43**: 153.
Friedel, J. (1954) *Adv. Phys.* **3**: 446.
Friedel, J. (1958) *Nuovo Cim. Suppl.* **2**: 287.
Kanamori, J., Terakura, K. and Yamada, K. (1969) *Prog. Theor. Phys.* **41**: 1426.
Kondo, J. (1969) *Solid State Phys.* **23**: 183. [Note added by the translators: The original article is J. Kondo (1964) *Prog. Theor. Phys.* **32**: 37.]
Kume, K. (1967) *J. Phys. Soc. Japan* **23**: 1226.
Linde, J. O. (1931) *Ann. Physik* **10**: 52.
Linde, J. O. (1932) *Ann. Physik* **15**: 219.
Shiba, H. (1973) *Prog. Theor. Phys.* **50**: 1797.
Yamada, K. (1975) *Prog. Theor. Phys.* **53**: 970.
Yosida, K. and Yamada, K. (1975) *Prog. Theor. Phys.* **53**: 1286.

Further reading

In addition, there are some comprehensive reports on the problem of s–d interaction, for example, the monograph written by several authors: *Magnetism*, Vol. V, ed. H. Suhl (Salt Lake City, UT: Academic Press, 1973). See also:

Fischer, K. H. (1971) *Phys. State Solidi* **B47**: 11.
Grüner, G. and Zawadowski, A. (1978). In D. F. Brewer (ed.), *Progress in Low Temperature Physics*, Vol. VIIb (Amsterdam: North-Holland), p. 593.
Murao, T. (1982). In T. Matsubara (ed.), *The Structure and Properties of Matter* (Berlin: Springer-Verlag), p. 325.

6
The infrared divergence in metals

Let us investigate in more detail the origin of the logarithmic term that we found in the previous chapter. We examine the s–d Hamiltonian from this point of view. We first consider the two total wavefunctions of conduction electrons moving in two different local potentials. The overlap integral and matrix elements of these two wavefunctions contain logarithmic terms. In particular, the overlap integral vanishes at zero temperature in general. Second, we investigate the development of the wavefunction under a time-dependent perturbation where one potential changes into the other one abruptly at a particular time. This problem is closely related to the physics of the s–d interaction, since the potential changes quite abruptly as the spin-flips occur due to the s–d interaction. This problem is solved using the Nozières–de Dominicis method, and we apply these general considerations to the s–d Hamiltonian to obtain the perturbation expansion of the partition function and other physical quantities. The behavior of the localized spin at low temperatures is clarified using the scaling method applied to the s–d Hamiltonian.

6.1 The Anderson orthogonality theorem

This section is based mainly on Anderson (1967a,b).

In the previous chapter, we showed that the logarithmic singularity arises as a function of temperature in several physical quantities for systems governed by the s–d Hamiltonian due to flips of the localized spin. It is essential in this phenomenon that the localized spin has an internal degree of freedom. For the static impurity potential, the logarithmic singularity never appears. As stated in §5.9, the localized spin arises from the electron–electron interaction U, and is absent in systems represented by a one-body Hamiltonian. For example, in the system described in §5.6, in spite of the importance of spin fluctuations, no logarithmic divergences are caused by them. We investigate in more detail how this logarithmic singularity arises.

The s–d interaction plays two roles. One is to change the spin state of the localized spin from up to down and vice versa, and simultaneously the spin state of the conduction electrons. The other is to produce the spin-dependent local potential for conduction electrons. Hence the local potential for conduction electrons changes as the localized spin changes its state, and we must consider two wavefunctions for two different local potentials. This never occurs in the study of conduction electrons in a static potential. The matrix element and the overlap integral of these two wavefunctions contain the logarithmic singularity. Furthermore, the time dependence of these quantities should be addressed. This means that, after the sudden change of the local potential, the wavefunction of electrons will change into that of a static state corresponding to the new local potential. This time dependence is very important. We will investigate these subjects in more detail.

Let us consider the impurity potential

$$V_0 \sum_{kk'} a_{k}^{\dagger} a_{k'}. \tag{6.1}$$

We denote the wavefunction in this potential as Φ and that of free electrons as Φ_0. Using perturbation theory up to the second order, we obtain

$$\Phi = N \left[\Phi_0 + \sum_{k<k_F,\, k'>k_F} \frac{V_0}{\varepsilon_k - \varepsilon_{k'}} \Phi_0(\boldsymbol{k} \to \boldsymbol{k}') \right].$$

Here, $\Phi_0(\boldsymbol{k} \to \boldsymbol{k}')$ denotes the state where the electron with momentum $\boldsymbol{k}$ in the Fermi sea is excited to the electron with momentum $\boldsymbol{k}'$ above the Fermi energy. N is the normalization constant. The integral $\langle \Phi_0 | \Phi \rangle$ equals N, if Φ_0 is normalized, since each term (except Φ_0) in Φ is orthogonal to Φ_0. From the normalization condition of Φ, we have

$$1 = N^2 \left[1 + \sum_{k<k_F,\, k'>k_F} \frac{V_0^2}{(\varepsilon_k - \varepsilon_{k'})^2} \right].$$

Here, if the summation on the right-hand side is replaced by an integral, we see that it diverges. Thus N must vanish. This originates from the jump of the distribution function of electrons at the Fermi surface in metals. However, so long as we are concerned only with the static properties of the system, the normalization N never appears in the calculations of physical quantities.

The above discussion is within second-order perturbation theory. Anderson obtained the correct overlap integral in the limit of large electron numbers. In §5.2, we solved the Schrödinger equation of one electron in the impurity potential, which is more general than that in eq. (6.1). Hereafter, we set $l = 0$. It follows from eqs. (5.17) and (5.20) that

$$v_n(r) = \left(\frac{2}{R}\right)^{1/2} \frac{1}{r} \sin(k_n r + \delta_n) \quad (r > r_0) \tag{6.2}$$

is a solution for $r > r_0$, where k_n is determined by eq. (5.15); δ is a function of k_n, so δ is denoted as δ_n. Here, we assumed that the levels from $n = 1$ to the Fermi level n_F are occupied by electrons. In the case where there is no impurity potential, the wavefunction is, from eq. (5.13),

$$\phi_n(r) = \left(\frac{2}{R}\right)^{1/2} \frac{1}{r} \sin \kappa_n r. \tag{6.3}$$

(Equation (5.11) holds for every x if $l = 0$.) We again assumed that the levels from $n = 1$ to n_F are occupied by electrons. The overlap integral of wavefunctions in eqs. (6.2) and (6.3) is always less than 1. The overlap integral of the total wavefunctions, which is the product of each overlap integral, is much less than 1. The total wavefunctions are Slater determinants made from wavefunctions in eqs. (6.2) and (6.3). The total overlap integral is the determinant,

$$S = \det|S_{nm}|, \tag{6.4}$$

where S_{nm} $(n, m = 1, \ldots, n_F)$ are overlap integrals between ϕ_n and v_m:

$$\begin{aligned} S_{nm} &= \int_0^R r^2 \phi_n(r) v_m(r) dr \\ &\cong \frac{\sin \delta_m}{R(\kappa_n - k_m)} = \frac{\sin \delta_m}{(n - m)\pi + \delta_m}. \end{aligned} \tag{6.5}$$

Here we neglected the term which has $\kappa_n + k_m$ in the denominator since this term is small. We used the wavefunction in eq. (6.2) even for $0 < r < r_0$. It turns out that the error due to this is small. By using eq. (6.5), S in eq. (6.4) is written as

$$S = \left(\prod_{m=1}^{n_F} \frac{\sin \delta_m}{\pi}\right) \det \left| \frac{1}{n - m + \delta_m/\pi} \right|. \tag{6.6}$$

Here we use Cauchy's identity:

$$\det\left|\frac{1}{a_n+b_m}\right| = \frac{\prod_{n>m}(a_n-a_m)(b_n-b_m)}{\prod_{nm}(a_n+b_m)}. \tag{6.7}$$

This formula is proved by expanding the determinant and reducing the fractions to a common denominator. This yields a ratio of two polynomials in a and b. If we exchange two as or two bs, the determinant changes its sign. Thus the numerator contains the antisymmetrized forms of a and b, respectively. Now, since the order of the polynomial in the numerator is $N^2 - N$ where N is the order of the determinant, the remaining factor is a constant. This factor is easily proved to be 1, and then eq. (6.7) follows. By using Cauchy's identity, we have

$$S = \prod_m \frac{\sin\delta_m}{\delta_m} \prod_{n>m} \frac{1-\dfrac{\delta_n-\delta_m}{\pi(n-m)}}{\left[1+\dfrac{\delta_m}{\pi(n-m)}\right]\left[1-\dfrac{\delta_n}{\pi(n-m)}\right]}.$$

$\log S$ is expanded as

$$\log S \cong \sum_m \log\frac{\sin\delta_m}{\delta_m} + \frac{1}{2\pi^2}\sum_{n>m}\frac{{\delta_n}^2+{\delta_m}^2}{(n-m)^2} - \frac{1}{2\pi^2}\sum_{n>m}\frac{(\delta_n-\delta_m)^2}{(n-m)^2}.$$

The last term can be neglected since it remains small when $n \sim m$. By using the identity

$$\frac{\sin x}{x} = \prod_{n=1}^{\infty}\left(1-\frac{x^2}{\pi^2 n^2}\right),$$

we obtain

$$\begin{aligned}\log S &= -\frac{1}{2\pi^2}\sum_{n=1}^{n_F}{\delta_n}^2\left(\sum_{m=n}^{\infty}\frac{1}{m^2}+\sum_{m=n_F-n+1}^{\infty}\frac{1}{m^2}\right)\\ &= -\frac{1}{2\pi^2}\sum_{n=1}^{n_F}{\delta_n}^2\left(\frac{1}{n}+\frac{1}{n_F-n+1}\right).\end{aligned}$$

Since δ_n vanishes near $n = 1$ and is finite near $n = n_F$, the $1/n$ term gives a small contribution. This is because when the denominator is small, $n \sim 1 \ll n_F$, and the numerator ${\delta_n}^2$ is also small. We neglect this term and obtain

$$\log S \cong -\frac{1}{2\pi^2}{\delta_F}^2 \sum_{n=1}^{n_F} \frac{1}{n_F - n + 1} \cong -\frac{1}{2\pi^2}{\delta_F}^2 \log n_F, \tag{6.8}$$

where $\delta_F = \delta_{n=n_F}$. The above approximations have no effect on the coefficient of $\log n_F$ in eq. (6.8). When the system size is increased, $n_F \to \infty$, and eq. (6.8) gives the correct asymptotic form in this limit. It follows from eq. (6.8) that the overlap integral S approaches 0 as $n_F \to \infty$. This is known as the Anderson orthogonality theorem. If we consider two wavefunctions in two different potentials, δ_F in eq. (6.8) is replaced by the difference of phase shifts at the Fermi energy. Hence two wavefunctions with different phase shifts at the Fermi energy are orthogonal to each other.

6.2 Mahan's problem

The conduction electrons in a metal experience different local potentials as the localized spin changes its state due to the s–d interaction. Thus the time dependence should be examined in the study of the s–d interaction. Suppose that, until time $t = 0$, the system is in the unperturbed state and a local potential is abruptly switched on at time $t = 0$. We denote the wavefunction at $t = 0$ as ϕ, and then the wavefunction for $t > 0$ is

$$e^{-iHt/\hbar}\phi.$$

Here, H is the Hamiltonian including the local potential. If the local potential vanishes for $t > 0$, the wavefunction is $e^{-iH_0t/\hbar}\phi$ (where H_0 is the Hamiltonian without the local potential). The overlap integral,

$$\langle\phi|e^{iH_0t/\hbar}e^{-iHt/\hbar}|\phi\rangle = g(t), \tag{6.9}$$

is 1 at $t = 0$ and will approach 0 for large t according to eq. (6.8). Let us investigate the t dependence of $g(t)$.

To obtain the solution to $g(t)$ and apply it to the problem of the s–d interaction, we use the theory of the absorption and emission of soft X-rays by metals developed by Nozières and de Dominicis (1969; see also Roulet *et al.*, 1969; Nozières *et al.*, 1969). The X-ray is absorbed by metals by exciting an electron in the deep levels of an atom to above the Fermi energy. In the case of emission, there is a hole with a positive charge in the core, which results in a local potential for conduction electrons. Then, the conduction electrons are also excited, with energy transfer from X-rays and, as a result, the line shape of the X-ray absorption exhibits a singular behavior near the absorption edge. This problem was presented and examined by Mahan (1967a,b). In general, the absorption rate of photons with frequency ω in a system interacting with an electromagnetic field is represented by the interaction Hamiltonian H' as

$$\frac{2\pi}{\hbar}\sum_n |\langle 0|H'|n\rangle|^2 \delta(E_n - E_0 - \hbar\omega). \tag{6.10}$$

Here we assume that the initial state is the ground state $|0\rangle$. Equation (6.10) is then written as

$$\frac{2}{\hbar^2}\,\mathrm{Re}\int_0^\infty \langle 0|H'(t)H'|0\rangle e^{i\omega t}dt, \tag{6.11}$$

where Re refers to the real part, and $H'(t)$ is the Heisenberg representation of H'. To derive this formula, we perform integration by t, noting that

$$\langle 0|H'(t)H|0\rangle = \sum_n \langle 0|H'(t)|n\rangle\langle n|H'|0\rangle,$$

and

$$\mathrm{Re}\int_0^\infty e^{i\omega t}dt = \pi\delta(\omega).$$

When the electron in one of the deep levels is excited to the conduction band by X-rays, a positive charge in the core state and the conduction electrons interact with each other via the Coulomb interaction. The response of the conduction electrons shields the Coulomb interaction. There are two responses: one is fast and the other is slow. The former originates from the high-energy excitations and is called shielding due to plasma oscillation. As a result, long-range interaction is screened and becomes short-range. This short-range interaction binds just one electron near the hole in the core state, and this is demonstrated by calculation of the phase shift. Finally, localization of the electron takes place slowly, via interaction with lower excited states. In order to examine this behavior, Mahan considered the Hamiltonian

$$H = H_0 + \varepsilon_0 b^\dagger b + V_0 c^\dagger c b b^\dagger. \tag{6.12}$$

H_0 represents the kinetic energy of the conduction electrons, ε_0 is the energy level of the core state, and $b^\dagger$, b are its creation and annihilation operators, respectively. Here the spin indices are not shown explicitly and c is

$$c \equiv \sum_k a_k. \tag{6.13}$$

The last term in eq. (6.12) agrees with the potential in eq. (6.1) when there are no electrons in the core, that is, $bb^\dagger = 1$, and vanishes when there is an electron in the core, $b^\dagger b = 1$. c is the annihilation operator of the conduction electrons that are localized near the origin. The term that excites the core electron to the conduction band through interaction with the electromagnetic field is

$$H' \propto c^\dagger b + b^\dagger c.$$

In the ground state, the core level is occupied by an electron, and thus H' in eq. (6.11) means $c^\dagger b$ and we substitute $b^\dagger c$ for $H'(t)$. Hence the matrix element is

$$\langle 0|e^{iHt/\hbar}\, b^\dagger c\, e^{-iHt/\hbar}\, c^\dagger b|0\rangle. \tag{6.14}$$

We substitute $H = H_0 + V_0 c^\dagger c$ for $e^{-iHt/\hbar}$ in the middle because there are no electrons in the core, and $H = H_0 + \varepsilon_0$ for $e^{iHt/\hbar}$ at the left of this expectation value. Then the matrix element in eq. (6.14) is

$$e^{i\varepsilon_0 t/\hbar}\langle 0|e^{iH_0 t/\hbar}\, c\, e^{-(H_0+V_0 c^\dagger c)t/\hbar}\, c^\dagger|0\rangle. \tag{6.15}$$

This quantity is similar to that in eq. (6.9). Equation (6.15) represents an overlap integral of a couple of wavefunctions: one is the wavefunction where we put an electron near the origin in the potential V_0 at $t = 0$ and it is annihilated at t, and the other is the wavefunction of the non-interacting conduction electrons. The expressions in eqs. (6.9) and (6.15) indicate the time dependence of the response of conduction electrons under local perturbation, and the absorption rate of X-rays is represented by their Fourier transforms.

In conduction electron systems, low-energy excitations, where the electrons just below the Fermi energy are excited to above the Fermi energy, are quite important. This means that the time dependence of eqs. (6.9) and (6.15) for large $t \gg \hbar/\varepsilon_F$ is significant. The exact expressions for eqs. (6.9) and (6.15) for large t were obtained by Nozières and Dominicis.

The same problem, in a mathematical sense, occurs in the theory of s–d interaction. Even in the calculation of time-independent quantities such as the partition function, the localized spin frequently changes its direction in the intermediate states in the perturbation theory, and the conduction electrons experience time-dependent potential. Then we obtain a similar expression to eq. (6.15) for the partition function, where we use imaginary time instead of real time. This is a kind of thermal Green's function. Thus we describe the thermal Green's function in relation to the present problem and derive the solution of Nozières–Dominicis by using the thermal Green's function.

6.3 The thermal Green's function

This section is mainly based on Abrikosov *et al.* (1963).

Now let us consider the conduction electrons, neglecting the spin:

$$H_0 = \sum \varepsilon_k a_k{}^\dagger a_k. \tag{6.16}$$

We define the interaction representation for c in eq. (6.13) as

$$c_0(u) \equiv e^{uH_0} c e^{-uH_0}. \tag{6.17}$$

Here u is in the range $0 < u < \beta (= 1/k_B T)$ (which will be clarified later). The Green's function $G_0(u - v)$ is defined as

$$G_0(u - v) = \langle \mathrm{Tu}\, c_0(u) c_0{}^\dagger(v) \rangle_0, \tag{6.18}$$

where $\langle A \rangle_0$ indicates the thermal expectation value of A in a system with Hamiltonian H_0:

$$\langle A_0 \rangle = \frac{\mathrm{Tr}(e^{-\beta H_0} A)}{\mathrm{Tr}(e^{-\beta H_0})}.$$

The symbol Tu denotes the ordering operator with the sign determined by the exchange of fermionic operators:

$$\mathrm{Tu}\, c_0(u) c_0{}^\dagger(v) = \begin{cases} c_0(u) c_0{}^\dagger(v) & u > v, \\ -c_0{}^\dagger(v) c_0(u) & u < v. \end{cases}$$

We can show that the quantity in eq. (6.18) is a function of $u - v$ by substituting eq. (6.17) into eq. (6.18) and using

$$\mathrm{Tr}\,(AB) = \mathrm{Tr}\,(BA). \tag{6.19}$$

u and v are sometimes called time. Since u and v are in the range 0 to β, $u - v$ in eq. (6.18) is between $-\beta$ and β. By using eq. (6.19), we can show that

$$G_0(u + \beta) = -G_0(u)$$

for $-\beta < u < 0$. By expanding the range of the argument of G_0, we obtain G_0 as a periodic function with period 2β.

Now substituting eq. (6.13) into eq. (6.18) and using $\langle a_k{}^\dagger a_{k'} \rangle_0 = \delta_{kk'} f_k$ (where f_k is the Fermi distribution function for the state $\boldsymbol{k}$), we have

$$G_0(u) = \sum_k e^{-u\varepsilon_k}(1 - f_k) \quad (0 < u < \beta). \tag{6.20}$$

We can transform the summation over $\boldsymbol{k}$ into an integral using the density of states ρ. Since ρ can be regarded as being approximately constant, we can perform the integration by setting the limits of integration to $\pm\infty$. We obtain

$$G_0(u) = \frac{\pi T \rho}{\sin(\pi T u)}. \tag{6.21}$$

Hereafter we set $k_B = 1$. The same formula is obtained for $-\beta < u < 0$. This formula is, however, not exact for $|u| \lesssim D^{-1}$ (D is the bandwidth or a quantity of the order of the Fermi energy) since we cannot set the limits of integration to $\pm\infty$. To investigate the behavior of $G_0(u)$ in the region $|u| \lesssim D^{-1}$, we integrate $G_0(u)$ in the narrow region $(-\tau, \tau)$ including the origin. By using eq. (6.20) for $u > 0$ and the corresponding equation for $u < 0$, we have

$$\int_{-\tau}^{\tau} G_0(u) du = \mathrm{P} \int \frac{\rho(\varepsilon)}{\varepsilon} d\varepsilon \equiv \pi \rho \tan\theta,$$

where P stands for the principal part of the integral. From these results, $G_0(u)$ is written as

$$G_0(u) = \pi T \rho \left[\mathrm{P}\left(\frac{1}{\sin \pi T u} \right) + \pi \tan\theta \delta(\sin \pi T u) \right]. \tag{6.22}$$

Although this formula is not exact near $u = 0$, the integration in the narrow range including $u = 0$ gives the correct result. Now we introduce the complex function $\Phi_0(z)$:

$$\Phi_0(z) = \frac{\pi T \rho e^{i\theta}}{\cos\theta} \frac{1}{\sin \pi T z}. \tag{6.23}$$

Then $G_0(u)$ is given by

$$G_0(u) = \mathrm{Re}\Phi_0(u + i\delta), \tag{6.24}$$

where δ is a positive infinitesimal.

It is rather impressive that the function $1/\sin \pi T z$ appears here. At $T = 0$, this function tends to z^{-1}. The problems of analyticity associated with this function are often solved using Cauchy's integral theorem. The function $1/\sin \pi T z$ at $T \neq 0$ has the following properties, corresponding to Cauchy's integral theorem, which we employ frequently in what follows. If $\Psi(z)$ satisfies

$$\left. \begin{array}{l} \Psi(z) \text{ is holomorphic in the upper-half plane} \\ \Psi(z + \beta) = -\Psi(z) \end{array} \right\}, \tag{6.25}$$

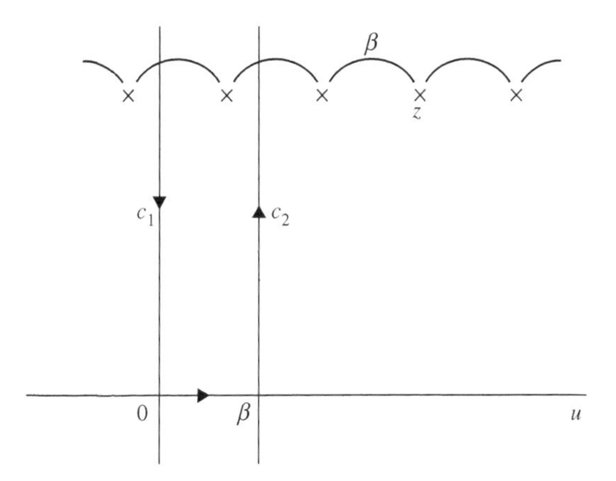

Figure 6.1 Integration contours c_1 and c_2 in the complex u-plane. The symbol × indicates the point of singularity of the integrand in eq. (6.26).

the following relation holds for z in the upper-half plane:

$$\int_0^\beta \frac{\Psi(u+i\delta)}{\sin \pi T(z-u)} du = -\frac{2i}{T}\Psi(z). \tag{6.26}$$

To prove this formula, we consider the integration along paths c_1 and c_2 as shown in Fig. 6.1. The resulting integral vanishes by virtue of eq. (6.25). Since the integrand vanishes in the limit $|z| \to \infty$, we can take a round path that includes c_1, $0 \to \beta$, c_2 and $|z| \to \infty$. Equation (6.26) is then proved by evaluating the residue of the pole with the help of eq. (6.25). When we put $z = u + i\delta$ and compare the real and imaginary parts of both sides of eq. (6.26), we obtain the following relations:

$$\text{Im}\,\Psi(u+i\delta) = T\int_0^\beta \text{P}\left(\frac{1}{\sin \pi T(u-u')}\right) \text{Re}\,\Psi(u'+i\delta)du', \tag{6.27}$$

$$\text{Re}\,\Psi(u+i\delta) = -T\int_0^\beta \text{P}\left(\frac{1}{\sin \pi T(u-u')}\right) \text{Im}\,\Psi(u'+i\delta)du'. \tag{6.28}$$

These correspond to the Kramers–Kronig relations. If we set $\Psi(z) = \Phi_0(z-v)$, we obtain, from eq. (6.24),

$$\frac{\cos^2\theta}{\pi^2\rho^2}\int_0^\beta G_0(u'-u)G_0(u'-v)du' = \pi T\delta(\sin \pi T(u-v)), \tag{6.29}$$

which is useful when deriving important equations later.

6.4 Thermal Green's functions in the presence of local potentials

We have investigated the Green's function of electrons at the origin in the free electron system in the previous section. In this section, let us examine the system in the presence of local potentials. The Hamiltonian is

$$H_\pm = H_0 \pm \frac{J}{2} c^\dagger c. \tag{6.30}$$

Although we may write $\pm J/2$ as V, we leave the notation as it is for later convenience. Now let us define

$$\left.\begin{aligned} c_\pm(u) &= e^{uH_\pm} c e^{-uH_\pm} \\ \langle \cdots \rangle_\pm &= \frac{\mathrm{Tr}(e^{-\beta H_\pm} \cdots)}{\mathrm{Tr}(e^{-\beta H_\pm})} \\ G_\pm(u-v) &= \langle \mathrm{Tu}\, c_\pm(u) c_\pm^\dagger(v) \rangle_\pm \end{aligned}\right\}. \tag{6.31}$$

$G_\pm$ is written as the average over the non-interacting system H_0, following standard methods found in many textbooks on quantum field theory and statistical physics (Abrikosov *et al.*, 1963).We can write

$$G_\pm(u-v) = \frac{\langle \mathrm{Tu}\, c_0(u) c_0{}^\dagger(v) S_\pm{}^0 \rangle_0}{\langle S_\pm{}^0 \rangle_0}, \tag{6.32}$$

where

$$S_\pm{}^0 = \mathrm{Tu} \exp\left(\mp \frac{J}{2} \int_0^\beta c_0{}^\dagger(u) c_0(u) du \right). \tag{6.33}$$

Similarly, following the methods described in these texts, we can show that $G_\pm$ satisfies

$$G_\pm(u-v) = G_0(u-v) \mp \frac{J}{2} \int_0^\beta G_0(u-u') G_\pm(u'-v) du'. \tag{6.34}$$

This equation is easily solved by the Fourier transformation. However, here we use the property of $G_0(u)$ shown in eq. (6.29) to obtain the solution. We can also apply this method directly to a problem that will be considered later. The inhomogeneous term of the equation, $G_0(u-v)$, consists of the principal part and a delta function, as shown in eq. (6.22). If we can neglect the principal part, the equation is easily solved. Therefore, we transform eq. (6.34) into an alternative form with the help of eq. (6.29). After integrating eq. (6.34) multiplied by $G_0(u-u')$ with respect to u from 0 to β, we obtain

$$\tan(\theta-\delta_\pm)G_\pm(u-v) - T\int_0^\beta \mathrm{P}\left(\frac{1}{\sin\pi T(u-u')}\right)G_\pm(u'-v)du' = \frac{\pi^2\rho T}{\cos^2\theta}\delta(\sin\pi T(u-v)). \tag{6.35}$$

$\delta_\pm$ is defined by

$$\tan(\theta-\delta_\pm) = \tan\theta \pm \frac{J}{2}\frac{\pi\rho}{\cos^2\theta}, \tag{6.36}$$

and so we have

$$\tan\delta_\pm = \mp\frac{\dfrac{\pi J\rho}{2}}{1\pm\dfrac{\pi J\rho}{2}\tan\theta}. \tag{6.37}$$

$\delta_\pm$ defined by this formula agrees with the phase shift in the scattering by the potential $\pm(J/2)c^\dagger c$, which is not proved here. When $\delta_\pm=0$, $G_\pm$ of course coincides with G_0. In this case, we can prove that G_0 satisfies eq. (6.35) as follows. We remember that G_0 is given by the real part of Φ_0 in eq. (6.24). Then, for $u\neq v$, eq. (6.35) is written as $\tan\theta\ \mathrm{Re}\Phi_0(u-v+i\delta)=\mathrm{Im}\Phi_0(u-v+i\delta)$ by virtue of eq. (6.27). This equation holds since the phase angle of $\Phi_0(u-v+i\delta)$ is θ. It is worth while noting that the residue of the pole of $\Phi_0(z)$ at $z=0$ agrees with the coefficient of the delta function on the right-hand side of eq. (6.35). Thus eq. (6.35) is easily solved even for $\delta_\pm\neq 0$. Let us define

$$\Phi_\pm(z) = \frac{\pi T\rho\cos(\theta-\delta_\pm)}{\cos^2\theta}\frac{e^{i(\theta-\delta_\pm)}}{\sin\pi Tz}, \tag{6.38}$$

then the solution is

$$G_\pm(u-v) = \mathrm{Re}\ \Phi_\pm(u-v+i\delta). \tag{6.39}$$

We can show similarly that $G_\pm$ satisfies eq. (6.35). It is also easy to prove the relation that is analogous to eq. (6.29):

$$\frac{\cos^4\theta}{\pi^2\rho^2\cos^2(\theta-\delta_\pm)}\int_0^\beta G_\pm(u'-u)G_\pm(u'-v)du' = \pi T\delta(\sin\pi T(u-v)). \tag{6.40}$$

6.5 The partition function in the s–d problem

Now let us return to the s–d interaction. The method mentioned above enables us to evaluate quantities such as that of eq. (6.15). First we examine how such a quantity appears in the theory of s–d interaction. Following Anderson and Yuval (1969), the s–d Hamiltonian is decomposed as

$$H = \sum_{k\sigma} \varepsilon_k a_{k\sigma}{}^{\dagger} a_{k\sigma} - J S_z (c_{\uparrow}{}^{\dagger} c_{\uparrow} - c_{\downarrow}{}^{\dagger} c_{\downarrow}), \tag{6.41}$$

$$H' = -J_{\perp}(S_+ c_{\downarrow}{}^{\dagger} c_{\uparrow} + S_- c_{\uparrow}{}^{\dagger} c_{\downarrow}), \tag{6.42}$$

where $c_\sigma = \sum_k a_{k\sigma}$. The total Hamiltonian is $H + H'$. In general, J and $J_\perp$ are different. As we mentioned earlier, the $J_\perp$ term changes the direction of spin and the J term indicates the spin-dependent potential for electrons. We expand the partition function in terms of H' using the decomposition (Appendix F):

$$\begin{aligned} Z &= \mathrm{Tr}(e^{-\beta(H+H')}) \\ &= \mathrm{Tr}\left[e^{-\beta H} \left\{ 1 + \iint_{\beta > u_1 > v_1 > 0} H'(u_1) H'(v_1) du_1 dv_1 + \cdots \right\} \right]. \end{aligned} \tag{6.43}$$

Here, concerning the trace operation, the summation with respect to the localized spin should be performed first. For example, in the H'^2 term, if the localized spin state is $\downarrow$, we take the S_+ term in eq. (6.42) for $H'(v_1)$ and S_- term for $H'(u_1)$. Then we must consider

$$\mathrm{Tr}[e^{-\beta H} e^{u_1 H} S_- c_{\uparrow}{}^{\dagger} c_{\downarrow} e^{-u_1 H} e^{v_1 H} S_+ c_{\downarrow}{}^{\dagger} c_{\uparrow} e^{-v_1 H}].$$

For the up-spin state $\uparrow$ to the left of S_+, H in eq. (6.41) reads H_-, and for the down-spin state $\downarrow$ H is H_+, where $H_\pm$ are defined in eq. (6.30), with the appropriate spin index. To the left of S_-, H_- is replaced by H_+ for the spin-down state. Hence, omitting the spin indices for the Hamiltonian, the above term is written as

$$\begin{aligned} &\mathrm{Tr}[e^{-\beta H_+} e^{u_1 H_+} c_{\uparrow}{}^{\dagger} e^{-(u_1 - v_1) H_-} c_{\uparrow} e^{-v_1 H_+}] \\ &\quad \times \mathrm{Tr}[e^{-\beta H_-} e^{u_1 H_-} c_{\downarrow} e^{-(u_1 - v_1) H_+} c_{\downarrow}{}^{\dagger} e^{-v_1 H_-}]. \end{aligned}$$

This is analogous to the matrix element in eq. (6.15), except that we have now imaginary time and the Hamiltonian is changed from $H_\pm$ to $H_\mp$ in the interval from v_1 to u_1. Now we define $S_\pm(u, v)$ as

$$S_\pm(u, v) = e^{uH_\pm} e^{-(u-v)H_\mp} e^{-vH_\pm}. \tag{6.44}$$

Then the second factor in the above expression can be written as $\mathrm{Tr}(e^{-\beta H_-})\langle c_{\downarrow -}(u_1)S_-(u_1,v_1)c_{\downarrow -}{}^\dagger(v_1)\rangle_-$, for instance, where $c_{\downarrow -}(u)=e^{uH_-}c_\downarrow e^{-uH_-}$ and $\langle\cdots\rangle_-$ are as defined in eq. (6.31). As a result, omitting the spin indices for simplicity, we obtain

$$\begin{aligned}Z &= 2\mathrm{Tr}(e^{-\beta H_+})\mathrm{Tr}(e^{-\beta H_-})\\&\times\Big[1+(-J_\perp)^2\iint_{\beta>u_1>v_1>0}\langle c_+{}^\dagger(u_1)S_+(u_1,v_1)c_+(v_1)\rangle_+\\&\times\langle c_-(u_1)S_-(u_1,v_1)c_-{}^\dagger(v_1)\rangle_- du_1dv_1+\cdots\Big].\end{aligned}\qquad(6.45)$$

In the fourth order, the term with four + indices

$$\langle c_+{}^\dagger(u_1)S_+(u_1,v_1)c_+(v_1)c_+{}^\dagger(u_2)S_+(u_2,v_2)c_+(v_2)\rangle_+$$

appears, and $c_-S_-c_-{}^\dagger$ appears twice in the term with − indices. The integration region is restricted to $\beta>u_1>v_1>u_2>v_2>0$. The higher-order terms can be deduced in the same way.

The above expectation values differ from the usual Green's functions in that they contain $S_\pm(u_i,v_i)$. The inclusion of $S_\pm$ indicates that the Hamiltonian changes from $H_\pm$ to $H_\mp$ for the intermediate region $v_i<u<u_i$ due to flips of the localized spin. If we set $H_\pm-H_\mp=V_\pm$, $S_\pm(u,v)$ is expanded as

$$S_\pm(u,v)=1+\int_{u>u_1>v}V_\pm(u_1)du_1+\iint_{u>u_1>u_2>v}V_\pm(u_1)V_\pm(u_2)du_1du_2+\cdots,$$

where $V_\pm(u)=e^{uH_\pm}V_\pm e^{-uH_\pm}$.

Proof: The derivative of this equation with respect to u leads to

$$\frac{\partial S_\pm(u,v)}{\partial u}=V_\pm(u)S_\pm(u,v).$$

This coincides with the derivative of eq. (6.44). Since this equation holds for $u=v$, the proof is concluded for general u and v.

This equation is equivalent to

$$S_\pm(u,v)=\mathrm{Tu}\exp\left(\int_v^u V_\pm(u_1)du_1\right).\qquad(6.46)$$

Since $S_\pm(u,v)$ appears n times in the $2n$th-order perturbation expansion in $J_\perp$, we define

$$S_\pm=S_\pm(u_1,v_1)\cdots S_\pm(u_n,v_n).\qquad(6.47)$$

Let U be the region in time where the Hamiltonian is changed into $H_{\mp}$. That is, $u \in U$ indicates $u_1 > u > v_1$ or $u_2 > u > v_2$, etc. If we define a function $U(u)$ by

$$U(u) = \begin{cases} 1 & u \in U, \\ 0 & u \notin U, \end{cases} \tag{6.48}$$

we easily obtain

$$S_{\pm} = \text{Tu} \exp\left(\int_0^{\beta} V_{\pm}(u)U(u)du \right). \tag{6.49}$$

By employing this formula, the quantity with the '$-$' index is expressed as $\langle \text{Tu}\, c_-(u_1)c_-^{\dagger}(v_1) \cdots c_-(u_n)c_-^{\dagger}(v_n)S_- \rangle_-$ in the 2nth-order of $J_{\perp}$.

This kind of expectation value is written as a sum of the contractions of pairs of c and $c^{\dagger}$ using Wick's theorem. Here we define the Green's function:

$$\mathcal{G}_{\pm}(u, v) = \frac{\langle \text{Tu}\, c_{\pm}(u)c_{\pm}^{\dagger}(v)S_{\pm} \rangle_{\pm}}{\langle \text{Tu}\, S_{\pm} \rangle_{\pm}}. \tag{6.50}$$

This is a function of u_i and v_i as well as u and v. If the Hamiltonian is always $H_{\pm}$ without the change in the potential, we have $U(u) \equiv 0$ and $\mathcal{G}_{\pm}$ reduces to $G_{\pm}$ in eq. (6.31). If the Hamiltonian is always $H_{\mp}$, then $U(u) \equiv 1$ and $\mathcal{G}_{\pm}$ reduces to $G_{\mp}$, since $\mathcal{G}_{\pm}$ is the Green's function for the system of the Hamiltonian $H_{\pm} - V_{\pm}$. By Wick's theorem, $\mathcal{G}_{\pm}$ is written as

$$\langle \text{Tu}\, c_-(u_1)c_-^{\dagger}(v_1) \cdots S_- \rangle = \langle \text{Tu}\, S_- \rangle_- \begin{vmatrix} \mathcal{G}_-(u_1, v_1) & \mathcal{G}_-(u_1, v_2) \cdots \\ \mathcal{G}_-(u_2, v_1) & \mathcal{G}_-(u_2, v_2) \\ \vdots & \end{vmatrix}. \tag{6.51}$$

$\langle \text{Tu}\, c_+^{\dagger}(u_1)c_+(v_1) \cdots S_+ \rangle_+$ is obtained in the same way by replacing $-$ by $+$ and exchanging u_i and v_i, with an overall sign factor $(-1)^n$.

6.6 The Nozières–de Dominicis solution

The results of this section are mainly based on Yuval and Anderson (1970), which gives a more concise derivation than Nozières and de Dominicis (1969).

The problem is now reduced to getting a solution for $\mathcal{G}_{\pm}$ in eq. (6.50). $\mathcal{G}_{\pm}$ contains $S_{\pm}$, which is different from the usual Green's functions, as we discussed previously. $S_{\pm}$ represents the effect of the perturbation which is applicable for

$u \in U$. If we expand $S_\pm$ in eq. (6.49) in terms of $V_\pm$, we obtain the following integral equation:

$$\mathcal{G}_\pm(u,v) = G_\pm(u-v) \pm J \int_0^\beta G_\pm(u-u')U(u')\mathcal{G}_\pm(u',v)du', \tag{6.52}$$

where $V_\pm = \pm Jc^\dagger c$ from eq. (6.30). This is analogous to eq. (6.34), which is satisfied by $G_\pm$ in eq. (6.32). We have only to take into account that the unperturbed Green's function G_0 is replaced by $G_\pm$ and the time-dependent perturbation $V_\pm(u)U(u)$ is included.

Equation (6.52) is solved in a similar manner to that used to solve eq. (6.34). For this purpose, we transform this equation to a familiar form by using the relation of eq. (6.40). Let us multiply eq. (6.52) by $G_\pm(u-u')$ and integrate from 0 to β with respect to u; then we have

$$\begin{aligned}\tan[\theta - \delta_\pm(u)]\mathcal{G}_\pm(u,v) - T \int_0^\beta \mathrm{P}\left(\frac{1}{\sin \pi T(u-u')}\right) \mathcal{G}_\pm(u',v)du' \\ = \frac{\pi^2 T \rho}{\cos^2 \theta}\,\delta(\sin \pi T(u-v)),\end{aligned} \tag{6.53}$$

where

$$\delta_\pm(u) = \begin{cases} \delta_\mp & u \in U, \\ \delta_\pm & u \notin U. \end{cases} \tag{6.54}$$

$\delta_\pm(u)$ is a function that takes the value $\delta_\pm$ when the Hamiltonian is $H_\pm$ and $\delta_\mp$ when the Hamiltonian is $H_\mp$. Otherwise eq. (6.53) coincides with eq. (6.35). If $\mathcal{G}_\pm(u)$ is written as Re $\Psi(u+i\delta)$ for some function $\Psi(z)$ which satisfies eq. (6.25), the phase of $\Psi(u+i\delta)$ should have a discontinuity as in eq. (6.54). Let us consider the following function:

$$X_\pm(z) = \prod_i \left[\frac{\sin \pi T(z-u_i)}{\sin \pi T(z-v_i)}\right]^{(\delta_\pm - \delta_\mp)/\pi}. \tag{6.55}$$

Now, let $v \in U$. The phase of $X_\pm(u+i\delta)/X_\pm(v+i\delta)$ then equals 0 when $u \in U$ and $\delta_\pm - \delta_\mp$ when $u \notin U$. We then define

$$\Psi_\pm(z,v) = \frac{\Phi_\mp(z-v)X_\pm(z)}{X_\pm(v+i\delta)}, \tag{6.56}$$

which satisfies eq. (6.25) so that $\mathcal{G}_\pm$ is written as

$$\mathcal{G}_\pm(u,v) = \mathrm{Re}\, \Psi_\pm(u+i\delta, v). \tag{6.57}$$

Since the phase of $\Psi_\pm(u+i\delta, v)$ is $\theta - \delta_\pm(u)$, $\mathcal{G}_\pm(u,v)$ in eq. (6.57) is a solution to eq. (6.53) for $u \neq v$. For $z = v$, the residue of the pole agrees with that of $\Phi_\mp(z-v)$, and is consistent with the δ function in eq. (6.53), which is similar to the solution to eq. (6.35). Now we have shown that $\mathcal{G}_\pm(u,v)$ satisfies eq. (6.53). If we define

$$Y(u) = \prod_i \left| \frac{\sin \pi T(u-u_i)}{\sin \pi T(u-v_i)} \right|, \tag{6.58}$$

it follows from eqs. (6.56), (6.57) and (6.39) that

$$\mathcal{G}_\pm(u,v) = G_\mp(u-v) \left[\frac{Y(u)}{Y(v)} \right]^{(\delta_\pm - \delta_\mp)/\pi}, \tag{6.59}$$

for $u \in U$ and $v \in U$. This has singularities at $u = u_i$ and $u = v_i$. This is because of an approximation adopted for $G_0(u)$ in eq. (6.21). $G_0(u)$ defined in this way is not exact for $|u| \lesssim D^{-1}$ and thus we must substitute $u - u_i \cong D^{-1}$ for $u \to u_i$. As is obvious from eq. (6.51), we must further evaluate $\langle \mathrm{Tu}\, S_\pm \rangle_\pm$ to obtain the partition function. Let us put $g_\pm = \langle \mathrm{Tu}\, S_\pm \rangle_\pm$. As shown in Appendix G, we obtain

$$\log g_+ g_- = \frac{(\delta_+ - \delta_-)^2}{\pi^2} \log A, \tag{6.60}$$

where

$$A = \prod_i \left[\frac{Y(u_i)}{Y(v_i)} \right]. \tag{6.61}$$

6.7 Calculation of the partition function

This section is mainly based on Anderson *et al.* (1970).

To summarize our foregoing results, in the expansion of the partition function in the s–d problem in terms of $J_\perp$, each term of the expansion is given as the integral of the product of eq. (6.51) and the corresponding equation with '+' indices. The integration is with respect to $u_1 v_1 u_2 v_2 \cdots u_n v_n$. The functions $\mathcal{G}_\pm$ and $g_\pm$, which are needed to calculate the partition function, are given in eqs. (6.59) and (6.60), respectively. $G_\pm(u)$ is given by eq. (6.39), but we may express it using eq. (6.38) as

$$G_\pm(u) = \frac{\pi T \rho \cos^2(\theta - \delta_\pm)}{\cos^2 \theta} \left[\mathrm{P}\left(\frac{1}{\sin \pi T u} \right) + \pi \tan(\theta - \delta_\pm) \delta(\sin \pi T u) \right]. \tag{6.62}$$

Hereafter we neglect the δ function since the singularity at $u=0$ will be investigated separately later. First, we must calculate the determinant of $\mathcal{G}_\pm$ in eq. (6.51). The coefficient $[Y(u)/Y(v)]^{(\delta_\pm-\delta_\mp)/\pi}$ in eq. (6.59) is a common factor and this leads to an overall factor $A^{(\delta_\pm-\delta_\mp)/\pi}$ in the determinant. We then only have to evaluate the determinant of $1/\sin\pi T(u_n-v_m)$, which, by virtue of Cauchy's identity in eq. (6.7), is given as

$$\det\left|\frac{1}{\sin\pi T(u_n-v_m)}\right| = \frac{\prod\limits_{n>m}\sin\pi T|u_n-u_m|\cdot\sin\pi T|v_n-v_m|}{\prod\limits_{n,m}\sin\pi T|u_n-v_m|}, \tag{6.63}$$

where we have used $\sin(u_n-v_m)=e^{-i(u_n+v_m)}(e^{2iu_n}-e^{2iv_m})/2i$. Since the determinant in eq. (6.63) is positive, the absolute value is used for each element.

Second, we note that A contains $Y(u_i)$, in which we must substitute $u=u_i$ for $\sin\pi T(u-u_i)$. As we mentioned earlier, in order to avoid singularities, $\sin\pi T(u-u_i)$ should be replaced by $\sin\pi T\tau\sim\pi T\tau$, where τ is a quantity of the order of the inverse of the bandwidth. As τ is also of the order of the density of states ρ, ρ is written as τ from now on. Now, from eq. (6.58), A equals the square of eq. (6.63) times $(\pi T\tau)^{2n}$. As a result, the square brackets in eq. (6.45) reduce to

$$\sum_n (J_\perp)^{2n}\int\cdots\int_{\beta>u_1>v_1>\cdots>v_n>0} e^{(2-\varepsilon)V}du_1\cdots dv_n, \tag{6.64}$$

where

$$V=\sum_{n>m}\left(\log\frac{\sin\pi T|u_n-u_m|}{\pi T\tau}+\log\frac{\sin\pi T|v_n-v_m|}{\pi T\tau}\right)$$
$$-\sum_{n,m}\log\frac{\sin\pi T|u_n-v_m|}{\pi T\tau} \tag{6.65}$$

and

$$\varepsilon=\left(\frac{4}{\pi}\right)(\delta_+-\delta_-)-\frac{2}{\pi^2}(\delta_+-\delta_-)^2. \tag{6.66}$$

Here V diverges at $u_n=u_m$, which is, as we discussed earlier, due to the approximation used for G_0 in eq. (6.21). If $|u_n-u_m|$ is less than τ, neither eq. (6.21) nor eq. (6.62) can be used. To take this into account, we exclude the region where any of $|u_n-u_m|$, $|v_n-v_m|$ and $|u_n-v_m|$ is less than τ from the integration.

In the integral of eq. (6.64), the parameters $u_1,\ldots,u_n$ and $v_1,\ldots,v_n$ take values between 0 and β. Let us sort them into ascending order, that is, as

v_n, u_n, v_{n-1}, u_{n-1}, etc., and label the result as β_1, β_2, ..., β_{2n}, so that $\beta_1 = v_n < \beta_2 = u_n \cdots < \beta_{2n} = u_1$. Let us define

$$v(x) \equiv \log \frac{\sin \pi T|x|}{\pi T \tau}.$$

We then obtain

$$V = \sum_{i>j} (-1)^{i-j} v(\beta_i - \beta_j). \tag{6.67}$$

The summation in eq. (6.64) can be regarded as the partition function of classical particles as follows: on a line segment of length β we have $2n$ particles with alternating charges, i.e., there are n positively charged particles and n negatively charged particles. Repulsive interaction $-v(\beta_i - \beta_j)$ takes effect between particles with the same charge, and attractive interaction $v(\beta_i - \beta_j)$ between particles with the opposite charges. The particles cannot get closer than length τ from each other. The temperature of the system is $1/(2 - \varepsilon)$, and the chemical potential is $\mu = (2 - \varepsilon)^{-1} \log J_\perp$. We can approach a system of this kind from a new viewpoint. Before we do that, let us derive the formula for susceptibility. We perform a similar calculation with the magnetic term $-2\mu_B H S_z$ added to the Hamiltonian in eq. (6.41). Since S_z changes from $\pm\frac{1}{2}$ to $\mp\frac{1}{2}$ at β_i, we have $\mu_B H(\beta_1 - \beta_2 + \cdots)$ in the exponent. We obtain F from the relation $Z = e^{-F/T}$ and then the susceptibility of the localized spin is derived from $\chi = -(\partial^2 F/\partial H^2)_{H=0}$. The result is

$$\chi = {\mu_B}^2 T \langle {\mu_n}^2 \rangle, \tag{6.68}$$

where

$$\mu_n = \begin{cases} \beta - 2(\beta_{2n} - \cdots + \beta_2 - \beta_1) & n > 0, \\ \beta & n = 0. \end{cases} \tag{6.69}$$

Here $\langle\ \rangle$ indicates the expectation value in a system with the partition function given by eq. (6.64); μ_n represents the area of the shaded region in Fig. 6.2, and just corresponds to the z-component of the localized spin.

In the system of classical particles, we have two characteristic quantities: one is the mean number of particles; the other is the correlation between particles. As $J_\perp$ is increased, the number of particles will increase since the chemical potential rises. This indicates that the higher-order terms are important if $J_\perp$ is large. If the temperature is high, β is small and the number of particles between 0 and β is small. This means that only the first few terms in the perturbation expansion are important at high temperatures. At high enough temperatures, the

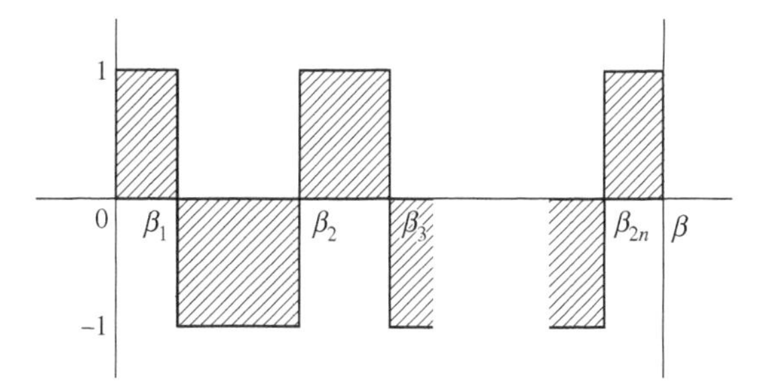

Figure 6.2 Graph of μ_n. The value of μ_n is defined by the area of the shaded region (taking into account the sign). The localized spin changes direction at β_i so that μ_n is the mean of S_z between 0 and β.

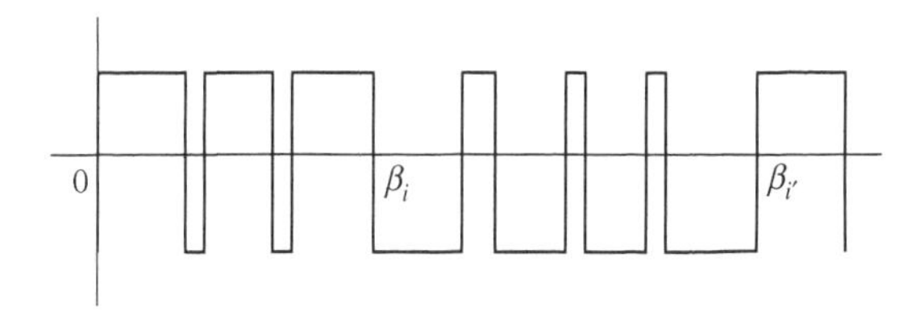

Figure 6.3 Spin-flips due to isolated particles β_i and $\beta_{i'}$. The localized spin is approximated as the spin-up state in the region between 0 and β_i, and the spin-down state between β_i and $\beta_{i'}$. The spin flips at a point where there is an isolated particle.

first term $n = 0$ is significant, and we have $\langle {\mu_n}^2 \rangle = \beta^2$, so $\chi \propto T^{-1}$. That is, the localized spin cannot change its direction in the extremely small interval of time. For large β, particles start to appear, and the localized spin flips repeatedly. It is now important to understand the electronic state with high density of particles at low temperatures with large β. First, we note that the mean value of the particle numbers is $\langle 2n \rangle = (2-\varepsilon)^{-1} \partial \log Z / \partial \mu \sim -\beta \partial E_g / \partial \log J_\perp$ since $Z \sim e^{-E_g/T}$. E_g is the ground-state energy, and is almost equal to $-{J_\perp}^2 \tau$ in magnitude. Hence $\langle 2n \rangle \sim \beta {J_\perp}^2 \tau$. The mean distance between particles $l_0 \sim 1/{J_\perp}^2 \tau$. Second, let us consider the particle correlations. Two particles with opposite charges have a tendency to form a pair state due to the attractive interaction, particularly for the case $\varepsilon < 0$ $(J > 0)$. Hence, in the region where there are several pairs in line as shown in Fig. 6.3, the localized spin is almost fixed in one direction, and gives a large contribution to ${\mu_n}^2$. However, the spin flips at a point where there is an isolated particle. Even in this case there may be a tendency for isolated particles to form pairs on the large scale. This indicates the possibility that localized spin is fixed in some direction on the whole for large β. If so, $\langle {\mu_n}^2 \rangle$ is proportional to β^2 as β is increased. There may be the other case, however, that on a large enough scale, the spin-up and spin-down states are equally distributed and we have $\langle {\mu_n}^2 \rangle \sim \beta$. In this case χ is independent of temperature.

6.8 A scaling approach

To investigate the points mentioned above in more detail, Anderson *et al.* (1970) employed a scaling method. The power series in eq. (6.64) depends only on ε, $(J_\perp \tau)^2$ and $T\tau$, although it is a function of $J_\perp$, β, ε and τ. Let us set $\eta \equiv |J_\perp \tau|$. The partition function is then $Z(\tau T, \varepsilon, \eta)$. From now on, $|\varepsilon|$ and η are assumed to be sufficiently small compared to 1. Then $\varepsilon \sim -4J_\perp \tau$ from eq. (6.66). In the integral of eq. (6.64), the β_i cannot be closer than τ from each other. We divide the integral into two parts: in one part, the β_i are more than $\tau + d\tau$ apart from each other; and in the other part, the distance is between τ and $\tau + d\tau$. In the latter case, in addition to having one pair separated by a distance between τ and $\tau + d\tau$, we also have the case where two pairs are between τ and $\tau + d\tau$ apart, and three pairs are between τ and $\tau + d\tau$ apart, and so on. Hence these contributions can be neglected if $d\tau$ is infinitesimally small. Now let us fix β_{i-1} and β_{i+2}, and integrate with respect to β_i and β_{i+1} (between β_{i-1} and β_{i+2}) assuming that the distance between β_i and β_{i+1} ranges between τ and $\tau + d\tau$. This procedure results in a new interaction between each of β_{i-1}, β_{i+2} and the other βs. As shown in Appendix I, we obtain

$$Z(\tau T, \varepsilon, \eta) = \exp({J_\perp}^2 \beta d\tau) Z(\tilde{\tau} T, \tilde{\varepsilon}, \tilde{\eta}). \tag{6.70}$$

Here

$$\tilde{\tau} = \tau + d\tau, \tag{6.71}$$

$$\tilde{\varepsilon} = \varepsilon + 4(2 - \varepsilon)\eta^2 d(\log \tau), \tag{6.72}$$

$$\tilde{\eta} = \eta + \frac{\varepsilon \eta}{2} d(\log \tau). \tag{6.73}$$

Apart from the first factor in eq. (6.70), the problem with parameters τ, ε and η is equivalent to that with parameters $\tilde{\tau}$, $\tilde{\varepsilon}$ and $\tilde{\eta}$. After integrating out the pairs separated by between τ and $\tau + d\tau$, we have remaining particles with parameters $\tilde{\varepsilon}$ and $\tilde{\eta}$. Note that the condition $\eta \ll 1$ is needed to derive eq. (6.70). The susceptibility is invariant as long as $\beta \gg \tilde{\eta}$, and can be calculated using eq. (6.68) with new parameters. We can transform the system to a new system with the parameters differing by a finite amount from the original ones by the repeated use of the transformation (scaling) in eqs. (6.71)–(6.73). Although the initial value of η is large, for example $|4J\rho \log(k_B T/D)| > 1$, for which the perturbation expansion cannot be applied, we can use the perturbation theory if $|4J\rho \log(k_B T/D)| < 1$ holds in the transformed system.

To see the nature of the transformation, we eliminate $d\tau$ in eqs. (6.72) and (6.73) (assuming $|\varepsilon| \ll 1$) so that we have $\varepsilon d\varepsilon = 16\eta d\eta$. This leads to

$$\tilde{\varepsilon}^2 - 16\tilde{\eta}^2 = \varepsilon^2 - 16\eta^2. \tag{6.74}$$

$\tilde{\varepsilon}$ and $\tilde{\eta}$ are on a hyperbola in the $\tilde{\varepsilon}-\tilde{\eta}$ plane. Let us examine the case of the isotropic exchange interaction $J = J_\perp$. Since $\eta = |J\tau|$ and $\varepsilon = -4J\tau$, we have $\varepsilon = -4\eta$ for $J > 0$ and $\varepsilon = 4\eta$ for $J < 0$. Then, since $(2/\varepsilon^2)d\varepsilon = d\tau/\tau$ holds from eq. (6.72), we obtain

$$\log\tilde{\tau} - \log\tau = \frac{2}{\varepsilon} - \frac{2}{\tilde{\varepsilon}}. \tag{6.75}$$

First, let us examine the case $J > 0$. In this case, as $\tilde{\tau}$ is increased, the absolute value of $\tilde{\varepsilon}(<0)$ is reduced and thus $\tilde{\eta}$ is also reduced. Let us assume that T is small enough compared to $\tau^{-1}e^{2/\varepsilon}$ at first. In this case the perturbative expansion cannot be applied since the higher-order terms are larger than the lower-order terms. However, if the relation

$$\beta \gg \tilde{\tau} \gg \left(\frac{\tau^{-1}e^{2/\varepsilon}}{T}\right)^{1/2}\tau$$

holds as $\tilde{\tau}$ is increased, we can employ the perturbation theory since $\tilde{\tau}^{-1}e^{2/\varepsilon} \ll T$ is derived from the second inequality and eq. (6.75). The first inequality is needed to guarantee the use of the scaling method. If $T \ll \tau^{-1}e^{2/\varepsilon}$, we have $\beta \gg (\tau^{-1}e^{2/\varepsilon}/T)^{1/2}\tau$, and thus there exists $\tilde{\tau}$ satisfying the above inequality. After the scaling transformation, $|\tilde{\varepsilon}|$ is extremely small, and the perturbative effect is also small. This means that the susceptibility equals that of the isolated spin. In the case $J > 0$, the number of pairs is reduced as pairs are eliminated repeatedly by the scaling transformation. As a result, the system reaches a state where the localized spin is always pointed in one direction.

Second, let us consider the case $J < 0$. Since $\tilde{\varepsilon}$ increases as $\tilde{\tau}$ increases, the perturbation theory is useless at low temperatures. In the end, $\tilde{\varepsilon}$ reaches $\tilde{\varepsilon} \sim 1$ at last in the scaling transformation. In this case, eq. (6.75) should be understood only as a qualitative relation, since the assumption used to derive eqs. (6.71)–(6.73) does not hold. We can, however, understand the behavior of the susceptibility near $T = 0$, as follows. Let us assume that we reach $\tilde{\varepsilon} = 1$ exactly. Then we have $\tilde{\tau} = \tau e^{2/\varepsilon - 2}$ from eq. (6.75), which is denoted as $1/eT_K$

here.[§] Since $\tilde{\eta} = 1/4$,

$$|\tilde{J}| = \frac{1}{4\tilde{\tau}} = \frac{e}{4} T_K. \tag{6.76}$$

When $\varepsilon = 1$, the system with the partition function in eq. (6.64) is equivalent to the model given below. Let us consider the Anderson model for spinless fermions:

$$H_0 = \sum \varepsilon_k a_k{}^\dagger a_k + \varepsilon_d b^\dagger b, \tag{6.77}$$

$$H' = |\tilde{J}|(b^\dagger c + c^\dagger b). \tag{6.78}$$

Here b indicates the operator for the localized orbital, and c is defined in eq. (6.13). Let us assume that the density of states of the conduction band ε_k is $\tilde{\tau}$. We define the 'susceptibility' χ' of localized particles in this system:

$$\chi' = -\left.\frac{d}{d\varepsilon_d}\langle b^\dagger b\rangle\right|_{\varepsilon_d = 0}. \tag{6.79}$$

The case $\varepsilon_d = 0$ is examined here. We can show that (Appendix J)

$$\chi' = \frac{T}{4}\langle \mu_n{}^2\rangle, \tag{6.80}$$

where $\langle \mu_n{}^2\rangle$ is given by eq. (6.69), setting $\varepsilon = 1$ and $J_\perp = \tilde{J}$. Since the model of eqs. (6.77) and (6.78) was solved exactly in §5.6, χ' at $T = 0$ is also obtained as $1/\pi\Delta$, where $\Delta = \pi \tilde{J}^2 \tilde{\tau}$. Hence eq. (6.80) is written as $1/\pi\Delta = (T/4)\langle \mu_n{}^2\rangle$, and we obtain $\chi = 4\mu_B{}^2/\pi\Delta$ by substituting $\langle \mu_n{}^2\rangle$ into eq. (6.68). This leads to $\chi \sim \mu_B{}^2/T_K$ because $\Delta \sim T_K$ from eq. (6.76).

This result can be interpreted as follows: although τ is increased by eliminating pairs in the scaling transformation for $J < 0$, the number of pairs to be eliminated is never reduced, and spin flips occur continuously. Schotte and Schotte (1971) investigated eqs. (6.64) and (6.68) by using a Monte Carlo method for $2n$ classical particles with fixed n, which is determined as a main contribution at some temperature. The Monte Carlo procedure proceeds by changing the positions of particles in small amounts and recalculating the Boltzmann weight in eq. (6.64) for $2n$ particles scattered randomly in the region ranging from 0 to β. Within numerical errors, the susceptibility follows the Curie law at high temperatures and tends to a constant at low temperatures, as described above.

[§] Note by the translators: the quantity T_K introduced here is commonly known as the Kondo temperature.

References and further reading

Abrikosov, A. A., Gorkov, L. P., and Dzyaloshinski, I. E. (1963) *Methods of Quantum Field Theory in Statistical Physics*, trans. R. A. Silverman (Upper Saddle River, NJ: Prentice Hall).
Anderson, P. W. (1967a) *Phys. Rev. Lett.* **18**: 1049.
Anderson, P. W. (1967b) *Phys. Rev.* **164**: 352.
Anderson, P. W. and Yuval, G. (1969) *Phys. Rev. Lett.* **23**: 89.
Anderson, P. W., Yuval, G. and Hamann, D. R. (1970) *Phys. Rev.* B**1**: 4464.
Mahan, G. D. (1967a) *Phys. Rev.* **153**: 882.
Mahan, G. D. (1967b) *Phys. Rev.* **163**: 612.
Nozières, P. and de Dominicis, C. T. (1969) *Phys. Rev.* **178**: 1097.
Nozières, P., Gavoret, J. and Roulet, B. (1969) *Phys. Rev.* **178**: 1084.
Roulet, B., Gavoret, J. and Nozières, P. (1969) *Phys. Rev.* **178**: 1072.
Schotte K. D., and Schotte, U. (1971) *Phys. Rev.* B**4**: 2228.
Yuval, G. and Anderson, P. W. (1970) *Phys. Rev.* B**1**: 1522.

Note added by the translators:

We suggest G. D. Mahan: *Many-Particle Physics* (New York, Kluwer Academic/Plenum, 2000) as a useful general reference for this chapter.

7
Wilson's theory

Wilson (1975) introduced the following method, which is distinct from the perturbative calculation where an expansion in terms of J is used. In his method, one starts from a two-body system composed of a localized spin and a conduction electron located at the same point. One then proceeds to expand the system by gradually taking into account electrons that are in the nearest neighborhood. As one goes further away from the localized spin, the states that are far away from the Fermi surface are taken out of consideration. That is, when using this method, the states around the Fermi surface are described more and more precisely as the system grows larger in size, and this corresponds to lowering the temperature.

As the system grows larger, the number of basis states increases rapidly. However, consideration of around a thousand excited states near the ground state is sufficient, and therefore the size of the matrix that needs to be diagonalized remains small enough for numerical simulation on the computer. In this way, for a reasonable value of J, when the system size is increased in a series of about 100 steps, Wilson obtained a state which was considered to be at sufficiently low temperature.

Both susceptibility and specific heat were obtained as a continuous function of temperature from high to low temperatures. There is just one scaling parameter, which is denoted by T_K. Furthermore, at low temperature, these physical quantities can be described as functions of J and, in particular, a constant factor of order unity was calculated, within the precision of the numerical calculation. This method of Wilson's is called the renormalization group method.

7.1 Wilson's Hamiltonian

The form of the s–d interaction in eq. (5.90) is not suitable for the treatment in this chapter. In the representation of the Hamiltonian in eq. (5.90), all the plane waves are equivalent. That is, all the waves are equivalent regardless of how far away they are from the localized spin in space and how far away their energies are from the Fermi energy. However, the wave that is localized at the origin (which is the position of the localized spin) in space must be the most important. This wave is represented by eq. (6.13), which indicates that the wave localized at the origin is composed of plane waves with all of the wavenumbers. This is consistent with the uncertainty principle. Far away from the origin, in contrast, only those waves whose wavenumbers are near the Fermi surface are important. Any superposition of plane waves whose wavenumbers are near the Fermi surface has a large spatial extension. Therefore, in the following, we consider those waves which are well localized near the origin and have large energy broadening and, as one moves away from the origin, spread out in space and have small energy broadening.

For this purpose, we express the Hamiltonian in eq. (5.90) in the spherical representation by using eq. (5.53). The states in the spherical representation are labeled by indices klm. If the s–d interaction is defined in a spherical symmetric form as in eq. (5.90), only the states with $l = m = 0$ should be taken into account, and thus we have only k as the parameter.

Let us rewrite k as $k_F + k$, so that k takes a value in the range $-k_F$ to k_F. The energy is then represented as $\hbar^2 {k_F}^2/2m + \hbar^2 k_F k/2m$, where we used a linear approximation for the energy dispersion. We will omit the first term since this term is a constant. We replace k with k/k_F, and the annihilation operator of the electron is denoted as $a_{k\sigma}$. Now, eq. (5.90) is transformed into:

$$B\left\{\int_{-1}^{1} k \sum_{\sigma} {a_{k\sigma}}^{\dagger} a_{k\sigma} dk \; - J\rho \left[\left({A_\uparrow}^{\dagger} A_\uparrow - {A_\downarrow}^{\dagger} A_\downarrow\right) S_z \right.\right.$$
$$\left.\left. + \; {A_\uparrow}^{\dagger} A_\downarrow S_- + {A_\downarrow}^{\dagger} A_\uparrow S_+\right]\right\}, \tag{7.1}$$

where

$$A_\sigma = \int_{-1}^{1} a_{k\sigma} dk, \tag{7.2}$$

$$B = \frac{R\hbar^2 {k_F}^3}{\pi m}. \tag{7.3}$$

The V term has been neglected (see Appendix K). Here, R is the radius of the system. The wavenumber k is in the range -1 to 1. We will represent the Hamiltonian in terms of discrete variables. The wavenumbers k were defined as discrete variables in eq. (5.12). In this definition, however, the wavenumbers

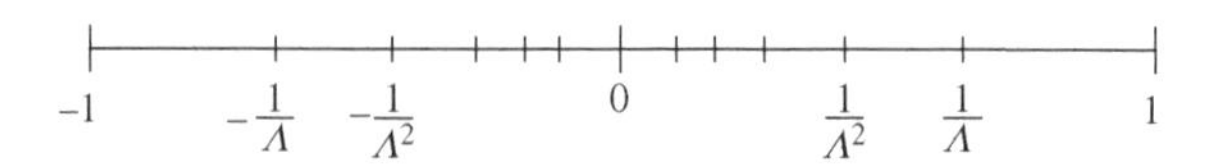

Figure 7.1 Logarithmic discretization of the k-space.

k near the Fermi surface and those away from the Fermi surface are equivalent. The best method of discretization for our problem is shown in Fig. 7.1. Here, Λ is a real number which is larger than 1. ($\Lambda \sim 2-3$ in practical calculations.) We can perform this transformation consistently as follows. In general, when we have a complete orthonormal system $\{\phi_n(k)\}$ defined in the range $-1 \leqq k \leqq 1$, we can transform this set to another complete orthonormal system via

$$a_n = \int_{-1}^{1} \phi_n^*(k) a_k dk.$$

The inverse transformation is

$$a_k = \sum_n \phi_n(k) a_n. \tag{7.4}$$

Now, we define a set of functions that are complete in a certain domain and vanish in the other domains. In the domain $\Lambda^{-m-1} < k < \Lambda^{-m}$, we define

$$\phi_{ml(+)}(k) = (\Lambda^{-m} - \Lambda^{-m-1})^{-1/2} e^{i\omega_m kl}, \tag{7.5}$$

$$\omega_m = \frac{2\pi}{\Lambda^{-m} - \Lambda^{-m-1}}. \tag{7.6}$$

These functions, for all integers l, form a complete orthonormal system. Similarly, we define a function $\phi_{ml(-)}(k)$ which takes the same value as in eq. (7.5) in the domain $-\Lambda^{-m} < k < -\Lambda^{-m-1}$ and vanishes in the other domains. Then, $\{\phi_n(k)\}$, for $n = ml(s)$ ($m = 0, 1, \ldots$; l is an integer; s is positive or negative), forms a complete orthonormal system in the range $-1 \leqq k \leqq 1$.

To represent A_σ in eq. (7.2) in terms of of a_n, we integrate eq. (7.4) with respect to k. We then only have the $l = 0$ terms. If we define the annihilation operators for $m0(+)$ and $m0(-)$ as a_m and b_m, respectively, we obtain

$$A = \int_{-1}^{1} a_k dk = (1 - \Lambda^{-1})^{1/2} \sum_{m=0}^{\infty} \Lambda^{-m/2}(a_m + b_m). \tag{7.7}$$

By substituting eq. (7.4) into the first term of eq. (7.1), the kinetic term in eq. (7.1) is written as

$$\int_{-1}^{1} k\phi_{n'}{}^{*}(k)\phi_n(k)dk.$$

This integral vanishes if indices $m(s)$ and $m'(s')$ are different, and is finite if $m(s)$ and $m'(s')$ coincide, in general even for different l and l'. We can substitute $k-k_m$ for k if $l \neq l'$ (owing to the orthogonality of $\phi_n(k)$) in the integral, where k_m is an arbitrary constant. If we set $k_m = (\Lambda^{-m} + \Lambda^{-m-1})/2$, $k - k_m$ becomes extremely small, especially for large m. Hence, hereafter we neglect the terms with $l \neq l'$. Since only the electrons with $l = 0$ interact with the localized spin, we consider only the terms with $l = l' = 0$ in the kinetic term. Thus, by eq. (7.1),

$$\begin{aligned} H_K = &\sum_m \frac{1}{2}(\Lambda^{-m} + \Lambda^{-m-1}) \sum_\sigma (a_{m\sigma}{}^{\dagger} a_{m\sigma} - b_{m\sigma}{}^{\dagger} b_{m\sigma}) \\ &-J\rho[(A_\uparrow{}^{\dagger}A_\uparrow - A_\downarrow{}^{\dagger}A_\downarrow)S_z + A_\uparrow{}^{\dagger}A_\downarrow S_- + A_\downarrow{}^{\dagger}A_\uparrow S_+], \end{aligned} \tag{7.8}$$

where we have neglected the factor B and have taken the spins into account. To consider only the $l = 0$ terms is equivalent to representing operators for all k in one domain by a single operator. The mean kinetic energy is given by the kinetic energy at the midpoint of the domain, and the real-space behavior of the wave that corresponds to this operator is represented by

$$\int \phi_n(k)\psi_k(r)dk,$$

which spreads more as m becomes greater. Here, $\psi_k(r)$ is given by eq. (K. 1), page 240, with $l = m = 0$.

It is quite natural to divide the domain into domains on the logarithmic scale, as shown in Fig. 7.1, for our purposes. Our treatment goes as follows. First, we diagonalize the sum of the s–d interaction and the largest kinetic term in the Hamiltonian. Second, we add the second largest kinetic term and diagonalize the new sum, where the basis states are the eigenfunctions obtained in the first step. Third, we add the next largest kinetic term to the second sum and perform diagonalization in a similar way. After repeating these steps further, the eigenfunction so obtained contains more information on the states near the Fermi surface. We may say that these perturbative steps, for instance, correspond to lowering the temperature. In order to perform this program completely, we need to transform the Hamiltonian again to a more suitable form. Note that the operators A interact with the localized spin, and they include terms with all values of m from small to large numbers as in eq. (7.7). Thus, we cannot use A as the interaction part and $m=0$ for the kinetic energy term in the initial Hamiltonian. Let us transform a_m and b_m to new variables f_n:

$$f_n = \sum_m (u_{nm} a_m + v_{nm} b_m). \tag{7.9}$$

First, f_0 is defined by

$$f_0 = \frac{A}{\sqrt{2}}. \tag{7.10}$$

Then, we have

$$u_{0m} = v_{0m} = \left[\frac{1-\Lambda^{-1}}{2}\right]^{1/2} \Lambda^{-m/2}. \tag{7.11}$$

The factor $1/\sqrt{2}$ is due to the normalization condition $f_0 f_0{}^\dagger + f_0{}^\dagger f_0 = 1$. Here, f_1 is defined so as to be orthogonal to f_0 and to satisfy the following condition, expressed for general terms f_n. If $\{f_n\}$ is a complete system, u and v are unitary matrices. Further, if both u and v are real, the inverse transformations are

$$\left.\begin{aligned} a_m &= \sum_n u_{nm} f_n \\ b_m &= \sum_n v_{nm} f_n \end{aligned}\right\}. \tag{7.12}$$

Our purpose is to transform the first term of eq. (7.8) into

$$\sum_n \varepsilon_n (f_n{}^\dagger f_{n+1} + f_{n+1}{}^\dagger f_n). \tag{7.13}$$

The spin indices have been omitted. Of course, if we could transform this term to $\sum_n \varepsilon_n f_n{}^\dagger f_n$, the task would be easier. This is, however, impossible, and thus the above form is the simplest. We substitute eq. (7.12) for a_m and b_m in the first term of eq. (7.8), and obtain

$$\sum_{mn} \frac{1}{2}(\Lambda^{-m} + \Lambda^{-m-1})(a_m{}^\dagger u_{nm} - b_m{}^\dagger v_{nm}) f_n.$$

Let us consider the f_0 term. The coefficient of f_0 should be $\varepsilon_0 f_1{}^\dagger$ according to eq. (7.13). Using the condition $f_1 f_1{}^\dagger + f_1{}^\dagger f_1 = 1$ (where u_{0m} and v_{0m} have been defined), ε_0 and $f_1{}^\dagger$ are determined as

$$\varepsilon_0 = \frac{1+\Lambda^{-1}}{2}\left(\frac{1-\Lambda^{-1}}{1-\Lambda^{-3}}\right)^{1/2}, \tag{7.14}$$

$$f_1^\dagger = \left[\frac{1-\Lambda^{-3}}{2}\right]^{1/2} \sum_m \Lambda^{-3m/2}(a_m^\dagger - b_m^\dagger). \tag{7.15}$$

The coefficient of f_1 must be equal to $\varepsilon_0 f_0^\dagger + \varepsilon_1 f_2^\dagger$, by eq. (7.13). This yields ε_1 and $f_2^\dagger$. This procedure can be generalized to all n, and we obtain

$$\varepsilon_n = \frac{[(1+\Lambda^{-1})/2]\Lambda^{-n/2}(1-\Lambda^{-n-1})}{[(1-\Lambda^{-2n-1})(1-\Lambda^{-2n-3})]^{1/2}}. \tag{7.16}$$

$\{f_n\}$, determined by this method, forms a complete set. If n is sufficiently large, ε_n is given by

$$\varepsilon_n = \left[\frac{1+\Lambda^{-1}}{2}\right]\Lambda^{-n/2}. \tag{7.17}$$

We consider this formula to be appropriate for all n in the problem. We use this expression in eq. (7.13). Then, the Hamiltonian is given in place of H_K by substituting eq. (7.10) into the s–d interaction term in eq. (7.8):

$$H = \sum_{n=0}^{\infty} \Lambda^{-n/2} \sum_\sigma (f_{n\sigma}^\dagger f_{n+1\sigma} + f_{n+1\sigma}^\dagger f_{n\sigma})$$

$$-2\tilde{J}[(f_{0\uparrow}^\dagger f_{0\uparrow} - f_{0\downarrow}^\dagger f_{0\downarrow})S_z + f_{0\uparrow}^\dagger f_{0\downarrow} S_- + f_{0\downarrow}^\dagger f_{0\uparrow} S_+], \tag{7.18}$$

$$\tilde{J} = \frac{2J\rho}{1+\Lambda^{-1}}, \tag{7.19}$$

where the constant factor $(1+\Lambda^{-1})/2$ is omitted.

The meaning of this Hamiltonian is as follows. The localized spin is located at the origin of a straight line, and the atomic orbitals are placed to its right on this line. The localized spin has an exchange interaction only with the leftmost atomic orbital. There are transfer integrals amongst neighboring orbitals, and these become smaller as we move further away from the origin. The energy of the atomic orbitals equals the chemical potential, which is set to zero.

This Hamiltonian is suitable for performing the method described above. First, we obtain the eigenfunctions for the system of the localized spin and f_0. Second, we consider the system of the localized spin, f_0 and f_1 with the perturbation $f_0^\dagger f_1 + f_1^\dagger f_0$, where the basis functions are the eigenfunctions obtained in the first step. We repeat this procedure for larger n, by treating $\Lambda^{-n/2}(f_n^\dagger f_{n+1} + f_{n+1}^\dagger f_n)$ as the perturbation. It is important that the perturbations are diminished as n is increased in this process. This completely differs from the conventional method where we use an expansion in terms of $\tilde{J}$. When n is large, we need to use computers. The solution for approximately $n = 100$

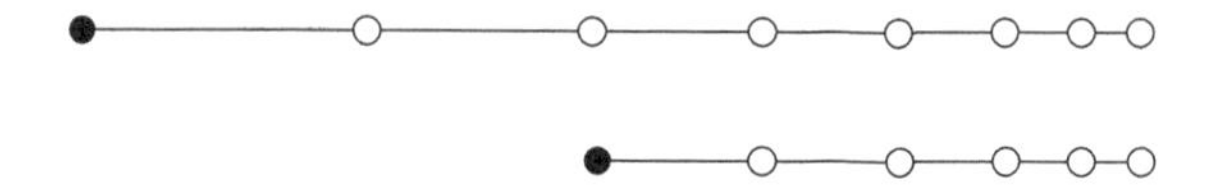

Figure 7.2 The Hamiltonian H_N. The solid circle and open circles denote the localized spin and the electronic orbitals, respectively. The solid circle has an exchange interaction with its neighbor on the right, which is an open circle. Between the nearest-neighbor open circles, there are hopping integrals, whose magnitudes are proportional to the distance between them. The upper figure is for $N+1=7$, and the lower figure is for $N+1=5$. Regardless of the value of N, the distance between the open circles at the rightmost end is unity. When N is increased, these figures stretch to the left.

can be regarded as the final solution for practical purposes. In this sense, we can say that the problem is exactly solved.

For this purpose, we define a set of Hamiltonians H_N as follows:

$$H_N = \Lambda^{(N-1)/2}\left\{\sum_{n=0}^{N-1}\Lambda^{-n/2}\sum_{\sigma}(f_{n\sigma}{}^{\dagger}f_{n+1\sigma}+f_{n+1\sigma}{}^{\dagger}f_{n\sigma})\right.$$
$$\left.-2\tilde{J}[(f_{0\uparrow}{}^{\dagger}f_{0\uparrow}-f_{0\downarrow}{}^{\dagger}f_{0\downarrow})S_z+f_{0\uparrow}{}^{\dagger}f_{0\downarrow}S_-+f_{0\downarrow}{}^{\dagger}f_{0\uparrow}S_+]\right\}. \tag{7.20}$$

Here, the infinite series of n in H is truncated at $n=N$. This Hamiltonian contains the factor $\Lambda^{(N-1)/2}$, so that the last term $f_{N-1}{}^{\dagger}f_N+f_N{}^{\dagger}f_{N-1}$ is of order unity regardless of the value of N (see Fig. 7.2). The excited states that are near the ground state are always important. We arranged the coefficients so that we can compare the results easily for different N. H_N is also written as

$$H_{N+1} = \Lambda^{1/2}H_N+\sum_{\sigma}(f_{N\sigma}{}^{\dagger}f_{N+1\sigma}+f_{N+1\sigma}{}^{\dagger}f_{N\sigma}). \tag{7.21}$$

7.2 Perturbative expansions

First, we consider the case $\tilde{J}=0$. We neglect the spin indices in this section. In this case, H_N is a quadratic form in f_n. Using the linear transformation

$$g_l = \sum_n M_{nl}f_n, \tag{7.22}$$

H_N is diagonalized as follows:

$$H_N(\tilde{J}=0)=\sum_l \eta_l g_l^\dagger g_l. \tag{7.23}$$

The kinetic term was diagonalized earlier by using a_m and b_m. g_l are not equal to a_m and b_m since there are only $N+1$ copies of f_n. $(N+1)\times(N+1)$ matrices can be easily diagonalized by numerical methods. The eigenvalues η_l have the following properties. When $N+1$ is even, all eigenvalues appear in pairs of equal absolute values and opposite signs. When $N+1$ is odd, one eigenvalue is zero and the others appear in pairs. This is due to the particle–hole symmetry. (H_N is invariant under the transformation $f_n \leftrightarrow (-)^n f_n^\dagger$.) Since the chemical potential is set to zero, the ground state is the state where the eigenstates with negative eigenvalues are occupied with electrons of both spins. For negative η_l, we define the creation operators of holes:

$$h_l^\dagger = g_l. \tag{7.24}$$

Then the excitation energy of holes is positive because we have $\eta_l g_l^\dagger g_l = \eta_l - \eta_l h_l^\dagger h_l$. The Hamiltonian H_N, after diagonalization, is written as

$$\left.\begin{aligned} H_N(\tilde{J}=0) &= \sum_{l=1}^{(N+1)/2} \eta_l (g_l^\dagger g_l + h_l^\dagger h_l) && (N \text{ odd})\\ H_N(\tilde{J}=0) &= \eta'_0 g_0^\dagger g_0 + \sum_{l=1}^{N/2} \eta'_l (g_l^\dagger g_l + h_l^\dagger h_l) && (N \text{ even}) \end{aligned}\right\}. \tag{7.25}$$

using only positive η_l. Here $\eta'_0=0$, and we neglected the constant term.

According to numerical calculation, if N is larger than approximately 20, η_l and η'_l are independent of N unless l is very close to N. For example, for $\Lambda=2$, they are given by

$$\left.\begin{aligned} &\eta_1 = 0.6555, \quad \eta_2 = 1.976, \quad \ldots, \quad \eta_l = 2^{l-1}\\ &\eta'_0 = 0, \quad \eta'_1 = 1.297, \quad \eta'_2 = 2.827, \quad \ldots, \quad \eta'_l = 2^{l-\frac{1}{2}} \end{aligned}\right\}. \tag{7.26}$$

For general Λ, we have $\eta_l = \Lambda^{l-1}$ and $\eta'_l = \Lambda^{l-\frac{1}{2}}$ unless l is near 1. The inverse transformations of eq. (7.22) are (for odd N and $\Lambda=2$)

$$\left.\begin{aligned} f_0 &= 2^{-(N-1)/4} \sum_{l=1}^{(N+1)/2} \alpha_l (g_l + h_l^\dagger) \\ f_1 &= 2^{-3(N-1)/4} \sum_{l=1}^{(N+1)/2} \gamma_l (g_l - h_l^\dagger) \end{aligned}\right\}, \tag{7.27}$$

$$\left.\begin{aligned} &\alpha_1 = 0.588, \quad \alpha_2 = 0.629, \quad \ldots, \quad \alpha_l = \alpha 2^{(l-1)/2} \quad \alpha = 0.4307 \\ &\gamma_1 = 0.386, \quad \gamma_2 = 1.243, \quad \ldots, \quad \gamma_l = \gamma 2^{3(l-1)/2} \quad \gamma = 0.4307 \end{aligned}\right\}. \tag{7.28}$$

For general Λ, we have $\Lambda^{-(N-1)/4}$ instead of $2^{-(N-1)/4}$ and so on.

The first term of eq. (7.20) has been diagonalized. Let us now carry out a calculation based on conventional perturbation theory in terms of $\tilde{J}$. We calculate the energy shift ΔE_G of the ground state. When N is odd, there are neither electrons nor holes in the ground state, and we have localized spin. This state is represented by, for instance, the spin-up state. If we substitute eq. (7.27) into the $\tilde{J}$ term in eq. (7.20), we obtain many terms. Among them, the terms that have the form $g_l^\dagger h_{l'}^\dagger$ create electron–hole pairs and they contribute to the second-order perturbation. After simple calculations, we obtain

$$\Delta E_G = -\frac{3\tilde{J}^2}{2} \sum_{ll'} \frac{\alpha_l^2 \alpha_{l'}^2}{\eta_l + \eta_{l'}}$$

up to second order. This quantity increases rapidly with the increase in N, but is not so important by itself. Next, we calculate the energy shift of the first excited state, ΔE_1. In the first excited state, we have one extra electron (or hole) corresponding to η_l which couples with the localized spin to form a singlet state $S = 0$. The energy of this state is higher than that of the ground state by an energy that corresponds to η_1. This is given by $(3\tilde{J}/2)\alpha_1^2$ in the first order of perturbation, and we have also corrections due to the second- and higher-order perturbation. After some calculations, z_N, defined by

$$z_N \equiv \Delta E_G - \Delta E_1, \tag{7.29}$$

is given by

$$z_N = -\frac{3}{2}\alpha_1^2 \tilde{J} + \frac{3}{4}\alpha_1^2 \tilde{J}^2 \left(\sum_l \frac{\alpha_l^2}{\eta_l + \eta_1} + 3 \sum_{l \neq 1} \frac{\alpha_l^2}{\eta_l - \eta_1} \right) \tag{7.30}$$

up to second order in $\tilde{J}$. Here, according to eqs. (7.26) and (7.28), the second-order terms are of order N. In general, the nth-order terms are of order N^{n-1}.

This is because the perturbative term includes spin operators. If the perturbative term is of the form $f_0{}^\dagger f_0$ etc., the higher-order terms are always of order unity. Thus, in this method, the divergence appears as N or powers of N, instead of the usual logarithmic divergence. This result originates from the discretization scheme shown in Fig. 7.1, and is natural.

z_N, defined by eq. (7.29), plays a significant role in the following. The central issue here is to investigate the behavior of z_N when N is increased. In this section, we examine this by using perturbation theory in terms of $\tilde{J}$. By numerical calculations, we obtain

$$z_N = 1.5J' + 27.2808J'^2 + 469.3577J'^3 + 7661.9091J'^4 + \cdots \quad (7.31)$$

for $\Lambda = 2$ and $N = 31$. Here, $J' = -\alpha_1{}^2\tilde{J}$. We can see that the higher-order coefficients become larger. Similarly, we obtain z_N for $N = 29$ by expanding in terms of J'. After J' is eliminated from these equations, we obtain

$$z_{N+2} = z_N + 0.71436z_N{}^2 + 0.14220z_N{}^3 - 2.9214z_N{}^4 \quad (N \text{ odd}). \quad (7.32)$$

This equation is for $N = 29$, and we can indeed confirm by numerical calculations that the coefficients are independent of N if N is not small. Since the coefficients in eq. (7.32) are clearly independent of $\tilde{J}$, eq. (7.32) can be regarded as a universal relation (where $\Lambda = 2$).

By the successive use of eq. (7.32), we can calculate $z_{N'}$ if z_N is known for some N. We convert eq. (7.32) to a differential equation, and then integrate it to obtain the formula which relates z_N to $z_{N'}$ (see Appendix L):

$$\Psi(z_N) + \tfrac{1}{2}N = \Psi(z_{N'}) + \tfrac{1}{2}N' \quad (N \text{ odd}), \quad (7.33)$$

$$\Psi(z) \equiv (0.71436z)^{-1} - 0.72135 \log(0.71436z) - 7.7310(0.71436z). \quad (7.34)$$

We can obtain z_N for arbitrary N from eq. (7.33); z_N increases as N increases. This equation is irrelevant if z_N is of order unity. This is because our discussions are based on the perturbative expansion, eq. (7.31), where J' is assumed to be sufficiently smaller than 1.

Equation (7.33) indicates that $\Psi(z_N) + \frac{1}{2}N$ is independent of N. This quantity, however, depends on $J' = -\alpha_1{}^2\tilde{J}$. We can clarify the dependence on J' by substituting eq. (7.31) and $N = 31$ into $\Psi(z_N) + \frac{1}{2}N$. Later on, we will need to know the functional form of $\Psi(z_N) + \frac{1}{2}N$ as a function of $2J\rho$, which was defined in Chapter 5. For this purpose, eq. (7.19) is inappropriate, since some approximations are used in this transformation. To obtain the correct functional form of $\Psi(z_N) + \frac{1}{2}N$, we calculate the susceptibility by both the conventional and present methods and extrapolate the two results. This leads to the following result (see Appendix M):

$$\left[\Psi(z_N) + \frac{1}{2}N\right] \cdot \log 2 = -\frac{0.5}{J\rho} - 0.30873 - 0.5 \log |2J\rho| - \log \frac{\tilde{D}}{A}, \quad (7.35)$$

$$\tilde{D} = 0.6001D(1 + O(J\rho)),$$
$$A = \frac{1 + \Lambda^{-1}}{2}B. \quad (7.36)$$

Here, B is given by eq. (7.3).

7.3 Numerical calculations: scaling

Next, we discuss the treatment of the Hamiltonian in eq. (7.20) without using perturbative calculations. There are $N+1$ orbitals in eq. (7.20), and each orbital has four states (i.e., the occupation number of each orbital is 0, 1, or 2; we have spin-up and down states when the occupation number is 1). Since the localized spin has two spin states, the total basis set is composed of 2^{2N+3} wavefunctions. Let us suppose that both the eigenfunctions and eigenvalues of this system are given and the matrix elements of f_N and $f_N{}^\dagger$ between these eigenfunctions are also computed. Now, we have all the information to solve H_{N+1} on the basis of eq. (7.21). For each eigenstates $|k\rangle$ of H_N, we have four states that form a basis for H_{N+1},

$$|k\rangle, \quad f_{N+1\uparrow}{}^\dagger|k\rangle, \quad f_{N+1\downarrow}{}^\dagger|k\rangle, \quad f_{N+1\uparrow}{}^\dagger f_{N+1\downarrow}{}^\dagger|k\rangle. \quad (7.37)$$

$\Lambda^{1/2}H_N$ in eq. (7.21) is diagonalized with respect to these bases, and its diagonal elements are $\Lambda^{1/2}E_N(k)$, where $E_N(k)$ is the energy of $|k\rangle$. The matrix elements of $f_N{}^\dagger f_{N+1} + f_{N+1}{}^\dagger f_N$ are calculated by using matrix elements $\langle k'|f_N|k\rangle$ and others. Hence, we can solve H_{N+1}. It follows in general that the matrix elements $\langle k|f_N{}^\dagger f_{N+1} + f_{N+1}{}^\dagger f_N|k'\rangle$ are of order unity if $E_N(k)$ is close to $E_N(k')$; otherwise they are extremely small. Now, we can proceed as follows. We keep about 1000 of the lowest eigenstates of H_N, and calculate 4000 eigenstates of H_{N+1} given by eq. (7.37). We then keep 1000 lowest eigenstates and calculate the matrix elements of f_{N+1} and $f_{N+1}{}^\dagger$ between these states. We now have all the necessary information to set up the matrix for H_{N+2}. We can increase N by successive use of this procedure. In actual computations, we only need to diagonalize an approximately 100×100 matrix since the diagonalization

Table 7.1 *The first and second excitation energies for* $\tilde{J} = -0.024$ *and* $\Lambda = 2$.

N	20	22	108	110	130	180
E_1	0.0314	0.0321	0.313	0.363	0.6541	0.6555
E_2	0.0419	0.0428	0.446	0.529	1.3055	1.3110

is performed in the subspace with fixed quantum numbers such as the total number of electrons, the magnitude of the total spin, and the z-component of the total spin. According to numerical calculations performed in this way, the energy of the first excited state is $\lesssim 1$ above the ground state, and the first 1000 energy levels of H_N are inside the energy range of up to $\sim$6–10.

The first and second excitation energies, denoted E_1 and E_2, respectively, are very important. The behaviors of these quantities as a function of N are completely different, depending on whether N is odd or even, like the case $\tilde{J} = 0$. For example, we show the results for $\tilde{J} = -0.024$ and $\Lambda = 2$ for even N in Table 7.1. Since the present value of $\tilde{J}$ is sufficiently smaller than unity, we can roughly estimate these excitation energies within the perturbative calculation if N is small. For even N, both the first and second excitation energies vanish for $\tilde{J} = 0$. This is consistent with Table 7.1. On the other hand, if N is sufficiently large, E_1 and E_2 coincide with the corresponding excitation energies at $\tilde{J} = -\infty$. When $\tilde{J} = -\infty$, the localized spin and the f_0 electron form a singlet, and we need an infinite amount of energy to break this singlet. This state is thus separated from the others, and we have only free electrons, denoted $f_1, \ldots, f_N$, in the Hamiltonian. The first excited state is given by $g_1^\dagger$ and the second excited state is given by $g_1{}^\dagger h_1{}^\dagger$ etc. Their energies are 0.6555 and 0.6555 × 2, respectively, from eq. (7.26). The higher excited states coincide completely with those for $\tilde{J} = -\infty$, and the total number of electrons, the magnitude of total spin and the z-component of total spin are also consistent. Hence we have obtained an important result here that the behavior for $\tilde{J} = -\infty$ can be obtained even for small $\tilde{J}$ if N is sufficiently large.

Another important result is the scaling property. In general, when the absolute value of $\tilde{J}$ is large, the $\tilde{J} = -\infty$ limit is reached quickly even for small N. It is shown that the results for two different $\tilde{J}$, such as E_1 and E_2, match very well when N is shifted by a constant (see Table 7.2) if N is not small. To be more precise, the solution for $\tilde{J}$ and N is equivalent to that for appropriately chosen $\tilde{J}_l$ and $N + l$. When N is increased further, one solution coincides with the other after the shift of N. This is easily demonstrated as follows. Let us suppose that

Table 7.2 *If N is not small, the variation of $\tilde{J}$ is equivalent to that of N. The first three columns from the left show the two excitation energies for $\tilde{J} = -0.055016$. The three columns to the right show those for $\tilde{J} = -0.02424$. When N is shifted by 56, the two results completely coincide (unless N is small).*

$\tilde{J} = -0.055016$			$\tilde{J} = -0.02424$		
N	E_1	E_2	N	E_1	E_2
6	0.075257	0.10067	62	0.07261	0.09714
8	0.07777	0.10407	64	0.07662	0.10255
10	0.08156	0.10919	66	0.08109	0.10858
20	0.11367	0.15275	76	0.11377	0.15290
30	0.18440	0.25076	86	0.18450	0.25092
40	0.38734	0.56332	96	0.38737	0.56338
50	0.68168	1.31203	106	0.68166	1.31199

we have two chains with different lengths as in Fig. 7.2, and the solutions at the rightmost ends are the same. To be more exact, we assume that the eigenvalues and eigenfunctions of the two chains, whose excitation energies are of order unity, are the same. We add an extra orbital and a perturbation of order unity at the rightmost end. Then, clearly, we obtain the same result for both chains. This scaling never holds for short chains where the effect of the leftmost end is not negligible. Thus, in the regime where the scaling property holds, all the properties (in the region where the energy is of order unity) are determined by a single parameter. Let us set z_N in eq. (7.29) as the parameter of the system. This is written as

$$z_N = \eta_1 - E_{1N}, \tag{7.38}$$

where E_{1N} is the first excitation energy of H_N. The dependence of z_N on $\tilde{J}$ and N is shown in Fig. 7.3 for odd N. Except in the region of small z_N, the curves for two different $\tilde{J}$ coincide after lateral translation. In the region of sufficiently small z_N, we expect z_N to satisfy eqs. (7.32) and (7.33). This has been confirmed with sufficient precision.

As discussed previously, a state with energy of order unity will approach the solution for $\tilde{J} = -\infty$ as N is increased. We define H^* as the limiting form of H_N in the limit $\tilde{J} = -\infty$ and $N = \infty$. For even N, this is written as

$$H^* = \sum_{l=1}^{\infty} \eta_l \sum_{\sigma} (g_{l\sigma}{}^{\dagger} g_{l\sigma} + h_{l\sigma}{}^{\dagger} h_{l\sigma}). \tag{7.39}$$

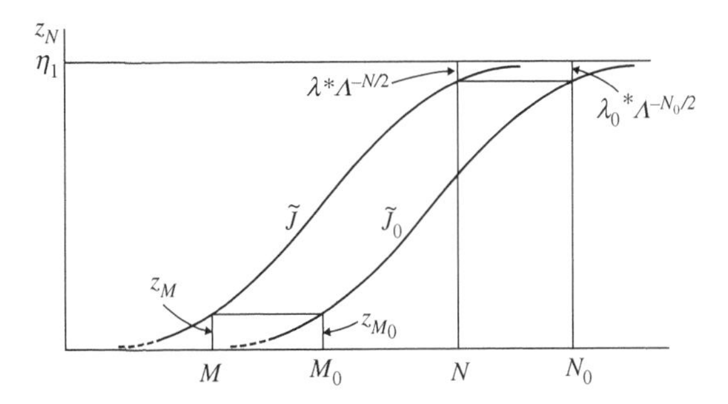

Figure 7.3 The z_N curves for two different values of $\tilde{J}$ ($|\tilde{J}| > |\tilde{J}_0|$). Except in the region of very small z_N (dashed lines), the two curves coincide after lateral translation. Thus, $M - M_0 = N - N_0$.

Now, let us consider the effective potential that represents the difference between the spectra of H_N for large N and those of H^*. This is defined as

$$H_N = H^* + \frac{\lambda^* \Lambda^{(N-1)/2}}{2\sqrt{2}\alpha_1 \gamma_1} \sum_\sigma (f_{0\sigma}{}^\dagger f_{1\sigma} + f_{1\sigma}{}^\dagger f_{0\sigma}) + \frac{\omega^* \Lambda^{(N-1)/2}}{2\sqrt{2}\alpha_1{}^4} (f_{0\uparrow}{}^\dagger f_{0\uparrow} + f_{0\downarrow}{}^\dagger f_{0\downarrow} - 1)^2. \tag{7.40}$$

The coefficients of the last two terms are defined for later convenience, and λ^* and ω^* are parameters. The N dependence of each of these two terms is $\Lambda^{-(N-1)/2}$ as shown by substituting eqs. (7.27) etc. into them. (We should be careful with the second term.) This dependence needs to be rather slow so that this potential is effective for large N. The terms of higher order than or equal to f_2 are excluded. Moreover, the effective potential must have particle–hole symmetry (i.e., must be invariant under the transformation $g_l \leftrightarrow h_l$) as must H_N. The form of eq. (7.40) is determined from these conditions.

Next, we need to confirm that this form indeed represents H_N for large N. We substitute eq. (7.27) for H_N in eq. (7.40) to obtain

$$H_N = H^* + \lambda^* \Lambda^{-N/2} \sum_l \frac{\alpha_l \gamma_l}{\alpha_1 \gamma_1} \sum_\sigma (g_{l\sigma}{}^\dagger g_{l\sigma} + h_{l\sigma}{}^\dagger h_{l\sigma}) + \omega^* \Lambda^{-N/2} \sum_{l_1 \cdots l_4} \frac{\alpha_{l_1}\alpha_{l_2}\alpha_{l_3}\alpha_{l_4}}{2\alpha_1^4} \sum_\sigma g_{l_1\sigma}{}^\dagger g_{l_2-\sigma}{}^\dagger g_{l_3-\sigma} g_{l_4\sigma} + \cdots. \tag{7.41}$$

Here we neglect the constant terms, non-diagonal terms, and some of the diagonal terms. In order to determine the two parameters, we examine the two excited states. For example, let us consider $g_{1\uparrow}{}^{\dagger}|0\rangle$ and $(g_{1\uparrow}{}^{\dagger}h_{1\downarrow}{}^{\dagger} - g_{1\downarrow}{}^{\dagger}h_{1\uparrow}{}^{\dagger})/\sqrt{2}|0\rangle$; $|0\rangle$ is the ground state of H^*, where we have neither electrons nor holes ($g_l|0\rangle = h_l|0\rangle = 0$). In the first-order perturbation of eq. (7.41), the energies of these states are, respectively,

$$\eta_1 + \lambda^* \Lambda^{-N/2}, \quad 2\eta_1 + (2\lambda^* + \omega^*)\Lambda^{-N/2}. \tag{7.42}$$

On the other hand, we can evaluate the excitation energies for these two states from the numerical results for H_N. Then we can determine λ^* and ω^* consistently. For example, we obtain $\lambda^* = -7.015 \times 10^{-4} \times 2^{66}$ and $\omega^* = 13.850 \times 10^{-4} \times 2^{66}$ for $\Lambda = 2$, $\tilde{J} = -0.024$ and $N = 132$. Using these values, let us determine the other excitation energies. For instance, the energy of the state $g_{1\uparrow}{}^{\dagger}g_{2\uparrow}{}^{\dagger}|0\rangle$ is $\eta_1 + \eta_2 + 4.444\lambda^*\Lambda^{-N/2}$, which agrees well with the results of numerical calculations if we use λ^* as estimated above. Other excitation energies are determined in a similar manner. Furthermore, we find that λ^* and ω^* estimated for $N = 144$ agree with the above values to within an error of less than 0.5%. Thus, when N is sufficiently large, the effective Hamiltonian in eq. (7.41) is a good approximation for H_N (within first-order perturbation). The analytical form of an effective Hamiltonian such as this is useful for treating the limit of large N.

λ^* and ω^* obtained in this way depend on $\tilde{J}$. However, as mentioned previously, H_N is completely determined by a single parameter z_N. Therefore, λ^* and ω^* are not independent of each other, and the ratio of them does not depend on $\tilde{J}$ (but depends on Λ). For $\Lambda = 2$, we have $\lambda^*/\omega^* = -0.507$.

It becomes important later on to know the form of λ^* and ω^* as a function of $\tilde{J}$. Let us achieve this as follows. For sufficiently large odd N, the first excitation energy is obtained as

$$z_N = \eta_1 + \lambda^* \Lambda^{-N/2}, \tag{7.43}$$

using the effective Hamiltonian in eq. (7.41). Let us suppose that we obtain $\lambda^* = \lambda_0{}^*$ for a certain value of $\tilde{J}_0$ by numerical calculation (see Fig. 7.3). We then consider the dependence of λ^* on $\tilde{J}$. As shown in Fig. 7.3, two z_N curves coincide after translation. If we determine N such that the two curves agree, we can obtain λ^* from the relation

$$\lambda^* \Lambda^{-N/2} = \lambda_0{}^* \Lambda^{-N_0/2}. \tag{7.44}$$

It is worth noting that $N - N_0$, shown in Fig. 7.3, can be obtained even for small N. Let us suppose that z_N is calculated as z_{M_0} for $\tilde{J}_0$ and M_0, as shown in Fig. 7.3. If z_M, calculated for $\tilde{J}$ and M, coincides with z_{M_0}, then λ^* can be obtained from $M - M_0 = N - N_0$. Here, according to eq. (7.35) (for $\Lambda = 2$), we have

$$\Psi(z_M) + \frac{1}{2}M = \frac{-\dfrac{0.5}{J\rho} - 0.30873 - 0.5\log|2J\rho| - \log\dfrac{\tilde{D}}{A}}{\log 2}.$$

$J\rho$, which is defined in Chapter 5, corresponds to $\tilde{J}$. Thus, using $z_M = z_{M_0}$, we obtain

$$\frac{1}{2}(M - M_0) = \frac{-\dfrac{0.5}{J\rho} - 0.30873 - 0.5\log|2J\rho| - \log\dfrac{\tilde{D}}{A}}{\log 2} - \Psi(z_{M_0}) - \frac{1}{2}M_0. \tag{7.45}$$

By using the formula $M - M_0 = N - N_0$, and eq. (7.44), we have

$$\lambda^* = \lambda_0{}^* \exp\left[-\frac{0.5}{J\rho} - 0.30873 - 0.5\log|2J\rho| - \log\frac{\tilde{D}}{A} - \log 2 \cdot \left(\Psi(z_{M_0}) + \frac{1}{2}M_0\right)\right]. \tag{7.46}$$

When $\tilde{J}_0 = -0.024$, $z_{M_0} = 0.038657$ for $M_0 = 39$. Hence, $\Psi(z_{M_0}) = 38.588$ from eq. (7.34). Since $\lambda_0{}^* = -7.015 \times 10^{-4} \times 2^{66}$, we obtain after substituting this value,

$$\lambda^* = -0.1241 \exp\left[-\frac{0.5}{J\rho} - 0.5\log|2J\rho| - \log\frac{\tilde{D}}{A}\right]. \tag{7.47}$$

This is an important result that will be used later.

7.4 Susceptibility and specific heat

Let us calculate susceptibility and specific heat on the basis of the above results. We consider how these quantities change due to the localized spin. We define H_{N0} by eliminating the localized spin from H_N. The g-factor of both the localized spin and the electrons is set to 2. We take into account the factor

$B \times (1 + \Lambda^{-1})/2$ when we go back to the original Hamiltonian from H_N. Then, the susceptibility χ is

$$\chi = \frac{4\mu_B{}^2}{k_B T} \lim_{N\to\infty} \left\{ \frac{\operatorname{Tr} \mathcal{S}_N{}^2 \exp[-\Lambda^{-(N-1)/2} H_N/\tau]}{\operatorname{Tr} \exp[-\Lambda^{-(N-1)/2} H_N/\tau]} - \frac{\operatorname{Tr} \mathcal{S}_{N0}{}^2 \exp[-\Lambda^{-(N-1)/2} H_{N0}/\tau]}{\operatorname{Tr} \exp[-\Lambda^{-(N-1)/2} H_{N0}/\tau]} \right\}. \tag{7.48}$$

Here we defined

$$\tau = \frac{k_B T}{A}. \tag{7.49}$$

(A was defined by eq. (7.36).) τ is approximately of order 2^{43} for $T = 1$ K and $R = 1$ cm. $\mathcal{S}_N$ is the sum of the z-components of the spin of $N + 1$ electrons and the localized spin, and $\mathcal{S}_{N0}$ is the same sum for $N + 1$ electrons. Tr in the second term indicates the trace operation excluding the localized spin. From now on, we set

$$\beta_N = \frac{\Lambda^{-(N-1)/2}}{\tau}. \tag{7.50}$$

As mentioned above, H_N approaches H^* for sufficiently large N. Therefore, one may perhaps think that we can substitute H^* for H_N in eq. (7.48). This is not, however, correct at high temperatures. The reason is as follows. In the evaluation of the susceptibility in eq. (7.48), the excited states with energies up to the order of $\Lambda^{(N-1)/2}\tau$ are important. However, this threshold energy $\Lambda^{(N-1)/2}\tau$ increases at high temperature, and we cannot replace H_N by H^* in this threshold region. Hence, we first examine the susceptibility in eq. (7.48) at sufficiently low temperature, and substitute H^* for H_N. We also substitute eq. (7.25) for H_{N0} and set $N \to \infty$. As discussed previously, for sufficiently large N, the eigenvalues and the properties of the eigenfunctions (for example, the z-component of the spin) of H_N are described by H^*. Thus, we can substitute the sum of the z-components of the spin of the electrons and holes for $\mathcal{S}_N$:

$$\mathcal{S}_N \to \mathcal{S} \equiv \sum_l \frac{1}{2}(g_{l\uparrow}{}^\dagger g_{l\uparrow} - g_{l\downarrow}{}^\dagger g_{l\downarrow}) - \sum_l \frac{1}{2}(h_{l\uparrow}{}^\dagger h_{l\uparrow} - h_{l\downarrow}{}^\dagger h_{l\downarrow}). \tag{7.51}$$

Similarly, $\mathcal{S}_{N0}$ is expressed only in terms of electrons and holes. Therefore, eq. (7.48) indicates the difference between the Pauli susceptibilities of two free electron systems. This difference is just due to whether N is odd or even. If N is even, H^* is given by eq. (7.39), and H_{N0} is given by the second equation of eq. (7.25). If we define $\sigma_l = \frac{1}{2}(g_{l\uparrow}{}^\dagger g_{l\uparrow} - g_{l\downarrow}{}^\dagger g_{l\downarrow})$, we obtain

$\langle \sigma_l{}^2 \rangle = (1/2)e^{-\beta_N \eta_l}/(1+e^{-\beta_N \eta_l})^2$, and $\langle \sigma_l \sigma_{l'} \rangle = 0$ for $l \neq l'$. Hence we have

$$\frac{k_B T \chi}{4\mu_B{}^2} = \lim_{N\to\infty}\left[\sum_{l=1}^{\infty}\frac{e^{-\beta_N \eta_l}}{(1+e^{-\beta_N \eta_l})^2} - \sum_{l=1}^{\infty}\frac{e^{-\beta_N \eta'_l}}{(1+e^{-\beta_N \eta'_l})^2} - \frac{1}{8}\right]. \quad (7.52)$$

The $1/8$ term comes from η'_0. When β_N is sufficiently smaller than unity, we have contributions only from the region where $\beta_N \eta_l$ (or $\beta_N \eta'_l$) is of order unity. (For small l, the two terms cancel each other. For $\beta_N \Lambda^l \gg 1$, they are exponentially suppressed.) Thus, we can substitute Λ^{l-1} and $\Lambda^{l-\frac{1}{2}}$ for η_l and η'_l, respectively. We can replace the lower limit $l = 1$ by $l = -\infty$. Hence

$$\frac{k_B T \chi}{4\mu_B{}^2} = \lim_{N\to\infty}\left[\sum_{l=-\infty}^{\infty}\frac{e^{-\beta_N \Lambda^{l-1}}}{(1+e^{-\beta_N \Lambda^{l-1}})^2} - \sum_{l=-\infty}^{\infty}\frac{e^{-\beta_N \Lambda^{l-\frac{1}{2}}}}{(1+e^{-\beta_N \Lambda^{l-\frac{1}{2}}})^2} - \frac{1}{8}\right]. \quad (7.53)$$

In the right-hand side of this equation, we can substitute β_N for $\beta_N \Lambda^n$ (since this simply corresponds to the shift of l to $l+n$). Therefore, β_N need not be so small. For example, β_N can take a value in the range 1 to Λ. In actual computations, the value of eq. (7.53) is of order 10^{-5} for $\Lambda = 2$ when β_N is in the range 1 to 2, and is of order 10^{-9} for $\Lambda = 1.4$. We can in fact prove that the above value approaches zero in the limit $\Lambda \to 1$.

We have shown that eq. (7.48) vanishes when we substitute H^* for H_N. As described above, this also holds at sufficiently low temperatures. Next, we consider the corrections to H^*, as in eq. (7.40) or eq. (7.41), up to the first-order perturbation. We expand the correction term in the exponent to obtain

$$\frac{k_B T \chi}{4\mu_B{}^2} = (7.52) - \lim_{N\to\infty} \beta_N (\langle \mathcal{S}^2 H'_N \rangle - \langle \mathcal{S}^2 \rangle \langle H'_N \rangle). \quad (7.54)$$

H'_N is the correction term, and $\langle \cdots \rangle$ is defined by

$$\langle \cdots \rangle = \frac{\mathrm{Tr}(e^{-\beta_N H^*} \cdots)}{\mathrm{Tr}(e^{-\beta_N H^*})}.$$

By using eqs. (7.41) and (7.51), we obtain

$$\frac{k_B T \chi}{4\mu_B{}^2} = \lim_{N\to\infty}\left\{-\Lambda^{-N/2}\beta_N \lambda^* \sum_l \frac{\alpha_l \gamma_l}{\alpha_1 \gamma_1}\frac{e^{-\beta_N \eta_l}(1-e^{-\beta_N \eta_l})}{(1+e^{-\beta_N \eta_l})^3}\right.$$
$$\left. +2\Lambda^{-N/2}\beta_N \omega^* \left[\sum_l \frac{\alpha_l{}^2}{\alpha_1{}^2}\frac{e^{-\beta_N \eta_l}}{(1+e^{-\beta_N \eta_l})^2}\right]^2\right\}. \quad (7.55)$$

We then set $\eta_l = \Lambda^{l-1}$ and use eq. (7.28). The summation with respect to l can be replaced with that from $-\infty$ to ∞. We define χ_1 and χ_2 as

$$\chi_1 = \beta_N{}^2 \log \Lambda \sum_{l=-\infty}^{\infty} \Lambda^{2l-2} \frac{e^{-\beta_N \Lambda^{l-1}}(1 - e^{-\beta_N \Lambda^{l-1}})}{(1 + e^{-\beta_N \Lambda^{l-1}})^3}, \tag{7.56}$$

$$\chi_2 = \beta_N{}^2 (\log \Lambda)^2 \left[\sum_{l=-\infty}^{\infty} \Lambda^{l-1} \frac{e^{-\beta_N \Lambda^{l-1}}}{(1 + e^{-\beta_N \Lambda^{l-1}})^2} \right]^2. \tag{7.57}$$

Then we have

$$\frac{k_B T \chi}{4\mu_B{}^2} = \lim_{N \to \infty} \left[-\frac{\Lambda^{-N/2}\lambda^*}{\beta_N \log \Lambda} \frac{\alpha\gamma}{\alpha_1 \gamma_1} \chi_1 + \frac{2\Lambda^{-N/2}\omega^*}{\beta_N (\log \Lambda)^2} \frac{\alpha^4}{\alpha_1{}^4} \chi_2 \right]. \tag{7.58}$$

Here, both χ_1 and χ_2 are invariant under the change $\beta_N \to \beta_N \Lambda^N$. We thus calculate χ_1 and χ_2 for β_N in the range 1 to Λ. We obtain $\chi_1 \sim 0.5$ and $\chi_2 \sim 0.25$ from numerical calculations. It can be proved that they are exactly 0.5 and 0.25, respectively, in the limit $\Lambda \to 1$. Finally, we obtain

$$\chi = \frac{4\mu_B{}^2}{A} \frac{0.5}{\sqrt{\Lambda} \log \Lambda} \left(-\lambda^* \frac{\alpha\gamma}{\alpha_1 \gamma_1} + \frac{\omega^*}{\log \Lambda} \frac{\alpha^4}{\alpha_1{}^4} \right). \tag{7.59}$$

By using λ^* in eq. (7.47) and $\lambda^*/\omega^* = -0.507$, we obtain

$$\chi = C \frac{4\mu_B{}^2}{T_K}, \tag{7.60}$$

where

$$T_K \equiv \tilde{D}\sqrt{|2J\rho|} e^{1/2J\rho}, \tag{7.61}$$

and C is determined as 0.1032 ± 0.0005 according to the detailed calculations by Wilson.

Next, let us calculate the specific heat at low temperature. In general, the specific heat c is defined as

$$c = k_B \frac{d}{dT} T^2 \frac{d}{dT} \log Z. \tag{7.62}$$

In order to obtain the contributions from the localized spin, we use the formula

$$\log Z = \lim_{N \to \infty} \left(\log \mathrm{Tr}\, e^{-\beta_N H_N} - \log \mathrm{Tr}\, e^{-\beta_N H_{N0}} \right). \tag{7.63}$$

Here, at sufficiently low temperature, we can again substitute eq. (7.41) for H_N; H_{N0} is given by eq. (7.25) in the limit $N \to \infty$. For even N, we have

$$\mathrm{Tr}\, e^{-\beta_N H_{N0}} = 4 \prod_{l=1}^{\infty} (1 + e^{-\beta_N \eta_l'})^4.$$

We expand the first term of eq. (7.63) in terms of H_N', where H_N' is defined by $H_N = H^* + H_N'$. Up to the first order of H_N', we obtain

$$\begin{aligned} \mathrm{Tr}\, e^{-\beta_N H_N} &= \prod_{l=1}^{\infty} (1 + e^{-\beta_N \eta_l})^4 \\ &\quad \times \left[1 - 4\lambda^* \frac{\beta_N}{\sqrt{\Lambda}} \sum_{l=1}^{\infty} \frac{\alpha_l \gamma_l}{\alpha_1 \gamma_1} \frac{e^{-\beta_N \eta_l}}{1 + e^{-\beta_N \eta_l}} \right]. \end{aligned}$$

The ω^* term does not contribute to the specific heat. We define the quantity I as follows:

$$I = \beta_N{}^2 \log \Lambda \sum_{l=1}^{\infty} \Lambda^{2l-2} \frac{e^{-\beta_N \eta_l}}{1 + e^{-\beta_N \eta_l}}. \tag{7.64}$$

We can show that this value approaches $\pi^2/12$ for $\Lambda \to 1$. Using this, we obtain

$$Z = \lim_{N \to \infty} \frac{1}{4} \prod_{l=1}^{\infty} \left(\frac{1 + e^{-\beta_N \eta_l}}{1 + e^{-\beta_N \eta_l'}} \right)^4 \cdot \left(1 - \frac{4\tau\lambda^*}{\sqrt{\Lambda} \log \Lambda} \frac{\alpha\gamma}{\alpha_1 \gamma_1} I \right).$$

The product over l yields 4, and hence

$$Z = 1 - \frac{\pi^2}{3} \frac{\lambda^*}{\sqrt{\Lambda} \log \Lambda} \frac{\alpha\gamma}{\alpha_1 \gamma_1} \frac{k_B T}{A}. \tag{7.65}$$

It follows from this that

$$c = -\frac{2\pi^2}{3} \frac{k_B{}^2 T}{A} \frac{\lambda^*}{\sqrt{\Lambda} \log \Lambda} \frac{\alpha\gamma}{\alpha_1 \gamma_1}. \tag{7.66}$$

By using eqs. (7.59) and (7.66), we obtain

$$r \equiv \frac{\chi T}{c} \frac{\pi^2}{3} \frac{k_B{}^2}{\mu_B{}^2} = 1 - \frac{\omega^*}{\lambda^*} \frac{1}{\log \Lambda} \frac{\gamma_1}{\gamma} \left(\frac{\alpha}{\alpha_1} \right)^3. \tag{7.67}$$

As discussed previously, ω^*/λ^* is independent of $\tilde{J}$ and depends only on Λ. Hence, the right-hand side of eq. (7.67) is dependent only on Λ. This value is, however, almost equal to 2 (within 0.1%) and is nearly independent of Λ in the range $2 \leqq \Lambda \leqq 3$, from the results obtained by the method described above. This result is consistent with the result of eq. (5.88).

We have obtained the susceptibility and specific heat of the localized spin at sufficiently low temperature. They are consistent with those predicted in Chapters 5 and 6. It is highly significant that they are obtained as a function of $J\rho$ exactly (although with some numerical error). Furthermore, it is remarkable that the behaviors of the susceptibility and specific heat are clarified from high to low temperatures: at high temperatures, the susceptibility is calculated perturbatively in terms of J as in eq. (M. 6) (see page 243), and at intermediate temperatures, it can be obtained by numerical calculations. The results of the numerical calculation are shown in Fig. 8.6.

References and further reading

Wilson, K. G. (1975) *Rev. Mod. Phys.* **47**: 773.

8

Exact solution to the s–d problem

Andrei and Wiegmann derived the exact solution to the s–d problem independently. If the s–d interaction is localized, only the radial degree of freedom needs to be considered, and the problem may be reduced to the Schrödinger equation in one-dimensional real space. When the solution is assumed to be in accord with Bethe's ansatz, the problem can be treated in exactly the same manner as in the one-dimensional Hubbard model, and the exact solution method used therein can be adapted as it is to the s–d problem. In this way, each of the various physical quantities can be represented by a single function all the way from high to low temperatures. This is a function of T/T_K, and its functional form is found to be consistent with the result of Wilson, insofar as they can be compared. This approach is mathematically powerful, and we may apply it to the case with a magnetic field, the case with $S > 1/2$, and the Anderson model. However, the focus of our attention in this chapter will be to discuss the initial analysis of Andrei (Andrei, 1980; Andrei and Lowenstein, 1981; Andrei *et al.*, 1983) and Wiegmann (Weigmann, 1981; Filyov *et al.*, 1981).

8.1 A one-dimensional model

At first, let us consider the movement of conduction electrons on a one-dimensional line. Even in the three-dimensional case, if the interaction is δ function-like, only s-wave scattering arises. As a result, only the radial dynamics is important, and this is essentially a one-dimensional problem.

Let us say that the electrons move on the line segment $-L/2 \leqq x \leqq L/2$, on which an impurity is placed at the origin. Out of the conduction electrons, those near the Fermi surface play a significant role, and we approximate their energy spectrum by a linear function of k (Fig. 8.1). Although there are two Fermi surfaces, one near k_F and the other near $-k_F$, we only consider the one

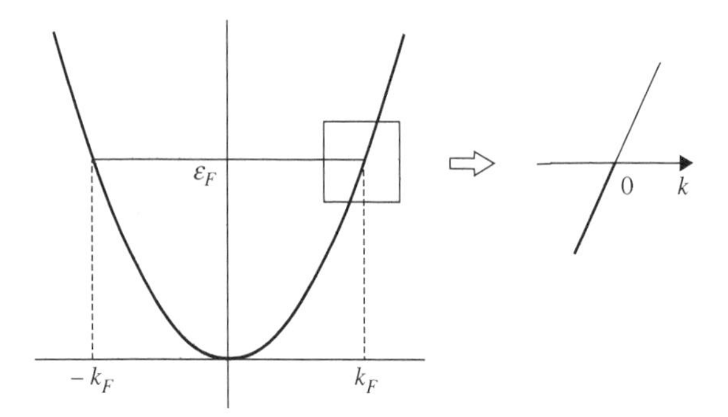

Figure 8.1 The energy spectrum of the one-dimensional model. We consider only the electrons with momenta near $+k_F$, and adopt a linear approximation for their spectrum. The origins of the wavenumber and energy are shifted to k_F and ε_F, respectively.

near k_F. We denote the wavenumber as $k + k_F$, where this k is small. Denoting the plane waves as e^{ikx} using this k, k_F has been removed from $-id/dx$. The kinetic energy of an electron, $\hbar^2 k^2/2m$, is approximated by $\varepsilon_F + \hbar v_F k$ using this new k. Here, v_F is the velocity of an electron at the Fermi level. Furthermore, the origin of energy is shifted to ε_F, so that the energy spectrum is now given by $\hbar v_F k$. This is also written as $-i\hbar v_F d/dx$.

The s–d interaction is represented by $J\boldsymbol{S} \cdot \boldsymbol{\sigma}\, \delta(x)$.[§] Using $\hbar v_F$ as the unit of energy, $J/\hbar v_F$ is written as J, and we obtain the Hamiltonian

$$\mathcal{H} = -i \sum_i \frac{\partial}{\partial x_i} + J \sum_i \boldsymbol{S} \cdot \boldsymbol{\sigma}_i\, \delta(x_i). \tag{8.1}$$

J is now dimensionless, and this greatly simplifies the calculations later on. $\boldsymbol{S}$ is the spin operator of the impurity, and $\boldsymbol{\sigma}_i$ is the spin operator of the ith electron.

Before going on to solving eq. (8.1), let us leave the problem of spin for a while, and consider a problem in which the kinetic energy is given by the first-order derivative and the potential energy is a δ function:

$$-i\psi' + J\delta(x)\psi = E\psi. \tag{8.2}$$

We represent the δ function as shown in Fig. 8.2, and take the limit $a \to 0$ as $Va = J$ remains constant. In region I in the figure, let us say $\psi = e^{ikx}$ so that $E = k$. In region II, we adopt $\psi = e^{i(E-V)x}$ so that it satisfies the above equation

[§] In this chapter, for the sake of convenience, we take the sign of J to be opposite to that used up to the last chapter. Hence our discussions mainly concern the case $J > 0$ in this chapter.

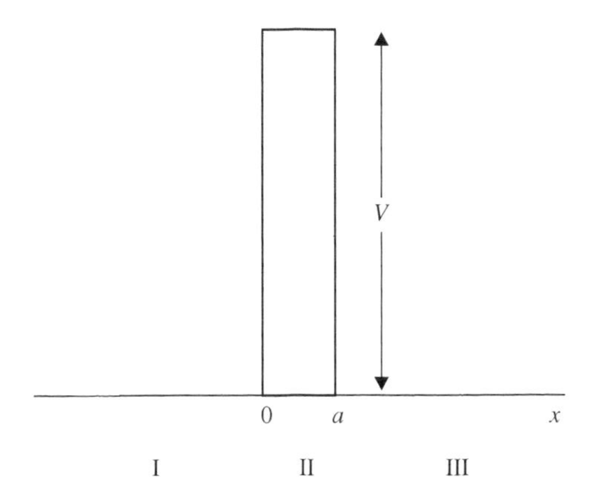

Figure 8.2 One-dimensional potential barrier. Keeping aV fixed and taking the limit $a \to 0$ corresponds to the potential $aV\delta(x)$.

and is continuous at the origin. In region III, we adopt $\psi = e^{ikx - iVa}$ so that it satisfies the above equation and is continuous at $x = a$. In short, the wave that moves in the form of e^{ikx} for negative x moves in the form of $e^{-iJ}e^{ikx}$ for positive x due to the potential $J\delta(x)$ at the origin. No reflection occurs: the energy of the wavenumber $-k$ differs from that of the wavenumber k.

Let us return to eq. (8.1) and consider the case with a single conduction electron. The impurity spin is defined to have $S = 1/2$. The wavefunction is a function of the space coordinate x_1 and the spin coordinate ζ_1 of the electron, and of the spin coordinate ζ_0 of the impurity spin. From the form of the interaction $JS \cdot \sigma_1$, the eigenfunction is an eigenfunction of the total spin, and it can be singlet or triplet. We hence define the form of the eigenfunction as

$$\Psi_s(x_1, \zeta_0, \zeta_1) = \psi_s(x_1) \frac{\alpha_0\beta_1 - \beta_0\alpha_1}{\sqrt{2}}, \tag{8.3}$$

or

$$\Psi_t(x_1, \zeta_0, \zeta_1) = \psi_t(x_1) \begin{cases} \alpha_0\alpha_1, \\ \dfrac{\alpha_0\beta_1 + \beta_0\alpha_1}{\sqrt{2}}, \\ \beta_0\beta_1. \end{cases} \tag{8.4}$$

Here, for example, $\alpha_0\beta_1$ denotes $\alpha(\zeta_0)\beta(\zeta_1)$. Equation (8.3) corresponds to the singlet state where $\boldsymbol{S} \cdot \boldsymbol{\sigma}_1$ takes the value $-3/4$, and eq. (8.4) corresponds to the triplet state where $\boldsymbol{S} \cdot \boldsymbol{\sigma}_1$ takes the value $1/4$. ψ_s therefore needs to satisfy

$$-i\psi_s{}' - \frac{3J}{4}\delta(x_1)\psi_s = E\psi_s,$$

whereas ψ_t needs to satisfy

$$-i\psi_t{}' + \frac{J}{4}\delta(x_1)\psi_t = E\psi_t.$$

Hence, by the above discussions, when the wavefunction for $x_1 > 0$ is defined to be equal to that for $x_1 < 0$ multiplied by $e^{i3J/4}$ (singlet) or $e^{-iJ/4}$ (triplet), it satisfies the above equations.

We would like to represent these factors in a unified form for both singlet and triplet. In order to do so, let us introduce the operator P_{01}. For an arbitrary function $f(\zeta_0, \zeta_1)$ of ζ_0 and ζ_1, $P_{01}f(\zeta_0, \zeta_1) = f(\zeta_1, \zeta_0)$. We then have $P_{01}\Psi_s = -\Psi_s$ and $P_{01}\Psi_t = \Psi_t$. Defining R_{10} as

$$R_{10} = e^{iJ/4}\left(\cos\frac{J}{2} - i\sin\frac{J}{2}P_{01}\right), \tag{8.5}$$

we see that the relation

$$\Psi(x_1 > 0) = R_{10}\Psi(x_1 < 0) \tag{8.6}$$

holds for both singlet and triplet, and so it also holds for any linear combination of them, i.e., for an arbitrary spin function. Since R_{10} includes the spin exchange operator P_{01}, the spin function for $x_1 > 0$ differs in general from that for $x_1 < 0$. Other than the problem of obtaining the eigenstates, when we consider the scattering of electrons by an impurity, for example, it is natural that the spin function differs before and after the scattering.

In order to determine the eigenvalues, we adopt the periodic boundary condition. From the condition that the wavefunction is equal at $x_1 = -L/2$ and $x_1 = L/2$, for the triplet for example,

$$e^{-ikL/2} = e^{-iJ/4}e^{ikL/2}$$

holds. We can hence determine the eigenvalues as $kL - J/4 = 2\pi n$ where n is an integer.

8.2 The three-body problem

Next, let us consider the case with the number of conduction electrons $N = 2$. As almost all problems that arise in the case of general N appear already in this case, let us discuss this case in some detail. We consider the following four regions:

$$\begin{aligned}
\text{I}: &\quad x_1, x_2 < 0, \\
\text{II}: &\quad x_2 < 0, x_1 > 0, \\
\text{III}: &\quad x_2 > 0, x_1 < 0, \\
\text{IV}: &\quad x_1, x_2 > 0.
\end{aligned}$$

Moving from I to II means that the first electron passes through the origin, and so the relation of eq. (8.6) holds between $\Psi(\text{II})$ and $\Psi(\text{I})$. Moving from II to IV means that the second electron passes through the origin, and so $\Psi(\text{IV})$ is related to $\Psi(\text{II})$ by R_{20}. Here, R_{20} is obtained by replacing P_{01} by P_{02} in eq. (8.5), where P_{02} is an operator that exchanges ζ_0 and ζ_2. By the above two statements, we see that the following must hold:

$$\Psi(\text{IV}) = R_{20}R_{10}\Psi(\text{I}).$$

On the other hand, we may pass III en route for IV, namely I→III→IV. In this case, x_2 first passes through the origin and then x_1, and so:

$$\Psi(\text{IV}) = R_{10}R_{20}\Psi(\text{I}).$$

Now, because $P_{02}P_{01}$ and $P_{01}P_{02}$ do not lead to the same result, these two are not equivalent in general. For example, let us say that electron 1 has up-spin, electron 2 has down-spin, and the impurity 0 has down-spin. The spin-flip of electron 1 can be followed by the spin-flip of electron 2, but spin-flip cannot occur if electron 2 comes first. This means that the electron scattering by localized spin cannot be regarded as a process that is independent of the other electrons, and that it should be treated as a many-body problem.

To avoid the difficulty described above, we need to treat the electronic spin function as being different between $x_1 > x_2$ and $x_1 < x_2$. We thus consider the following six regions:

$$\begin{aligned}
\text{I}: &\quad x_1 < x_2 < 0, \\
\text{II}: &\quad x_2 < x_1 < 0, \\
\text{III}: &\quad x_1 < 0 < x_2, \\
\text{IV}: &\quad x_2 < 0 < x_1, \\
\text{V}: &\quad 0 < x_1 < x_2, \\
\text{VI}: &\quad 0 < x_2 < x_1.
\end{aligned}$$

In the lth region, let us say that the wavefunction is given by

$$\Psi_l(x_1\, x_2\, \zeta_0\, \zeta_1\, \zeta_2) = a_l(\zeta_0\, \zeta_1\, \zeta_2)e^{i(k_1x_1+k_2x_2)}, \tag{8.7}$$

where we have omitted the commas that separate the arguments. In general, this equation does not satisfy the Pauli principle:

$$\Psi_l(x_2\, x_1\, \zeta_0\, \zeta_2\, \zeta_1) = -\Psi_l(x_1\, x_2\, \zeta_0\, \zeta_1\, \zeta_2).$$

However, if eq. (8.7) satisfies the wave equation, so does $\Psi_l(x_2\, x_1\, \zeta_0\, \zeta_2\, \zeta_1)$, with the same eigenvalue. Hence the difference of the two has the same eigenvalue and satisfies the Pauli principle. Now, let us determine a_l such that eq. (8.7) satisfies the wave equation. First considering I → III, since this means that x_2 passes through the origin, Ψ_{III} needs to be related to Ψ_{I} by R_{20}. Hence,

$$a_{\mathrm{III}}(\zeta_0\, \zeta_1\, \zeta_2) = R_{20}a_{\mathrm{I}}(\zeta_0\, \zeta_1\, \zeta_2), \tag{8.8}$$

and similarly,

$$a_{\mathrm{IV}} = R_{10}a_{\mathrm{II}}, \tag{8.9}$$

$$a_{\mathrm{V}} = R_{10}a_{\mathrm{III}}, \tag{8.10}$$

$$a_{\mathrm{VI}} = R_{20}a_{\mathrm{IV}}. \tag{8.11}$$

Now, for example, let us consider I → VI. There are two possible routes for this, the first being I→III→V→VI. In this case, by eqs. (8.8) and (8.10), we have

$$a_{\mathrm{V}} = R_{10}R_{20}a_{\mathrm{I}}.$$

However, since we have not yet fixed the relation between a_{V} and a_{VI}, a_{VI} cannot be related to a_{I} as yet. The other route is I→II→IV→VI and, by eqs. (8.9) and (8.11), we have

$$a_{\mathrm{VI}} = R_{20}R_{10}a_{\mathrm{II}}.$$

However, the relation between a_{II} and a_{I} has not been fixed yet. If we were to adopt $a_{\mathrm{I}} = a_{\mathrm{II}}$ and $a_{\mathrm{VI}} = a_{\mathrm{V}}$ as before, we would arrive at the contradictory relations $a_{\mathrm{VI}} = R_{20}R_{10}a_{\mathrm{I}}$ and $a_{\mathrm{VI}} = R_{10}R_{20}a_{\mathrm{I}}$. Let us momentarily adopt

$$a_{\mathrm{II}} = P_{12}a_{\mathrm{I}}, \tag{8.12}$$

$$a_{\mathrm{VI}} = P_{12}a_{\mathrm{V}}. \tag{8.13}$$

Here, P_{12} is an operator that exchanges ζ_1 and ζ_2. This yields

$$a_{\rm VI} = P_{12}R_{10}R_{20}a_{\rm I}$$

and

$$a_{\rm VI} = R_{20}R_{10}P_{12}a_{\rm I}.$$

These two equations are equivalent because we can easily derive the following operator identity:

$$P_{12}R_{10}R_{20} = R_{20}R_{10}P_{12}. \tag{8.14}$$

Equations (8.8)–(8.13) provide relations between adjacent regions such as I→II→IV→VI→V→III→I. Using these relations, we may arrive at an arbitrary region starting from a particular region, but the routes are manifold. For each such route, there is a corresponding product composed of P and R. The inverse operator, R^{-1}, of R, may be necessary. The products of operators are identical for all routes, and this is ensured by eq. (8.14). In this way, we have shown that one is able to satisfy the wave equation by adopting a wavefunction of the form eq. (8.7).

We have now shown that all spin functions a_l can be determined when one of them, for example $a_{\rm I}$, is given. In order to fix $a_{\rm I}$ and obtain the eigenvalues, let us adopt the periodic boundary condition for the wavefunction. We introduce the condition that the value of the function is the same at $x_1 = -L/2$ and $x_1 = L/2$. When $x_2 < 0$, these points are in regions I and IV, and so we obtain

$$a_{\rm I}e^{i(-k_1L/2+k_2x_2)} = a_{\rm IV}e^{i(k_1L/2+k_2x_2)},$$

i.e.,

$$e^{ik_1L}a_{\rm IV} = a_{\rm I}.$$

For $x_2 > 0$, these points are in regions III and VI, and we obtain

$$e^{ik_1L}a_{\rm VI} = a_{\rm III}.$$

Now, using $a_{\rm IV} = R_{10}P_{12}a_{\rm I}$, $a_{\rm III} = R_{20}a_{\rm I}$ and $a_{\rm VI} = R_{20}R_{10}P_{12}a_{\rm I}$, all spin functions can be written in terms of $a_{\rm I}$. The above two equations then both reduce to

$$e^{ik_1L}R_{10}P_{12}a_{\rm I} = a_{\rm I}. \tag{8.15}$$

By setting a similar condition for x_2, we obtain a similar equation in which k_1 in the above is replaced by k_2.

Let us consider, firstly, the case when all spin is up-spin, that is, $a_{\rm I} = \alpha_0\alpha_1\alpha_2$. In this case, since $a_{\rm I}$ is invariant under both P_{12} and P_{01}, we may adopt the replacement $P_{12} \to 1$ and $P_{01} \to 1$. Hence, we may adopt $R_{10} \to e^{-iJ/4}$ (see eq. (8.5)). Equation (8.15) then reduces to

$$e^{ik_iL - iJ/4} = 1 \quad (i = 1, 2),$$

and the eigenvalues are determined by

$$k_iL - \frac{J}{4} = 2\pi n_i \quad (n_i \text{ is an integer}).$$

The energy eigenvalue is $E = k_1 + k_2$.

Secondly, for the case of one down-spin, there are three independent spin functions: $\alpha_0\alpha_1\beta_2$, $\alpha_0\beta_1\alpha_2$ and $\beta_0\alpha_1\alpha_2$. Writing a_{I} as a linear combination of these three, the application of P_{12} or P_{01} on it yields another linear combination of the three. We can henceforth obtain linear equations with three unknowns from eq. (8.15), and determine k_i by solving these secular equations.

Now, there is a difficulty (Schulz, 1982) associated with the solution in the form of eq. (8.7). When we consider the case where the interaction J is 0, and the solution is given in the form of eq. (8.7), a_l must clearly be the same for all l. Since both R_{10} and R_{20} become 1 for $J = 0$, although eqs. (8.8)–(8.11) are all right, eqs. (8.12) and (8.13) are problematic. However, the solution that we require is the antisymmetrized version of eq. (8.7). So long as we work with such a function, it can be shown that the difficulty described above does not occur. For this purpose, let us first antisymmetrize eq. (8.7).

As we discussed before, antisymmetrizing a certain wavefunction $\Psi(x_1\, x_2)$ means forming $\Psi(x_1\, x_2) - \Psi(x_2\, x_1)$. Here, in the second term, the coordinate of the first electron is x_2 and that of the second electron is x_1. Thus, if x_1 and x_2 belong to region V, the second term is a wavefunction in region VI. Hereby, for example, the wavefunction that has been antisymmetrized in region V can be written as

$$\begin{aligned}\Psi_{\text{V}} &= a_{\text{V}}(\zeta_0\,\zeta_1\,\zeta_2)e^{i(k_1x_1+k_2x_2)} - a_{\text{VI}}(\zeta_0\,\zeta_2\,\zeta_1)e^{i(k_1x_2+k_2x_1)} \\ &= a_{\text{V}}(\zeta_0\,\zeta_1\,\zeta_2)\left(e^{i(k_1x_1+k_2x_2)} - e^{i(k_1x_2+k_2x_1)}\right) \\ &\equiv a_{\text{V}}(\zeta_0\,\zeta_1\,\zeta_2)D(x_1\,x_2). \end{aligned} \tag{8.16}$$

Here, we used eq. (8.13). The last equality defines $D(x_1\, x_2)$. In the same way, the wavefunction that has been antisymmetrized in region VI can be written as

$$\Psi_{\text{VI}} = a_{\text{V}}(\zeta_0\,\zeta_2\,\zeta_1)D(x_1\,x_2). \tag{8.17}$$

In order to combine the above two functions in regions V and VI, we introduce the step function $\theta(x)$:

$$\theta(x) = \begin{cases} 1 & x > 0, \\ 0 & x < 0. \end{cases} \tag{8.18}$$

Then, in the region V + VI, that is, if $0 < x_i < L/2$ $(i = 1, 2)$, the wavefunction can be written as

$$\Psi = \{a_{\rm V}(\zeta_0\,\zeta_1\,\zeta_2)\theta(x_2 - x_1) + a_{\rm V}(\zeta_0\,\zeta_2\,\zeta_1)\theta(x_1 - x_2)\}\,D(x_1\,x_2). \tag{8.19}$$

This should, of course, be the solution even for $J = 0$. On the other hand, since D is identically zero when $k_1 = k_2$ or $x_1 = x_2$, it seems as if the expected solution for $J = 0$ is not included. This problem is actually another form of the difficulty described above, and as long as we deal with the function that has been antisymmetrized in the region V + VI, eq. (8.19) can be shown to form a complete set, and so such a difficulty is in fact only superficial.

In order to demonstrate this, let us remember the Fourier series expansion. An arbitrary function defined in the range $-L/2 \leqq x \leqq L/2$, with, some restrictions of course, can be expanded in terms of $\sin 2\pi nx/L$ and $\cos 2\pi nx/L$. However, if we restrict the range to $0 \leqq x \leqq L/2$, the expansion may be in terms of $\sin 2\pi nx/L$ only, or $\cos 2\pi nx/L$ only. Functions that are finite at $x = 0$ can also be expanded in this way, that is, the series can tend to a finite value at $x = 0$. On the other hand, using some tricks, an arbitrary function in the range $0 \leqq x \leqq L/2$ can also be expanded in terms of $\sin(2n + 1)\pi x/L$.

Now, let us say that a function, $F(x_1\,x_2)$, of x_1 and x_2 is expanded as follows:

$$F(x_1\,x_2) = \sum a_{n_1 n_2} e^{i2\pi(n_1 x_1 + n_2 x_2)/L}. \tag{8.20}$$

Here, both x_1 and x_2 are in $(-L/2, L/2)$. Using $X = (x_1 + x_2)/2$ and $x = x_2 - x_1$, the function may be expanded as follows:

$$F(x_1\,x_2) = \sum b_n(x) e^{i2\pi nX/L}. \tag{8.21}$$

We then consider region V, that is, $0 < x_1 < x_2 < L/2$. Since $0 < x < L/2$, $b_n(x)$ can be expanded in terms of either $\sin 2m\pi x/L$ or $\sin(2m + 1)\pi x/L$. Choosing the former for even n and the latter for odd n, $b_n(x)$ can be written as

$$b_n(x) = \sum_{\substack{m \\ 2m-n>0}} c_{nm} \sin(2m - n)\frac{\pi x}{L}.$$

Substituting this into eq. (8.21), and denoting $m = n_1$, $n - m = n_2$, and $-c_{nm}/2i = a_{n_1 n_2}$, we find that $F(x_1\, x_2)$ can be expanded as

$$F(x_1\, x_2) = \sum_{n_1 > n_2} a_{n_1 n_2} \left[e^{i2\pi(n_1 x_1 + n_2 x_2)/L} - e^{i2\pi(n_2 x_1 + n_1 x_2)/L} \right]$$
$$(0 < x_1 < x_2 < L/2). \tag{8.22}$$

By defining $k_i = 2\pi n_i / L$, we find that the expression inside square brackets in eq. (8.22) is precisely $D(x_1\, x_2)$. Thus even when F is non-zero for $x_1 = x_2$, such an expansion is possible. Equation (8.22) can be written in the following form:

$$F(x_1\, x_2) = \sum_{n_1 n_2} a_{n_1 n_2} e^{i2\pi(n_1 x_1 + n_2 x_2)/L} \quad (0 < x_1 < x_2 < L/2), \tag{8.23}$$

where

$$a_{n_1 n_2} = -a_{n_2 n_1}. \tag{8.24}$$

We shall make use of this expression later.

Next, we may consider region VI. This is given merely by the exchange of x_1 and x_2, and since $D(x_2\, x_1) = -D(x_1\, x_2)$, F can also be expanded using $D(x_1\, x_2)$ in this region. However, in general, the expansion coefficients will be different from $a_{n_1 n_2}$. We therefore conclude that $\theta(x_2 - x_1) D(x_1\, x_2)$ and $\theta(x_1 - x_2) D(x_1\, x_2)$ form a complete set in the region V + VI.

For a given function $F(x_1\, x_2\, \zeta_1\, \zeta_2)$, with the spin coordinates, the expansion using

$$\left\{ a_{n_1 n_2}(\zeta_1\, \zeta_2)\theta(x_2 - x_1) + a'_{n_1 n_2}(\zeta_1\, \zeta_2)\theta(x_1 - x_2) \right\} D(x_1\, x_2) \tag{8.25}$$

is possible in region V + VI. Here a and a' are arbitrary spin functions. If F is antisymmetrized, it can be expanded using the antisymmetrized version of eq. (8.25), which reads

$$\begin{aligned} &\left[\left\{ a_{n_1 n_2}(\zeta_1\, \zeta_2) + a'_{n_1 n_2}(\zeta_2\, \zeta_1) \right\} \theta(x_2 - x_1) \right. \\ &\left. + \left\{ a_{n_1 n_2}(\zeta_2\, \zeta_1) + a'_{n_1 n_2}(\zeta_1\, \zeta_2) \right\} \theta(x_1 - x_2) \right] D(x_1\, x_2). \end{aligned}$$

When we compare this with eq. (8.19), we find that there is perfect correspondence. In other words, we have shown that eq. (8.19) forms a complete set, if we restrict our discussion to functions that are antisymmetrized in the region V + VI. While the above discussion demonstrates completeness in the region V + VI, similar discussions are possible in the other regions.

In eqs. (8.12) and (8.13), we have been somewhat loose in defining the relations between $a_{\rm I}$ and $a_{\rm II}$ and between $a_{\rm V}$ and $a_{\rm VI}$. On the other hand, we may repeat the discussions in exactly the same manner if we introduce a minus sign on the right-hand sides of both equations. The wavefunction then acquires a different form. For example, if we write eq. (8.16) using $D \equiv e^{i(k_1x_1+k_2x_2)} + e^{i(k_2x_1+k_1x_2)}$, we can still prove completeness in the same way. For this, we would use $\cos 2\pi mx/L$ etc. in the expansion of $b_n(x)$. The conclusion is unaffected.

Hereafter, we adopt the plus sign on the right-hand side of both eqs. (8.12) and (8.13); k_1 and k_2 are determined by eq. (8.15) etc. and we can use them to obtain the wavefunction as the antisymmetrized version of eq. (8.7). Here, as can be seen from eq. (8.19), the wavefunction for a certain pair of k_1 and k_2 differs only in sign from that in which k_1 and k_2 are swapped, and so they should not be considered as distinct. In addition, the case $k_1 = k_2$ is excluded since the wavefunction vanishes.

8.3 Symmetric groups

Let us review some simple results about symmetric groups. For natural numbers 1 to N, the operation that assigns $P_1, P_2, \ldots, P_N$ to $1, 2, \ldots, N$, respectively, is called the permutation, and is represented as

$$P = \begin{pmatrix} 1 & 2 & \cdots & N \\ P_1 & P_2 & \cdots & P_N \end{pmatrix}. \tag{8.26}$$

Here $(P_1\, P_2 \cdots P_N)$ is $(1\, 2 \cdots N)$ placed in some order. We shall denote this as $P = (P_1\, P_2 \cdots P_N)$ so long as there is no ambiguity.

There are altogether $N!$ distinct permutations. The set of these permutations is denoted as S_N, and is called the order N symmetric group. The product of two permutations means the net result of the permutation on the right followed by that on the left. For example, when $P = (1\,3\,2)$ and $Q = (2\,3\,1)$, we have $PQ = (3\,2\,1)$ and $QP = (2\,1\,3)$. Note the following:

(i) The product of two elements of S_N also belongs to S_N.
(ii) Associativity, holds, that is, $P(QR) = (PQ)R$.
(iii) The permutation that assigns 1, 2, ..., N to 1, 2, ..., N, respectively, is called the identity permutation and is represented as E. Then, for an arbitrary P,

$$PE = EP = P.$$

(iv) For P given by eq. (8.26), and denoting by P^{-1} the permutation that assigns 1, 2, ..., N to P_1, P_2, ..., P_N, respectively, we have $PP^{-1} = P^{-1}P = E$. Moreover, $(PQ)^{-1} = Q^{-1}P^{-1}$.

(v) Let us consider the product PQ (or QP). When Q is a fixed element of S_N and P is an arbitrary element of S_N, PQ forms a set with $N!$ elements, and this set is identical to S_N. Later on, we often have to deal with the summation of function $f(P)$ over all permutations P. There, we may make use of

$$\sum_P f(P) = \sum_P f(PQ) = \sum_P f(QP). \tag{8.27}$$

(vi) The permutation that assigns i to j, j to i, and leaves all other k unchanged, is called transposition and is represented by P_{ij}. We then have $P_{ij}P_{ji} = E$ and $P_{ij}{}^{-1} = P_{ij}$.

(vii) An arbitrary permutation can be represented as the product of a number of transpositions. There is no unique way of representation, but a permutation (called an even permutation) that can be represented as the product of an even number of transpositions can only be represented as products of even numbers of transpositions. Similarly for odd permutations.

(viii) For a given permutation $P = (\cdots i\, j\, k \cdots)$, we have $P_{ij}P = (\cdots j\, i\, k \cdots)$. Such transposition of neighboring pairs (i and j in this case) is called adjacent transposition. As i and k are neighbors in $P_{ij}P$, $P_{ik}P_{ij}P = (\cdots j\, k\, i \cdots)$ is also an adjacent transposition. One can start from arbitrary permutation P, and arrive at another arbitrary permutation P', by repeated application of adjacent transposition.

(ix) Hereafter, we shall denote the permutation of $N+1$ integers $(0\,1\,2 \cdots N)$ as Q, and the permutations of N integers $(1\,2 \cdots N)$ as P, R. In Q, 0 is discriminated from the other integers, that is, $(Q_1\, Q_2 \cdots 0 \cdots Q_N)$. Operating on Q by P means operating on $(Q_1\, Q_2 \cdots Q_N)$ by P, keeping the position of 0 as it is.

8.4 The N-electron problem

Let us now consider the N-electron problem. For $\mathcal{H}$ given by eq. (8.1), we would like to solve

$$\mathcal{H}\Psi(x_1 \cdots x_N\, \zeta_0\, \zeta_1 \cdots \zeta_N) = E\Psi. \tag{8.28}$$

In the same way as for the $N = 2$ case, we introduce regions of the electrons' coordinates. For a given permutation $Q = (Q_1\, Q_2 \cdots 0 \cdots Q_N)$, the region Q is defined as follows:

$$-\frac{L}{2} < x_{Q_1} < x_{Q_2} < \cdots < 0 < \cdots < x_{Q_N} < \frac{L}{2}.$$

The wavefunction in this region is defined as

$$\Psi = a_Q(\zeta_0\,\zeta_1 \cdots \zeta_N)e^{i(k_1x_1+k_2x_2+\cdots+k_Nx_N)}. \tag{8.29}$$

It is then clear that

$$E = \sum k_i. \tag{8.30}$$

This wavefunction does not satisfy the Pauli principle. However, one may first obtain a solution in the form of eq. (8.29) and then antisymmetrize it. For a region $Q = (Q_1 \cdots Q_i\, 0 \cdots Q_N)$ and its neighboring region $Q' = (Q_1 \cdots 0\, Q_i \cdots Q_N)$, let us consider the relation between a_Q and $a_{Q'}$. Since $Q \to Q'$ means that x_{Q_i} passes through the origin, in exactly the same way as in the case of $N = 2$ the relation

$$a_{Q'} = R_{Q_i0}a_Q \tag{8.31}$$

needs to hold. Here, R_{Q_i0} is obtained by replacing P_{01} by P_{0Q_i} in eq. (8.5). P_{0Q_i} is the transposition operator for ζ_0 and ζ_{Q_i}. For $Q = (\cdots 0\, Q_i \cdots)$ and $Q' = (\cdots Q_i\, 0 \cdots)$, the relation

$$a_{Q'} = {R_{Q_i0}}^{-1}a_Q \tag{8.32}$$

of course needs to hold.

Next, let us consider the region $Q = (\cdots Q_i\, Q_{i+1} \cdots)$ and a neighboring region $Q' = (\cdots Q_{i+1}\, Q_i \cdots)$. Here, neither Q_i nor Q_{i+1} is 0. The relation between a_Q and $a_{Q'}$ cannot be determined by the wave equation. In the same way as in eqs. (8.12) and (8.13), we adopt

$$a_{Q'} = P_{Q_iQ_{i+1}}a_Q \tag{8.33}$$

with some degree of arbitrariness. Here, $P_{Q_iQ_{i+1}}$ is the transposition operator for ζ_{Q_i} and $\zeta_{Q_{i+1}}$.

Let us combine eqs. (8.31)–(8.33). Let us say that there is a certain region $Q = (\cdots i\, j \cdots)$. This includes the case where i or j is 0. In a neighboring region $Q' = (\cdots j\, i \cdots)$ (where all sequences other than i and j are the same), the spin function $a_{Q'}$ is given by

$$a_{Q'} = X_{ij}a_Q. \tag{8.34}$$

Here,

$$X_{ij} = \begin{cases} P_{ij} & i \neq 0, j \neq 0, \\ R_{i0} & j = 0, \\ R_{j0}{}^{-1} & i = 0. \end{cases} \tag{8.35}$$

As $Q' = P_{ij}Q$, eq. (8.34) can also be written as

$$a_{P_{ij}Q} = X_{ij}a_Q. \tag{8.36}$$

As stated in §8.3(viii), an arbitrary permutation Q can be transformed into another arbitrary permutation Q' by the repeated application of adjacent transposition on Q. Thus, an arbitrary $a_{Q'}$ can be obtained from a_Q by using eq. (8.36) repeatedly. However, the route from Q to Q' is not unique. Each route corresponds to a product of X_{ij}. Should they all give us an identical result? As in the $N = 2$ case, it can be verified that in fact they do. Corresponding to eq. (8.14), the operator identity

$$X_{jk}X_{ik}X_{ij} = X_{ij}X_{ik}X_{jk} \tag{8.37}$$

holds. From this, we see that the two routes from the region $(\cdots i\,j\,k \cdots)$ to the region $(\cdots k\,j\,i \cdots)$ give an identical result. Moreover, since

$$X_{ij}X_{ji} = 1, \tag{8.38}$$

and

$$X_{ij}X_{kl} = X_{kl}X_{ij} \quad (i\,j\,k\,l \text{ are all different}) \tag{8.39}$$

can be easily shown, we conclude that any route for $Q \to Q'$ gives an identical result. This conclusion has been brought about by the assumption of eq. (8.33). Even if we have a minus sign in the right-hand side of eq. (8.33), the conclusion remains the same, as we saw in the $N = 2$ case.

In this way, when a_Q is given in a certain region, $a_{Q'}$ can be consistently determined in all other regions. Let us define the standard region as $(0\,1\,2 \cdots N)$, and denote this as I. We shall express all a_Q in terms of a_{I}. a_{I} is determined by the boundary condition. Let us choose the periodic boundary condition. Now, let us say that a region $Q = (Q_1\,Q_2 \cdots 0 \cdots Q_N)$ is given, and $x_{Q_1} = -L/2$. In this case, eq. (8.29) becomes

$$a_Q e^{i(k_1x_1 + \cdots - (L/2)k_{Q_1} + \cdots)}.$$

Next, we keep all elements other than x_{Q_1} fixed, and set $x_{Q_1} = L/2$. This region is denoted as $\bar{Q}$, where $\bar{Q} = (Q_2 \cdots 0 \cdots Q_N\,Q_1)$. Equation (8.29) then becomes

$$a_{\bar{Q}} e^{i(k_1x_1 + \cdots + (L/2)k_{Q_1} + \cdots)}.$$

By the condition that these two are equal, we obtain

$$e^{ik_{Q_1}L}a_{\bar{Q}} = a_Q. \tag{8.40}$$

We apply a suitable X_{ij} to both sides of this equation and then would like to write a_Q and $a_{\bar{Q}}$ in terms of $a_{\rm I}$ using eq. (8.36). At first, we keep Q_1 apart and choose X_{ij} such that the sequence $Q_2 \cdots 0 \cdots Q_N$ is arranged in ascending order. (Q_1 is denoted by j hereafter.) We then obtain

$$e^{ik_jL}a_{\bar{Q}'} = a_{Q'}, \tag{8.41}$$

$$Q' = (j\,0\,1\,2 \cdots (j) \cdots N),$$
$$\bar{Q}' = (0\,1\,2 \cdots (j) \cdots N\,j),$$

where, (j) means the absence of j at that position. On the other hand, by eq. (8.36), we have

$$a_{Q'} = X_{0j}X_{1j} \cdots X_{j-1\,j}\, a_{\rm I},$$
$$a_{\bar{Q}'} = X_{jN} \cdots X_{j\,j+1}\, a_{\rm I}.$$

We substitute these into eq. (8.41), and use eq. (8.38) so that $a_{\rm I}$ appears in the left-hand side. This yields

$$e^{ik_jL}a_{\rm I} = X_{j+1\,j} \cdots X_{Nj}X_{0j}X_{1j} \cdots X_{j-1\,j}\, a_{\rm I}. \tag{8.42}$$

This is the eigenvalue equation. Since X_{0j} is the sum of a constant and P_{0j}, and all the others, X_{1j} etc., are equal to P_{1j} etc., we can easily see the result of the operations on the right-hand side. We find that the result is independent of j, and so j may be replaced by, for example, unity. Equation (8.42) then becomes

$$e^{ik_jL}a_{\rm I} = X_{21} \cdots X_{N1}X_{01}\, a_{\rm I}. \tag{8.43}$$

Since j is arbitrary, eq. (8.43) needs to hold for $j = 1, \ldots, N$. This corresponds to eq. (8.15). However, the definition of $a_{\rm I}$ slightly differs from that in eq. (8.15). When we determine $a_{\rm I}$ and k_j by solving eq. (8.43), eq. (8.29) is evaluated in all regions.

8.5 Antisymmetrization

Before solving eq. (8.43), let us antisymmetrize eq. (8.29) for given $a_{\rm I}$. For some function $\Psi(x_1 \cdots x_N\, \zeta_1 \cdots \zeta_N)$, let us define the permutation P as

$$P\Psi(x_1 \cdots x_N\, \zeta_1 \cdots \zeta_N) = \Psi(x_{P_1} \cdots x_{P_N}\, \zeta_{P_1} \cdots \zeta_{P_N}). \tag{8.44}$$

The antisymmetrized function is then represented as

$$\sum_P (-)^P P\Psi(x_1 \cdots x_N\, \zeta_1 \cdots \zeta_N).$$

The summation is over the $N!$ permutations P, and $(-)^P$ is $+1$ and -1 for even and odd permutations, respectively. In order to demonstrate its antisymmetry, we apply the permutation R:

$$\begin{aligned} R\sum_P (-)^P P\Psi &= \sum_P (-)^P RP\Psi = (-)^R \sum_P (-)^{RP} RP\Psi \\ &= (-)^R \sum_P (-)^P P\Psi. \end{aligned}$$

This is equal to the original function times $(-)^R$. This is precisely the statement of antisymmetry. Here, we used eq. (8.27).

Equation (8.29) is the wavefunction for region Q. We would like to express the wavefunction in a combined form for all regions. In order to do so, we define the step function θ_Q such that θ_Q is a function of $x_1, \ldots, x_N$, and yields 1 when these arguments are in region Q, and 0 otherwise. $P\theta_Q$ is defined as the operation on the arguments of θ_Q, $x_1, \ldots, x_N$, by the permutation shown in eq. (8.44). It is easy to see that $P\theta_Q = \theta_{PQ}$. Here, the meaning of the product of the permutation Q, composed of $N+1$ integers, and the permutation P, composed of N integers, is as given in §8.3(ix). Thus by eq. (8.29),

$$\sum_Q \theta_Q a_Q(\zeta_0\, \zeta_1 \cdots \zeta_N) e^{i(k_1 x_1 + \cdots + k_N x_N)}$$

is the wavefunction for all regions. When we antisymmetrize it, we obtain

$$\begin{aligned} \Psi &= \sum (-)^P P\left[\theta_Q a_Q e^{i(k_1 x_1 + \cdots)}\right] \\ &= \sum (-)^P \theta_{PQ} (P a_Q) e^{i(k_1 x_{P_1} + \cdots)} \\ &= \sum_Q \theta_Q \sum_P (-)^P (P a_{P^{-1}Q}) e^{i(k_1 x_{P_1} + \cdots)}. \end{aligned} \tag{8.45}$$

This equation is expressed as the sum of $N!$ plane waves, whose coefficients differ according to the regions. However, the wavenumbers, $k_1, \ldots, k_N$, are identical in all regions. Writing the wavefunction in this form is called Bethe's ansatz. In the above discussion, we obtained the wavefunction of Bethe's ansatz by antisymmetrizing eq. (8.29).

Equation (8.45) is a solution to eq. (8.28). Is it the only solution? In other words, does it form a complete set? This is the same problem as encountered in the case $N = 2$. In order to simplify this problem, let us consider only regions in which all x_i are located on the positive side. We denote such regions in general as Q_0. For these regions, eq. (8.36) becomes

$$a_{P_{ij}Q_0} = P_{ij} a_{Q_0}.$$

Hence, for an arbitrary P, we obtain

$$a_{PQ_0} = P a_{Q_0}.$$

Then, if the summation in eq. (8.45) is restricted to Q_0, we obtain

$$\begin{aligned} \Psi &= \sum_{Q_0} \theta_{Q_0} a_{Q_0} \sum_P (-)^P e^{i(k_1 x_{P_1} + \cdots)} \\ &= \sum_{Q_0} \theta_{Q_0} a_{Q_0} D(x_1 \cdots x_N), \end{aligned}$$

where

$$D(x_1 \cdots x_N) \equiv \begin{vmatrix} e^{ik_1 x_1} & e^{ik_1 x_2} & \cdots \\ e^{ik_2 x_1} & \ddots & \\ \vdots & & \ddots \end{vmatrix}.$$

This corresponds to eq. (8.19). D is a determinant composed of plane waves, and is zero when any two coordinates x or any two wavenumbers k are the same. However, remembering the $N = 2$ case, it is sufficient, for the proof of completeness, to show that $D(x_1 \cdots x_N)$ is complete in a certain region. Let us therefore adopt region I:

$$0 < x_1 < x_2 < \cdots < x_N < \frac{L}{2}. \tag{8.46}$$

An arbitrary function of $x_1 \cdots x_N$, $F(x_1 \cdots x_N)$, is expanded as

$$F(x_1 \cdots x_N) = \sum_{n_1 \cdots n_N} a_{n_1 \cdots n_N} e^{i2\pi(n_1 x_1 + \cdots)/L}. \tag{8.47}$$

If the arguments are restricted to the form of eq. (8.46), the coefficients may also have restrictions. As shown by eq. (8.24), the coefficients can be defined such that $a_{n_1 n_2 \cdots} = -a_{n_2 n_1 \cdots}$. In exactly the same way, we can define the coefficients such that they change signs when any two subscripts of $a_{n_1 n_2 \cdots}$ are exchanged. Thus, out of the $N!$ coefficients generated by permutations on the subscripts, only one is independent, and the others are equal to this coefficient times $+1$ or -1, respectively, for even or odd permutations.

Now, let us divide the summation in eq. (8.47) into the summation over the combinations of N integers and the summation over permutations. Carrying out the latter first, we obtain $D(x_1 x_2 \cdots)$ since we sum exponential functions times $+1$ or -1 respectively for even or odd permutations. Here, $k_i = 2\pi n_i / L$. Therefore, an arbitrary function $F(x_1 \cdots x_N)$ can be expanded as

$$F(x_1 \cdots x_N) = \sum_{n_1 > n_2 > \cdots > n_N} a_{n_1 \cdots n_N} D(x_1 \cdots x_N)$$

for the region of eq. (8.46). This proves the completeness. It is important, furthermore, that every k_i should be different. We shall make use of this fact later on. This stems from the definition of the right-hand side of eq. (8.33) with positive sign, and is unrelated to the Fermi statistics of electrons.

8.6 The eigenvalue problem

Returning to eq. (8.42) or eq. (8.43), if we solve this eigenvalue problem to determine k_j, the energy eigenvalue will be given by eq. (8.30). a_{I} is a function of the $N+1$ spin coordinates, and can in general be expanded using the α and β functions. The wavefunction corresponding to the state in which the i_1th, i_2th, and so on up to the i_Mth electrons are in the β (down-spin) state, and the rest are in the α (up-spin) state is denoted as $\phi_{i_1 i_2 \cdots i_M}$. When we operate X_{ij} on this, we obtain terms in which ζ_i and ζ_j are exchanged, plus terms that are unchanged. Since both kinds of terms leave M unchanged, a_{I} can be expanded, for constant M, as

$$a_{\mathrm{I}} = \sum_{0 \leqq i_1 < i_2 < \cdots < i_M \leqq N} a_{i_1 i_2 \cdots i_M} \phi_{i_1 i_2 \cdots i_M}. \tag{8.48}$$

We now try to solve the eigenvalue problem by substituting this into eq. (8.42). In order to apply the method of solution described in Appendix N (Yang, 1967), we shall slightly rearrange the equation. In place of X_{ij} defined by eq. (8.35), we define X'_{ij} as

$$X'_{ij} = \begin{cases} X_{ij} & i \neq 0, j \neq 0, \\ e^{-iJ/4} X_{0j} & i = 0, \\ e^{iJ/4} X_{i0} & j = 0. \end{cases}$$

It then follows that

$$X'_{ij} = \alpha_{ij} + \beta_{ij} P_{ij}. \tag{8.49}$$

Here, α and β are defined by

$$\alpha_{ij} = \frac{\lambda_i - \lambda_j}{\lambda_i - \lambda_j + ic}$$

and

$$\beta_{ij} = \frac{ic}{\lambda_i - \lambda_j + ic},$$

respectively, where λ and c are defined by $\lambda_0 = 1$, $\lambda_i = 0\,(i = 1, \ldots, N)$, and

$$c \equiv \tan \frac{J}{2}.$$

We can verify eq. (8.49) by the direct substitution of α and β. Equation (8.42) can then be written as

$$e^{ik_jL - iJ/4} a_{\mathrm{I}} = X'_{j+1\,j} \cdots X'_{Nj} X'_{0j} X'_{1j} \cdots X'_{j-1\,j} a_{\mathrm{I}}. \tag{8.50}$$

By identifying $N+1$, λ_i, and X'_{ij} in this equation with N, k_i, and X_{ij}, respectively, in Appendix N, we find that the two problems are identical. We shall therefore use the result of Appendix N in the following.

We define $e(x)$ by

$$e(x) = \frac{x+i}{x-i}.$$

When there are M down-spin electrons, we define M numbers, $\Lambda_1, \ldots, \Lambda_M$, by

$$e\left(\frac{\Lambda_\alpha - 1}{c'}\right) e\left(\frac{\Lambda_\alpha}{c'}\right)^N = -\prod_\beta^M e\left(\frac{\Lambda_\alpha - \Lambda_\beta}{c}\right) \quad (\alpha = 1, \ldots, M), \tag{8.51}$$

where

$$c' \equiv \frac{c}{2}.$$

The product on the right-hand side includes the $\beta = \alpha$ component. Using Λ_α as determined by eq. (8.51), k_j is given by

$$e^{ik_jL - iJ/4} = \prod_\alpha^M e\left(-\frac{\Lambda_\alpha}{c'}\right). \tag{8.52}$$

This is the result of Appendix N. An important point about this result is that the right-hand side of eq. (8.52) is independent of j. As discussed earlier, all of k_j, of which there are N, must differ from each other. Furthermore, since the right-hand side of eq. (8.52) is independent of j, we conclude that every k_j differs from each other by an integral multiple of $2\pi/L$.

Now, for the ground state, we take N k_js, starting from the lowest one and going upwards. However, there is no minimum k_j in our problem since the spectrum is approximated by a straight line, as shown in Fig. 8.1. Here, let us define a cut-off for k so that we have a minimum. The value of the cut-off does not make any difference so long as we are not interested in the value of the ground-state energy itself: we shall consider excitations by a magnetic field and thermal excitations.

Let us move on to the problem of determining Λ_α using eq. (8.51). Here Λ_α is taken to be real. As the absolute value of both sides of eq. (8.51) is 1, let us take their logarithm. We define $\theta(x)$ as

$$\theta(x) \equiv -2\tan^{-1}\frac{x}{c}, \tag{8.52a}$$

where

$$-\pi < \theta(x) < \pi.$$

Note that this is different from the step function defined by eq. (8.18). Since

$$e(x) = e^{-i[2\tan^{-1}x+\pi]}, \tag{8.53}$$

eq. (8.51) can be written as

$$\sum_\beta \theta(\Lambda_\alpha - \Lambda_\beta) - \theta(2\Lambda_\alpha - 2) - N\theta(2\Lambda_\alpha) = 2\pi J_\alpha \quad (\alpha = 1, \ldots, M). \tag{8.54}$$

Here J_α is a half-integer for odd $N - M$ and an integer for even $N - M$, but this is unimportant.

When M J_αs are fixed, the corresponding Λ_α are uniquely determined by them. Using Λ_α that have been obtained in such a manner, we define the following function:

$$h(\Lambda) \equiv \frac{\sum_\beta \theta(\Lambda - \Lambda_\beta) - \theta(2\Lambda - 2) - N\theta(2\Lambda)}{2\pi}. \tag{8.55}$$

$h(\Lambda)$ tends to $\mp(N + 1 - M)/2$ as $\Lambda \to \pm\infty$. Equation (8.54) thus has no solutions unless $|J_\alpha| \leqq (N + 1 - M)/2$. Now, as shown in Fig. 8.3, we make a plot of $h(\Lambda)$. When we take J_α as a point on the vertical axis, the corresponding

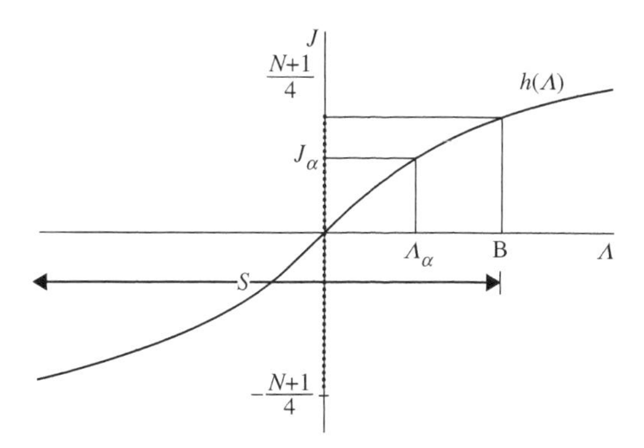

Figure 8.3 $h(\Lambda)$ vs Λ. On the vertical axis, M points are placed with equal spacing 1. There are hence $dh(\Lambda)$ points in an interval of length $dh(\Lambda)$ on the vertical axis.

point on the horizontal axis gives Λ_α. There are M points on the vertical axis, and there are M points on the horizontal axis corresponding to them. Now, let us take M J_αs that are placed side by side, that is,

$$J_{\alpha+1} = J_\alpha + 1.$$

In this case, let us denote by S the range on the horizontal axis that contains these points. When we take the large-N limit while keeping N/L and M/N fixed, the distribution of the points on the horizontal axis will become dense. We introduce the distribution function $\sigma(\Lambda)$, such that an interval of length $d\Lambda$ on the horizontal axis contains $\sigma(\Lambda)d\Lambda$ points; $\sigma(\Lambda)$ is proportional to N. The following is satisfied:

$$M = \int_S \sigma(\Lambda)d\Lambda. \tag{8.56}$$

8.7 The integral equation

Insofar as our problem is concerned, determining $\sigma(\Lambda)$ is sufficient. In order to do so, we set up an integral equation for $\sigma(\Lambda)$.

Let us denote the length of the interval on the vertical axis that corresponds to length $d\Lambda$ on the horizontal axis as $dh(\Lambda)$. In this interval, we have $dh(\Lambda)$ points, and this is equal to the number of points in the interval $d\Lambda$ on the horizontal axis. We thus have

$$\sigma(\Lambda) = \frac{dh(\Lambda)}{d\Lambda}. \tag{8.57}$$

Writing the sum with respect to β in eq. (8.55) as an integral by using $\sigma(\Lambda)$, and substituting the result into eq. (8.57) yields

$$\sigma(\Lambda) + \int_S K(\Lambda - \Lambda')\sigma(\Lambda')d\Lambda' = g(\Lambda), \tag{8.58}$$

where

$$g(\Lambda) \equiv \frac{1}{\pi}\frac{c'}{(\Lambda-1)^2 + c'^2} + \frac{N}{\pi}\frac{c'}{\Lambda^2 + c'^2},$$

$$K(\Lambda) \equiv \frac{1}{\pi}\frac{c}{\Lambda^2 + c^2}.$$

This is the integral equation.

We shall leave the solution to this equation until later. Let us express the energy in terms of $\sigma(\Lambda)$. Using eq. (8.52), we obtain

$$k_j L - \frac{J}{4} = 2\pi n_j - \sum_\beta [\pi + \theta(2\Lambda_\beta)] \quad (j = 1, \ldots, N), \tag{8.59}$$

where n_j are integers and are all distinct. We then have

$$\begin{aligned} E &= \sum k_j \\ &= \frac{2\pi}{L}\sum_j n_j + \frac{J}{4}D - D\sum_\beta [\pi + \theta(2\Lambda_\beta)], \end{aligned} \tag{8.60}$$

where

$$D \equiv \frac{N}{L}. \tag{8.60a}$$

As discussed before, we may define the ground state by $n_j = j - N$ $(j = 1, \ldots, N)$. The first term of E is equal to the energy of a system of free Fermi particles without the spin degree of freedom and confined within an interval of length L. This remains true at finite temperature. Denoting the third term of E as E_Λ, we obtain

$$E_\Lambda = -D\int_S [\pi + \theta(2\Lambda)]\sigma(\Lambda)d\Lambda. \tag{8.61}$$

Equation (8.58) is to be solved for Λ inside S. On the other hand, by substituting $\sigma(\Lambda)$, which has been obtained in this manner, into the integrand of eq. (8.58), $\sigma(\Lambda)$ can also be defined for Λ outside S. Hence integrating eq. (8.58) from $\Lambda = -\infty$ to ∞, we obtain

$$\int_{-\infty}^{\infty} \sigma(\Lambda)d\Lambda + \int_S \sigma(\Lambda)d\Lambda = N + 1.$$

From this result and eq. (8.56), we obtain

$$S_z = \tfrac{1}{2}(N+1) - M$$
$$= \tfrac{1}{2}\int_{\bar{S}} \sigma(\Lambda) d\Lambda. \tag{8.62}$$

Here, S_z refers to the value of S_z of the wavefunction that is given by eq. (8.48), and $\bar{S}$ refers to the region outside S for Λ.

8.8 The ground state

We may guess that, in the ground state, S should be taken as the whole of $\Lambda = -\infty$ to ∞. If so, from eq. (8.62), we have $S_z = 0$ and $M = \frac{1}{2}(N+1)$. From the discussion in Appendix B, the magnitude of spin S in eq. (8.48) is also zero. Denoting all quantities that represent the ground state with subscript 0, we obtain

$$\sigma_0(\Lambda) + \int_{-\infty}^{\infty} K(\Lambda - \Lambda')\sigma_0(\Lambda')d\Lambda' = g(\Lambda), \tag{8.63}$$

$$E_{\Lambda 0} = -D\int_{-\infty}^{\infty} \sigma_0(\Lambda)[\pi + \theta(2\Lambda)]d\Lambda. \tag{8.64}$$

Equation (8.63) can be solved easily by using Fourier transformation. However, in order to be able to solve eq. (8.58) later on, we adopt a more generic scheme. $K(\Lambda)$ is expressed as the Fourier integral:

$$K(\Lambda) = \frac{1}{2\pi}\int_{-\infty}^{\infty} e^{i\omega\Lambda - c|\omega|} d\omega.$$

Similarly, $R(\Lambda)$ is defined as follows:

$$R(\Lambda) \equiv \frac{1}{2\pi}\int_{-\infty}^{\infty} \frac{e^{i\omega\Lambda}}{1+e^{c|\omega|}} d\omega. \tag{8.64a}$$

We can then easily verify the following equation:

$$\int_{-\infty}^{\infty} R(\Lambda - \Lambda'')K(\Lambda'' - \Lambda')d\Lambda'' = K(\Lambda - \Lambda') - R(\Lambda - \Lambda'). \tag{8.65}$$

Note that both $R(\Lambda)$ and $K(\Lambda)$ are even functions of Λ. We then multiply eq. (8.63) by $R(\Lambda - \Lambda'')$ and integrate the result from $\Lambda = -\infty$ to $\Lambda = +\infty$.

Applying eq. (8.65) to the result yields an expression for σ_0:

$$\sigma_0(\Lambda) = g(\Lambda) - \int_{-\infty}^{\infty} R(\Lambda - \Lambda')g(\Lambda')d\Lambda'. \tag{8.66}$$

Evaluating the integral on the right-hand side yields

$$\sigma_0(\Lambda) = \frac{1}{2c}\left[N \operatorname{sech} \frac{\pi\Lambda}{c} + \operatorname{sech} \frac{\pi(\Lambda - 1)}{c}\right]. \tag{8.67}$$

8.9 Susceptibility

Next, we multiply eq. (8.58) by $R(\Lambda - \Lambda'')$ and integrate the result from $\Lambda = -\infty$ to $\Lambda = +\infty$. Using eqs. (8.65) and (8.66), we obtain

$$\sigma(\Lambda) = \sigma_0(\Lambda) + \int_{\bar{S}} R(\Lambda - \Lambda')\sigma(\Lambda')d\Lambda'. \tag{8.68}$$

Hereafter, we shall solve this for $\sigma(\Lambda)$. We multiply this by D and $\pi + \theta(2\Lambda)$, and integrate it from $\Lambda = -\infty$ to $\Lambda = +\infty$. Using eqs. (8.61) and (8.64), we then obtain

$$\Delta E \equiv E_\Lambda - E_{\Lambda 0} = \int_{\bar{S}} \sigma(\Lambda)U(\Lambda)d\Lambda, \tag{8.69}$$

$$U(\Lambda) \equiv D\int_{-\infty}^{\infty} \bar{R}(\Lambda - \Lambda')[\pi + \theta(2\Lambda')]d\Lambda',$$

$$\bar{R}(\Lambda) \equiv \delta(\Lambda) - R(\Lambda).$$

Here, D is as defined by eq. (8.60a). By carrying out the integration in $U(\Lambda)$, we obtain

$$U(\Lambda) = D\left(\frac{\pi}{2} - 2\tan^{-1}\tanh\frac{\pi\Lambda}{2c}\right)$$
$$\cong 2De^{-\pi\Lambda/c} \quad (\text{when } e^{-\pi\Lambda/c} \ll 1).$$

When the range from $-\infty$ to $+\infty$ is taken to be S, eq. (8.69) becomes 0. Otherwise, it is positive. We therefore see that the case $S = (-\infty, \infty)$ gives the ground state, and that $U(\Lambda)$ corresponds to the excitation energy. When Λ is large, the excitation energy is small. In the excited states with some empty spaces in the Λ axis, S_z is not 0, as can be seen from eq. (8.62). We expect, therefore, that one of these states arises as the ground state in a magnetic field.

Let us now take S as the range $\Lambda \leqq B$ as shown in Fig. 8.3. When we determine B in a way such that the sum of ΔE and $-2\mu H S_z$,

$$\Delta E - 2\mu H S_z = \int_{\bar{S}} \sigma(\Lambda)(2De^{-\pi\Lambda/c} - \mu H)d\Lambda, \tag{8.70}$$

is minimized, B is a function of H, and the magnetization is given by $2\mu S_z$. Here, $\mu = \mu_B/\hbar v_F$; $g = 2$ for both electrons and the localized spin.

In order to carry out this program, we need to solve eq. (8.68):

$$\sigma(\Lambda) = \sigma_0(\Lambda) + \int_B^\infty R(\Lambda - \Lambda')\sigma(\Lambda')d\Lambda'.$$

In terms of a new variable $x = \Lambda - B$ and a function $\sigma'(\Lambda) = \sigma(\Lambda + B)$,

$$\sigma'(x) = \sigma_0(x + B) + \int_0^\infty R(x - x')\sigma'(x')dx'. \tag{8.71}$$

Here, $x > 0$. If, in addition,

$$B > 1, \tag{8.72}$$

we obtain the expansion

$$\sigma_0(x + B) = \frac{N}{c}\sum_{n=0}^{\infty}(-)^n e^{-(2n+1)(\pi/c)(B+x)} + \frac{1}{c}\sum_{n=0}^{\infty}(-)^n e^{-(2n+1)(\pi/c)(B+x-1)}$$

using eq. (8.67). Hence we can solve eq. (8.71) as a linear combination of the solutions to

$$y(x) = \alpha e^{-\beta x} + \int_0^\infty R(x - x')y(x')dx' \quad (x > 0, \beta > 0). \tag{8.73}$$

The solution to eq. (8.73) is given by Appendix O. As described there, we first define $G_+(z)$ by the following equation:

$$G_+(z) = \frac{\sqrt{2\pi}\left(-\frac{iz}{e}\right)^{-iz}}{\Gamma\left(\frac{1}{2} - iz\right)}. \tag{8.74}$$

This is regular in the upper half of the z complex plane, and is non-zero. It is a multivalued function, and we consider the branch on which

$$G_+(i\omega) = \frac{\sqrt{2\pi}\left(\frac{\omega}{e}\right)^{\omega}}{\Gamma\left(\frac{1}{2}+\omega\right)} \quad (\omega > 0) \tag{8.75}$$

along the imaginary axis. Defining $G_-(z)$ as $G_-(z) = G_+(-z)$, this function is regular in the lower half of the z plane, and is non-zero. Now, from Appendix O, the solution to eq. (8.73) can be expressed in the form of the Fourier integral as follows:

$$\int_0^\infty y(x)e^{i\omega x}dx = \frac{\alpha}{\beta - i\omega} G_+\left(\frac{c\omega}{2\pi}\right) G_-\left(\frac{c\beta}{2\pi i}\right). \tag{8.76}$$

Using this, the solution to eq. (8.71) is also represented in the form of the Fourier integral:

$$\begin{aligned}
\Psi(\omega) &\equiv \int_0^\infty \sigma'(x)e^{i\omega x}dx \\
&= G_+\left(\frac{c\omega}{2\pi}\right)\sum_{n=0}^{\infty}\frac{(-)^n G_+\left(i\left(n+\frac{1}{2}\right)\right)}{(2n+1)\pi - ic\omega} \\
&\quad \times \left[Ne^{-(2n+1)(\pi/c)B} + e^{-(2n+1)(\pi/c)(B-1)}\right].
\end{aligned} \tag{8.77}$$

Using this Ψ, we can transform eq. (8.62) into

$$\begin{aligned}
2S_z = \Psi(0) &= \frac{1}{\sqrt{\pi}}\sum_{n=0}^{\infty}\frac{(-)^n}{n!}e^{-n-(1/2)}\left(n+\frac{1}{2}\right)^{n-(1/2)} \\
&\quad \times \left[Ne^{-(2n+1)(\pi/c)B} + e^{-(2n+1)(\pi/c)(B-1)}\right].
\end{aligned} \tag{8.78}$$

Here, we used eq. (8.75). ΔE in eq. (8.70) is expressed as

$$\Delta E = 2De^{-(\pi/c)B}\Psi\left(\frac{i\pi}{c}\right),$$

and is also expressed as a series. We differentiate these with respect to B to obtain

$$\frac{\partial}{\partial B}(\Delta E - 2\mu H S_z) = \left(2De^{-(\pi/c)B}\sqrt{\frac{\pi}{e}} - \sqrt{2}\mu H\right)\sum_n \cdots,$$

where a factor has been taken out of the series. In order that this factor vanishes, we must have

$$e^{-(\pi/c)B} = \sqrt{\frac{e}{2\pi}}\frac{\mu H}{D}. \tag{8.79}$$

This equation gives B as a function of H. Substituting this B into eq. (8.78), we obtain the magnetization as a function of H.

Let us first consider the $n=0$ term of the N terms in eq. (8.78). This yields $2S_z = \mu LH/\pi$, which is precisely the Pauli paramagnetic term for the case without localized spin. As for the $n \neq 0$ terms, they are of order $(\mu H/D)^{2n}$ compared with the $n=0$ term. Let us consider the case of a small magnetic field, so that these terms are negligible. In this case, the terms that do not contain N in eq. (8.78) would give the magnetization due to localized spin. These terms form a power series in $e^{-(\pi/c)(B-1)}$, and, by eq. (8.79), we obtain

$$e^{-(\pi/c)(B-1)} = \frac{\mu H}{T_1}, \tag{8.80}$$

where

$$T_1 \equiv \sqrt{\frac{2\pi}{e}}De^{-(\pi/c)}. \tag{8.81}$$

Hence the magnetization due to localized (impurity) spin is given by

$$2\mu S_z^{\text{imp}} = \frac{\mu}{\sqrt{\pi}}\sum_{n=0}^{\infty}\frac{(-)^n}{n!}e^{-n-(1/2)}\left(n+\frac{1}{2}\right)^{n-(1/2)}\left(\frac{\mu H}{T_1}\right)^{2n+1}. \tag{8.82}$$

In particular, when μH is small compared with T_1, we obtain

$$2\mu S_z^{\text{imp}} = \left(\frac{\mu^2}{\pi T_0}\right)H, \tag{8.83}$$

where

$$T_0 \equiv De^{-(\pi/c)} = \sqrt{\frac{e}{2\pi}}T_1. \tag{8.84}$$

Since eqs. (8.72) and (8.80) are valid, the expansion of eq. (8.82) holds when $\mu H < T_1$. When $\mu H > T_1$, by analytically continuing eq. (8.82) (see Appendix P), we obtain

$$2\mu S_z^{\text{imp}} = \mu\left[1 - \frac{1}{\pi^{3/2}}\int_0^\infty \sin\pi t\,\Gamma\left(\frac{1}{2}+t\right)e^t t^{-t-1}\left(\frac{T_1}{\mu H}\right)^{2t}dt\right]. \tag{8.85}$$

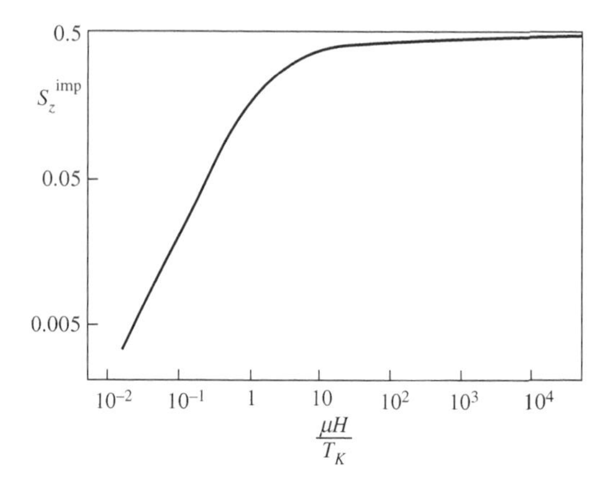

Figure 8.4 The relation between the magnetization of localized spin and the magnetic field. In low magnetic field, the magnetization behaves as $S_z^{\text{imp}} = 0.102676 \times 2\mu H/T_K$, and in high magnetic field, as $S_z^{\text{imp}} = \frac{1}{2} - [4\log(\mu H/T_K)]^{-1}$. From N. Andrei, K. Furuya and J. H. Lowenstein, *Rev. Mod. Phys.* **55** (1983) 331.

In particular, when $\mu H \gg T_1$, the vicinity of $t = 0$ makes the dominant contribution to the integral, and so we expand the integrand in the above equation as follows:

$$\int_0^\infty [1 - t\log t + (1 - C - 2\log 2)t]\, e^{-2t\log(\mu H/T_1)} dt.$$

$C = 0.5772\cdots$ is the Euler constant. Evaluating this integral yields

$$2\mu S_z^{\text{imp}} = \mu \left[1 - \frac{1}{2\log\dfrac{\mu H}{T_1}} + \frac{\log 2}{\left(2\log\dfrac{\mu H}{T_1}\right)^2} - \frac{\log\log\dfrac{\mu H}{T_1}}{\left(2\log\dfrac{\mu H}{T_1}\right)^2} + \cdots \right]. \tag{8.86}$$

This is an asymptotic expansion. These results are valid so long as $\mu H \ll D$. The result of a numerical calculation based on eqs. (8.82) and (8.85) is shown in Fig. 8.4.

8.10 Universality

This section is mainly based on Andrei and Lowenstein (1981).

The most important quantity in the last section is the ratio between T_0 and T_1. Here, T_0 appears in the expression for magnetization, eq. (8.83), which holds for $\mu H \ll T_1$, and T_1 appears in eq. (8.86), which holds for $\mu H \gg T_1$. However, we shall introduce a more strict definition for T_1. For some constant α of order unity, we substitute $T_1 = \alpha T_H$ into eq. (8.86). We then obtain an equation in which T_1 is replaced by T_H and $\log 2$ is replaced by $\log 2 - 2\log\alpha$. Let us adopt $\alpha = \sqrt{2}$ so that the latter vanishes. We then obtain

$$T_H = \sqrt{\frac{\pi}{e}}\, T_0. \tag{8.87}$$

In other words, using this T_H, we have

$$2\mu S_z^{\text{imp}} = \mu \left[1 - \frac{1}{2\log \dfrac{\mu H}{T_H}} - \frac{\log\log \dfrac{\mu H}{T_H}}{\left(2\log \dfrac{\mu H}{T_H}\right)^2} + \cdots \right], \tag{8.88}$$

in which the $\left(\log \dfrac{\mu H}{T_H}\right)^{-2}$ term is not present. Let us hence define T_H as the characteristic energy for the high magnetic field region, which makes the $\left(\log \dfrac{\mu H}{T_H}\right)^{-2}$ term vanish as shown above. We may then say that the ratio between T_0 and T_H is given by eq. (8.87), where T_0 is the quantity that appears in eq. (8.83), which represents the magnetization when $\mu H \ll T_1$, and T_H is the quantity that appears in eq. (8.88), which represents the magnetization when $\mu H \gg T_1$. This ratio characterizes the crossover from high to low magnetic field.

This quantity can be thought to be independent of the physics model. The present model is one-dimensional, so the movement of electrons is only in one direction, and the energy spectrum is linear. Moreover, in this model, the wavefunction is written in the form of Bethe's ansatz and is discontinuous at the boundaries, and we have set an artificial lower limit on the wavenumber k.

On the other hand, as in §5.9, one may use a model which is three-dimensional and has an exchange interaction that takes the form of a δ function. In that model, the wavefunction is expanded in terms of plane waves, and we set an upper limit to the wavenumber k. The density of states can be taken in various ways.

In each model, a quantity that is equivalent to T_0 may be given in terms of the parameters of the model. It is natural to think that all models that have the same T_0 have exactly the same magnetic field dependence so long as $\mu H \ll D$. This is called the universality hypothesis. We do not know at present how wide the class of models is for which this hypothesis is valid, but it seems certain that this hypothesis holds for a fairly wide class of models. Within this hypothesis, the models that have the same T_0 should also have the same T_H.

In the sense of the above discussion, we may say that T_H/T_0 is universal. One of the goals of the theory of s–d interaction is to calculate various universal parameters. There is one other universal parameter that characterizes the crossover from high to low temperatures. From eq. (8.83), the susceptibility at $T = 0$ is given by

$$\chi = \frac{\mu^2}{\pi T_0}. \tag{8.89}$$

When the susceptibility at high temperature is given by

$$\chi = \frac{\mu^2}{T}\left[1 - \frac{1}{\log\dfrac{T}{T_K}} - \frac{\log\log\dfrac{T}{T_K}}{2\left(\log\dfrac{T}{T_K}\right)^2} + \cdots\right], \tag{8.90}$$

the ratio between T_0 and T_K is such a quantity. Here, in eq. (8.90), as before, T_K is chosen so as to make the coefficient of $[\log(T/T_K)]^{-2}$ vanish. This equation is as obtained in §7.4. If the ratio can be derived from the present model, it will provide a useful check of universality. The direct calculation of this ratio by extending the exact solution to finite temperatures seems difficult, but we may derive the ratio in an indirect manner by using eq. (8.87).

In order to do so, we first note that the ratio between T_H in eq. (8.88) and T_K in eq. (8.90) is also expected to be universal. Then, since

$$\frac{T_0}{T_K} = \frac{T_0}{T_H}\frac{T_H}{T_K},$$

T_0/T_K follows from eq. (8.87) when we have obtained T_H/T_K. Since both eqs. (8.88) and (8.90) refer to the region in which perturbative expansion is valid, we can obtain T_H/T_K by perturbative expansion. If the universality hypothesis is correct, we may also obtain T_H/T_K on the basis of the perturbative expansion of the model discussed in Chapter 5. According to the model in §5.9, we have (eq. (F.9), page 234)

$$2S_z = 1 + J\rho - 2J^2\rho^2 \log\frac{D}{2\mu_B H} \tag{8.91}$$

when $\mu_B H \gg k_B T$ $(T = 0)$, and

$$\chi = \frac{\mu_B{}^2}{k_B T}\left(1 + 2J\rho - 4J^2\rho^2 \log\frac{0.6001D}{k_B T}\right) \tag{8.92}$$

(eq. (5.105)) when $T \gg T_K$. D is the cut-off energy of the conduction band, and $J\rho$ is the magnitude of the interaction. In eqs. (8.91) and (8.92), to the order $J^2\rho^2$, we have written not only the logarithmic term but also the constant term exactly. We now assume that S_z and χ can be expressed in the form of eqs. (8.88) and (8.90), respectively, for the model of §5.9. This assumption is nothing but a requirement of universality. We then compare eq. (8.88) with eq. (8.91), and eq. (8.90) with eq. (8.92), and equate the coefficients of $J^2\rho^2$ in them. This yields T_H and T_K. If we write

$$T_H = Ae^{1/2J\rho},$$
$$T_K = Be^{1/2J\rho},$$

we obtain $B/A = 1.2002$ and $B = 0.6001D\sqrt{|2J\rho|}$. Thus, this T_K is consistent with eq. (7.61). From this and eq. (8.87), we obtain

$$\frac{T_0}{T_K} = \sqrt{\frac{e}{\pi}}\frac{1}{1.2002} = 0.77503.$$

This is in good agreement with the value obtained by Wilson, $T_K/4\pi T_0 = 0.1032 \pm 0.0005$. As this level of agreement can hardly be coincidence, we conclude that universality holds for all three models in §5.9, Chapters 7 and 8.

8.11 The excited states

Next, we consider the case of finite temperature. The techniques at finite temperature are based on Takahashi (1971), in which the one-dimensional Heisenberg model and the one-dimensional system of Fermi particles are treated. The quantum numbers of this system are J_α in eq. (8.54) and n_j in eq. (8.59). The latter represents a system of N Fermi particles with no interaction nor the spin degree of freedom, and its behavior (for example, concerning the specific heat) at finite temperature can be directly derived from the results described in Chapter 3. In relation to the former, it has been found that $(N+1)/2$ J_αs, from $(N+1)/4$ to $-(N+1)/4$, are needed for the ground state, as shown in Fig. 8.3.

The ground state has $S_z = 0$ and $S = 0$, but there is no unique state with $S_z = 0$ and $S = 0$. In fact, the number of these states is equal to f_{N0}, given by eq. (B. 11). Now, how shall we obtain these states? J_α cannot be taken outside the range $(N+1)/4$ to $-(N+1)/4$: for such J_α, eq. (8.54) cannot have any solutions since $-\pi < \theta < \pi$. However, there are solutions for complex Λ, for which, of course, the complex conjugate also gives solutions. Thus, returning to eq. (8.51), out of the M Λs, let us say that two are of the form $\Lambda + ic' + i\delta$ and $\Lambda - ic' - i\delta$, and the rest, denoted as Λ_α, are real. Here δ is a sufficiently small real number. Equation (8.51) is now written as follows, omitting δs wherever appropriate:

$$\frac{\Lambda - 1 + ic}{\Lambda - 1}\left(\frac{\Lambda + ic}{\Lambda}\right)^N = \frac{c}{\delta}\prod_\alpha \frac{\Lambda - \Lambda_\alpha + 3ic'}{\Lambda - \Lambda_\alpha - ic'}, \tag{8.93}$$

$$\frac{\Lambda - 1}{\Lambda - 1 - ic}\left(\frac{\Lambda}{\Lambda - ic}\right)^N = \frac{\delta}{c}\prod_\alpha \frac{\Lambda - \Lambda_\alpha + ic'}{\Lambda - \Lambda_\alpha - 3ic'}, \tag{8.94}$$

$$e\left(\frac{\Lambda_\alpha - 1}{c'}\right) e\left(\frac{\Lambda_\alpha}{c'}\right)^N = -e\left(\frac{\Lambda_\alpha - \Lambda}{3c'}\right) e\left(\frac{\Lambda_\alpha - \Lambda}{c'}\right)\prod_\beta e\left(\frac{\Lambda_\alpha - \Lambda_\beta}{c}\right). \tag{8.95}$$

Multiplying the first two equations together yields

$$e\left(\frac{\Lambda - 1}{c}\right) e\left(\frac{\Lambda}{c}\right)^N = \prod_\beta e\left(\frac{\Lambda - \Lambda_\beta}{3c'}\right) e\left(\frac{\Lambda - \Lambda_\beta}{c'}\right). \tag{8.96}$$

We then determine Λ and Λ_α using eqs. (8.95) and (8.96), and substitute them into eqs. (8.93) or (8.94), to determine δ. If N is sufficiently large, we may expect that $|\delta| \ll 1$. The absolute values of both sides of eqs. (8.95) and (8.96) are equal to 1, and so we can solve them as before.

The number of pairs of complex solutions is not restricted to 1. Moreover, not only pairs but also triplets, quadruplets and in general n-plet solutions may exist. Here, $n = 1$ corresponds to a real solution. Now, when the number of down-spin electrons is given by M, let us say that the number of real solutions is given by M_1. These solutions are denoted as

$$\Lambda_\alpha^1 \quad \alpha = 1, \ldots, M_1.$$

The number of pairs of complex solutions is given by M_2, and we denote the solutions as

$$\left.\begin{array}{l} \Lambda_\alpha^2 + ic' \\ \Lambda_\alpha^2 - ic' \end{array}\right\} \quad \alpha = 1, \ldots, M_2,$$

where δ has been omitted. For general n, the number of n-plet solutions is given by M_n, and the solutions are denoted as

$$\left.\begin{array}{l} \Lambda_\alpha^n + (n-1)ic' \\ \Lambda_\alpha^n + (n-3)ic' \\ \vdots \\ \Lambda_\alpha^n - (n-1)ic' \end{array}\right\} \quad \alpha = 1, \ldots, M_n. \tag{8.97}$$

Here

$$M_1 + 2M_2 + 3M_3 + \cdots = M \tag{8.98}$$

needs to be satisfied. Although these are a hypothesis, it is generally thought, based on various circumstances, that these give an adequate description of the excited states.

Let us now move on to writing down the equations for Λ_α^n that correspond to eqs. (8.95) and (8.96), and solving them. The Λ_α in eq. (8.51) are now designated as $\Lambda_\alpha^n + lic'$ by the three indices, n, α and l. Equation (8.51) is then expressed as

$$e\left(\frac{\Lambda_\alpha^n - 1}{c'} + li\right) e\left(\frac{\Lambda_\alpha^n}{c'} + li\right)^N$$
$$= -\prod_{m=1}^{M_m} \prod_{\beta=1}^{m-1} \prod_{k=-(m-1)} e\left(\frac{\Lambda_\alpha^n - \Lambda_\beta^m}{c} + \frac{li}{2} + \frac{ki}{2}\right).$$

In the product on the right-hand side, k increases in steps of 2 (see eq. (8.97)). After multiplying out this product, the right-hand side yields

$$-\prod_{m\beta} e\left(\frac{\Lambda_\alpha^n - \Lambda_\beta^m + lic'}{(m-1)c'}\right) e\left(\frac{\Lambda_\alpha^n - \Lambda_\beta^m + lic'}{(m+1)c'}\right).$$

We then form the product with respect to l from $-(n-1)$ to $n-1$ in steps of 2, on both sides of the equation:

$$e\left(\frac{\Lambda_\alpha^n - 1}{nc'}\right) e\left(\frac{\Lambda_\alpha^n}{nc'}\right)^N = \prod_{m\beta}{}' E_{nm}\left(\frac{\Lambda_\alpha^n - \Lambda_\beta^m}{c'}\right), \tag{8.99}$$

where $'$ on the right-hand side refers to omitting the case where $m = n$ and $\beta = \alpha$ at the same time. $E_{nm}(x)$ is defined as

$$E_{nm}(x) \equiv e\left(\frac{x}{n+m}\right) e\left(\frac{x}{n+m-2}\right)^2 \cdots e\left(\frac{x}{|n-m|+2}\right)^2 e\left(\frac{x}{|n-m|}\right)$$

(omitting the last factor when $n = m$).

Using eq. (8.53), eq. (8.99) is transformed into

$$\theta\left(\frac{2\Lambda_\alpha^n - 2}{n}\right) + N\theta\left(\frac{2\Lambda_\alpha^n}{n}\right) + 2\pi J_\alpha^n = \sum_{m=1}\sum_{\beta=1}^{M_m}\theta_{nm}(\Lambda_\alpha^n - \Lambda_\beta^m). \quad (8.100)$$

Here θ_{nm} is defined by

$$\begin{aligned}\theta_{nm}(x) \equiv \theta\left(\frac{2x}{n+m}\right) &+ 2\theta\left(\frac{2x}{n+m-2}\right) + \cdots \\ &+ 2\theta\left(\frac{2x}{|n-m|+2}\right) + \theta\left(\frac{2x}{|n-m|}\right)\end{aligned}$$

(omitting the last term when $n = m$),

and $\theta(x)$ is defined by eq. (8.52a). J_α^n is an integer when $N - M_n$ is even, and a half-integer when it is odd.

When M_n J_α^ns are given for $n = 1, 2, \ldots$, the Λ_α^n are determined as the solutions to eq. (8.100). Since θ in eq. (8.100) tends to $\mp\pi$ in the limit $\Lambda_\alpha^n \to \pm\infty$, there are no solutions when the absolute value of J_α^n is large. By calculating the number of θs in θ_{nm}, we find that

$$|J_\alpha^n| \leqq \frac{1}{2}\left(N - \sum_{m=1} t_{nm}M_m\right), \quad (8.101)$$

where

$$t_{nm} = \begin{cases} 2\mathrm{Min}(n,m)^{\S} & n \neq m, \\ 2n-1 & n = m \end{cases}$$

needs to be satisfied in order that eq. (8.100) has a solution. For some given M, there are a number of sets of M_n $(n = 1, 2, \ldots)$ that satisfy eq. (8.98). For each such set of M_n $(n = 1, 2, \ldots)$, there are again a number of sets of J_α^n $(\alpha = 1, \ldots, M_n)$ that satisfy eq. (8.101). The number of all such sets can be calculated, and this is shown to be given by ${}_{N+1}\mathrm{C}_M - {}_{N+1}\mathrm{C}_{M-1}$. This is exactly the same as f_{N+1S} $(S = (N+1)/2 - M)$, which is given by eq. (B. 11). This is presumably not a coincidence, and is due to the fact that the complex Λ obtained in the above manner accommodates all excited states.

After Λ_α has been determined using eq. (8.100), the energy can be expressed as follows. From eq. (8.52), we have

$$e^{ik_jL - iJ/4} = \prod_{n\alpha l} e\left(-\frac{\Lambda_\alpha^n}{c'} - li\right) = \prod_{n\alpha} e\left(-\frac{\Lambda_\alpha^n}{nc'}\right).$$

§ Min(n, m) means the smaller of n and m.

From this, we may use eq. (8.53) to obtain

$$k_j L - \frac{J}{4} = 2\pi n_j - \sum_{n\alpha} \left[\theta\left(\frac{2\Lambda_\alpha^n}{n}\right) + \pi \right].$$

n_j ($j = 1, \ldots, N$) are distinct integers. In the presence of a magnetic field H, the energy becomes

$$E = \sum k_j - 2\mu H \left[\frac{1}{2}(N+1) - M \right]. \tag{8.102}$$

Since

$$M = \sum_n n M_n = \sum_{n\alpha} n,$$

we obtain

$$E = \frac{DJ}{4} + \frac{2\pi}{L} \sum_j n_j - D \sum_{n\alpha} g_n(\Lambda_\alpha^n) - \mu H(N+1), \tag{8.103}$$

$$g_n(\Lambda) = \theta\left(\frac{2\Lambda}{n}\right) + \pi - \frac{2\mu H}{D} n. \tag{8.104}$$

8.12 Free energy

We now seek the expression for the free energy of this system. Adopting the Hamiltonian of eq. (8.1), which is denoted as $\mathcal{H}$, we obtain

$$\begin{aligned} Z &= \operatorname{Tr} e^{-\beta(\mathcal{H} - 2\mu H S_z)} \\ &= \sum_{S=0}^{\frac{1}{2}(N+1)} \sum_{S_z=-S}^{S} \operatorname{Tr}_{SS_z} e^{-\beta(\mathcal{H} - 2\mu H S_z)} \\ &= \sum_S \frac{\sinh(2S+1)\beta\mu H}{\sinh \beta\mu H} \operatorname{Tr}_{SS} e^{-\beta\mathcal{H}} \quad (\beta = 1/k_B T). \end{aligned}$$

Here, Tr_{SS_z} refers to the summation over all states that have magnitude of spin S, and z-component of spin S_z. Tr_{SS} means the summation only over states with S_z equal to S. The wavefunctions of eq. (8.48) determined through the procedure described in Appendix N are such states. Since the magnitude of S

is of order N, we obtain

$$Z \cong \sum_S \mathrm{Tr}_{SS}\, e^{-\beta(\mathcal{H}-2\mu HS)}$$
$$= \sum_M \mathrm{Tr}_M\, e^{-\beta[-\mu H(N+1)+\mathcal{H}+2\mu HM]}.$$

We then use eq. (8.102) or (8.103) as the energy eigenvalue to obtain:

$$Z = Z_C Z_\Lambda, \tag{8.105}$$
$$Z_C = \mathrm{Tr}\, e^{-(2\pi\beta/L)\sum n_j}, \tag{8.106}$$
$$Z_\Lambda = \mathrm{Tr}\, e^{-\beta D\sum_{n\alpha} g_n(\Lambda_\alpha^n)+\beta\mu H(N+1)}. \tag{8.107}$$

Here, Z_C is the partition function of a Fermi particle system that does not have the spin degree of freedom, and can be easily calculated, Z_Λ is the part that is due to Λ_α^n, and Tr refers to the summation over all ways of giving the sets J_α^n. In practice, we only consider the case that maximizes Z_Λ.

Next, we move on to solving eq. (8.100). Here, we need the distribution function of Λ, and the integral equation for this function can be derived in the same way as in §8.7. However, at finite temperatures in general, J_α^n are not always given as a continuous sequence of integers (or half-integers), and so the concept of holes in the Λ distribution arises.

At first, we define $h_n(\Lambda)$ in terms of the Λ_α^n, which are obtained by solving eq. (8.100) for given J_α^n, as follows:

$$h_n(\Lambda) = \frac{\sum_{m\beta}\theta_{nm}(\Lambda - \Lambda_\beta^m) - \theta\left(\frac{2\Lambda-2}{n}\right) - N\theta\left(\frac{2\Lambda}{n}\right)}{2\pi}. \tag{8.108}$$

As shown in Fig. 8.5, when we make a plot of $h_n(\Lambda)$ and choose points J_α^n on the vertical axis, the values of corresponding Λ on the horizontal axis give Λ_α^n. Holes are those points on the horizontal axis that correspond to integers (or half-integers) on the vertical axis which have not been chosen as J_α^n. When N is large, the Λ_α^n and holes are densely distributed on the Λ axis. We then introduce the corresponding distribution functions as follows. We denote the number of the solutions Λ_α^n in the range $d\Lambda$ on the Λ axis as $\rho_n(\Lambda)d\Lambda$, and that of holes as $\tilde{\rho}_n(\Lambda)d\Lambda$. Then, in the same way as in eq. (8.57), we find that

$$\rho_n(\Lambda) + \tilde{\rho}_n(\Lambda) = \frac{dh_n(\Lambda)}{d\Lambda}. \tag{8.109}$$

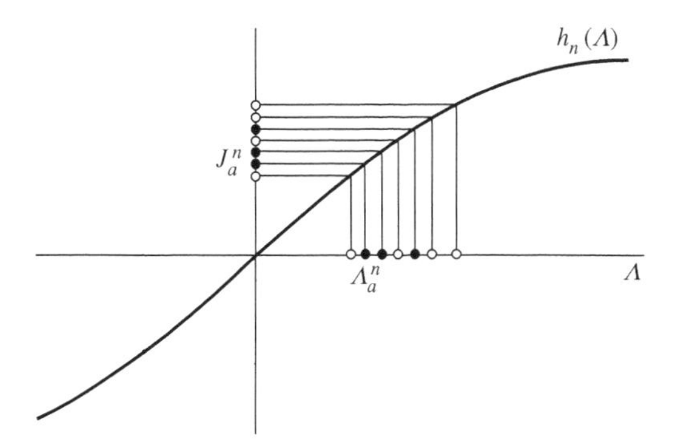

Figure 8.5 $h_n(\Lambda)$ vs Λ. On the vertical axis, points are placed with separation 1. We represent those points that are not adopted as J_α^n by open circles. The corresponding points on the Λ axis may be identified with holes.

We use ρ_m to express the summation over β in eq. (8.108) as an integral, and then differentiate it with respect to Λ. This yields

$$\frac{dh_n}{d\Lambda} = NF_n(\Lambda) + F_n(\Lambda - 1) - \sum_m \int B_{nm}(\Lambda - \Lambda')\rho_m(\Lambda')d\Lambda', \quad (8.110)$$

where

$$F_n(\Lambda) \equiv \frac{1}{\pi}\frac{nc'}{\Lambda^2 + (nc')^2}, \quad (8.110a)$$

and

$$B_{nm}(\Lambda) \equiv F_{|n-m|}(\Lambda) + 2F_{|n-m|+2}(\Lambda) + \cdots + 2F_{n+m-2}(\Lambda) + F_{n+m}(\Lambda)$$
$$\text{(the first term is 0 when } n = m\text{)}.$$

Here, although the first term of B_{nm} can also be taken as $\delta(\Lambda)$ when $n = m$, it is more correct to take this term to be zero, from the derivation of eq. (8.110). However, for the sake of convenience in the following, we define $A_{nm}(\Lambda)$, in which this term is taken as $\delta(\Lambda)$, as follows:

$$A_{nm}(\Lambda) = B_{nm}(\Lambda) + \delta_{nm}\delta(\Lambda). \quad (8.110b)$$

We substitute this equation and eq. (8.109) into eq. (8.110) and obtain

$$\tilde{\rho}_n(\Lambda) = NF_n(\Lambda) + F_n(\Lambda - 1) - \sum_m \int A_{nm}(\Lambda - \Lambda')\rho_m(\Lambda')d\Lambda'. \quad (8.111)$$

These form a system of coupled integral equations for ρ_n and $\tilde{\rho}_n$.

However, these equations are not sufficient for obtaining ρ_n and $\tilde{\rho}_n$. Remember that at first, in eq. (8.100), we obtained Λ_α^n only after giving J_α^n. Then, how should J_α^n be given? In order to answer this question, we should find the entropy and consider the free energy. Let us divide the vertical axis in Fig. 8.5 into small intervals. Although these intervals are small, let us say that the number of points in an interval is much larger than 1. For a given set of J_α^ns, let us say that n_i points in the ith interval belong to this set, and the remaining h_i points are empty. Out of the various sets of J_α^n, those for which n_i in all intervals are the same yield the same ρ_n and $\tilde{\rho}_n$ when eq. (8.100) is solved. The number of distinct combinations is then given by

$$W = \prod_i \frac{(n_i + h_i)!}{n_i!\, h_i!}.$$

Let us hence denote the sum of the last two terms in eq. (8.103) with the entropy term $-T \log W$ (here, this T denotes $k_B T/\hbar v_F$) as F_Λ. Using $n_i = \rho_n(\Lambda)d\Lambda$, $h_i = \tilde{\rho}_n(\Lambda)d\Lambda$, and Stirling's formula $n! \sim n^n e^{-n}$, we obtain

$$\begin{aligned} F_\Lambda = &-D \sum_n \int g_n(\Lambda)\rho_n(\Lambda)d\Lambda - \mu H(N+1) \\ &-T \sum_n \int \left[\rho_n(\Lambda) \log \left(1 + \frac{\tilde{\rho}_n(\Lambda)}{\rho_n(\Lambda)} \right) + \tilde{\rho}_n(\Lambda) \log \left(1 + \frac{\rho_n(\Lambda)}{\tilde{\rho}_n(\Lambda)} \right) \right] d\Lambda. \end{aligned} \tag{8.112}$$

F_Λ refers to the part of the free energy that involves Λ_α^n. By imposing the condition on ρ_n and $\tilde{\rho}_n$ that F_Λ be minimized, and using eq. (8.111), we can solve for ρ_n and $\tilde{\rho}_n$.

Let us first introduce the notation

$$\eta_n = \frac{\tilde{\rho}_n}{\rho_n}.$$

The variation of F_Λ can then be expressed as

$$\begin{aligned} \delta F_\Lambda = &-D \sum_n \int g_n(\Lambda)\delta\rho_n(\Lambda)d\Lambda \\ &-T \sum_n \int \Big[\delta\rho_n(\Lambda) \log (1 + \eta_n(\Lambda)) \\ &+ \delta\tilde{\rho}_n(\Lambda) \log \left(1 + \eta_n(\Lambda)^{-1} \right) \Big] d\Lambda. \end{aligned} \tag{8.113}$$

Next, we introduce the following notation for convolution:

$$A * B \equiv \int A(\Lambda - \Lambda')B(\Lambda')d\Lambda'. \tag{8.113a}$$

If $A(\Lambda)$ is an even function,

$$\int C(A * B)d\Lambda = \int B(A * C)d\Lambda. \tag{8.114}$$

Here, from eq. (8.111),

$$\delta\tilde{\rho}_n = -\sum_m A_{nm} * \delta\rho_m.$$

We substitute this into eq. (8.113) and apply eq. (8.114). We then set the coefficient of $\delta\rho_n$ equal to zero to obtain

$$Dg_n(\Lambda) + T\log(1+\eta_n) - T\sum_m A_{nm} * \log(1+\eta_m{}^{-1}) = 0. \tag{8.115}$$

Note that neither ρ nor $\tilde{\rho}$ appears in this equation. We substitute this equation into eq. (8.112) to obtain

$$F_\Lambda = -T\sum_n \int \log(1+\eta_n{}^{-1})[\tilde{\rho}_n + \sum_m A_{nm} * \rho_m]d\Lambda - \mu H(N+1).$$

On the other hand, using eq. (8.111), we see that

$$F_\Lambda = -T\sum_n \int \log(1+\eta_n{}^{-1})[NF_n(\Lambda) + F_n(\Lambda - 1)]d\Lambda - \mu H(N+1). \tag{8.116}$$

Again, this equation involves neither ρ nor $\tilde{\rho}$.

Altogether, we see that if we determine η_n by solving eq. (8.115) and substitute it into eq. (8.116), we obtain the free energy. As shown in Appendix Q, after suitable transformations of these two equations, we can phrase our problem as follows. We first define $s(\Lambda)$ by

$$s(\Lambda) = \frac{1}{2c}\operatorname{sech}\frac{\pi\Lambda}{c}, \tag{8.117}$$

and solve for η_n so that the following three equations are satisfied:

$$\log\eta_1 = s * \log(1+\eta_2) - \frac{2D}{T}\tan^{-1} e^{-\pi\Lambda/c}, \tag{8.118}$$

$$\log \eta_n = s * [\log(1 + \eta_{n+1}) + \log(1 + \eta_{n-1})] \quad (n \geqq 2), \tag{8.119}$$

$$\frac{2\mu H}{T} + F_{n+1} * \log(1 + \eta_n) - F_n * \log(1 + \eta_{n+1}) = 0 \quad (n \to \infty). \tag{8.120}$$

In terms of η_1 determined in this manner, F_Λ is obtained as

$$F_\Lambda = -T \int [Ns(\Lambda) + s(\Lambda - 1)] \log(1 + \eta_1) d\Lambda - \mu H(N + 1) + \text{const.} \tag{8.121}$$

Here, 'const.' refers to a constant which is independent of both the magnetic field and temperature.

8.13 Specific heat

It is difficult to solve eqs. (8.118)–(8.120) analytically. Here, let us examine their behavior at low temperature. When $D \gg T$, for $\Lambda \lesssim 0$, η_1 is highly suppressed and does not contribute to F_Λ. The second term on the right-hand side of eq. (8.118) can hence be written as

$$-\frac{2D}{T} e^{-\pi\Lambda/c} = -2e^{-\pi\Lambda/c + \log(D/T)}.$$

We therefore introduce a new variable ζ in place of Λ:

$$\zeta = \frac{\pi\Lambda}{c} - \log\frac{D}{T}.$$

Equations (8.118)–(8.121) are then written as

$$\log \eta_1(\zeta) = \frac{1}{2\pi} \int \operatorname{sech}(\zeta - \zeta') \log[1 + \eta_2(\zeta')] d\zeta' - 2e^{-\zeta}, \tag{8.122}$$

$$\log \eta_n(\zeta) = \frac{1}{2\pi} \int \operatorname{sech}(\zeta - \zeta') \log[1 + \eta_{n+1}(\zeta')] \times [1 + \eta_{n-1}(\zeta')] d\zeta' \quad (n \geqq 2), \tag{8.123}$$

$$\frac{2\mu H}{T} + \int \frac{2(n+1)}{4(\zeta - \zeta')^2 + \pi^2(n+1)^2} \log[1 + \eta_n(\zeta')] d\zeta' - \int \frac{2n}{4(\zeta - \zeta')^2 + \pi^2 n^2} \log[1 + \eta_{n+1}(\zeta')] d\zeta' = 0 \quad (n \to \infty), \tag{8.124}$$

$$F_\Lambda = -\frac{NT}{2\pi}\int \operatorname{sech}\left(\zeta + \log\frac{D}{T}\right)\log(1+\eta_1)d\zeta - N\mu H$$
$$-\frac{T}{2\pi}\int \operatorname{sech}\left(\zeta + \log\frac{T_0}{T}\right)\log(1+\eta_1)d\zeta - \mu H. \qquad (8.125)$$

Here, we made use of eq. (8.84). Looking at eqs. (8.122)–(8.124), we find that η_n are functions of only ζ and $2\mu H/T$, and the interaction parameter c does not appear. Thus the integral in the first line of eq. (8.125) is a function of $2\mu H/T$ and T/D, and that in the second line is a function of $2\mu H/T$ and T/T_0. Furthermore, it is significant that their functional form is identical. Another important result is that the interaction parameter c appears only via T_0.

Let us denote the integral in the first term as $H(2\mu H/T, T/D)$.

We now calculate the free energy F_C that corresponds to Z_C of eq. (8.106). The expression for the free energy of Fermi particles reads

$$F_C = -T\sum_j \log(1+e^{-\beta k_j}).$$

Using $k_j = 2\pi n_j/L$, we can transform this into the integral:

$$F_C = -\frac{LT}{2\pi}\int_{-K}^{\infty}\log(1+e^{-\beta k})dk \cong -\frac{L\pi T^2}{12} + \text{const.}$$

Here, $K = 2\pi N/L$ represents the cut-off for k. The last equation is valid for $K \gg T$.

Now adding together F_Λ and F_C, we obtain the total free energy:

$$F = -\frac{NT}{2\pi}H\left(\frac{2\mu H}{T}, \frac{T}{D}\right) - N\mu H - \frac{L\pi T^2}{12}$$
$$-\frac{T}{2\pi}H\left(\frac{2\mu H}{T}, \frac{T}{T_0}\right) - \mu H. \qquad (8.126)$$

As discussed earlier, the form of eqs. (8.122)–(8.124) is independent of whether the interaction parameter c is present or not. The functional form of H is therefore also independent of whether there is interaction or not. It is thus permissible to consider the sum of the first three terms in eq. (8.126) as the free energy in the case without localized spin, and the sum of the fourth and fifth terms as the contribution of the localized spin. The former is an elementary calculation, and the functional form of H can be found by comparing against it. This allows us to obtain the latter contribution. The former free energy is calculated as

$$F_0 = -\frac{LT}{2\pi}\int_{-K}^{\infty} \log(1+e^{-\beta(k+\mu H)})(1+e^{-\beta(k-\mu H)})dk$$
$$= -\frac{\pi LT^2}{6} - \frac{L\mu^2 H^2}{2\pi} + \text{const.} \quad (K \gg T, \mu H).$$

When we compare this with the first three terms in eq. (8.126), we obtain

$$TH\left(\frac{2\mu H}{T}, \frac{T}{D}\right) + 2\pi\mu H$$
$$= \frac{\pi^2}{6}\frac{T^2}{D} + \frac{\mu^2 H^2}{D} + \text{const.} \quad (D \gg T, \mu H),$$

and, from this, we obtain the free energy due to localized spin as

$$F^i = -\frac{\pi}{12}\frac{T^2}{T_0} - \frac{\mu^2 H^2}{2\pi T_0} \quad (T_0 \gg T, \mu H).$$

The susceptibility that follows from this is the same as eq. (8.89), and the specific heat is given by $c = \pi T/6T_0$.

In §5.8 and Chapter 7, we considered the ratio

$$r = \frac{\chi T}{c}\frac{\pi^2}{3\mu^2}.$$

From the above results, we have $r = 2$, and this is consistent with the earlier result. This quantity can also be considered to be universal. In §5.8, we saw that $r \to 2$ as $U \to \infty$. When U is large but finite, r is close to 2 but not exactly equal to 2. If we were to replace the model of §5.8 with the s–d model, r would be exactly equal to 2 as shown here. This difference may be attributed to the contributions that were not taken into account when the Anderson model was transformed into the s–d model.

We thus see that, by transforming the s–d problem into the problem of a one-dimensional system, and applying the exact solution method for the one-dimensional system, which has previously been established, we are able to solve this problem exactly. In particular, an analytic solution, eq. (8.82) or eq. (8.85), has been obtained regarding the H-dependence at $T = 0$. As for the temperature dependence, the specific heat and susceptibility at an arbitrary temperature can be obtained just by numerically solving the coupled integral equations.

One of the important results in this chapter and the preceding chapter is universality. That is to say, the results in these two chapters suggest that the

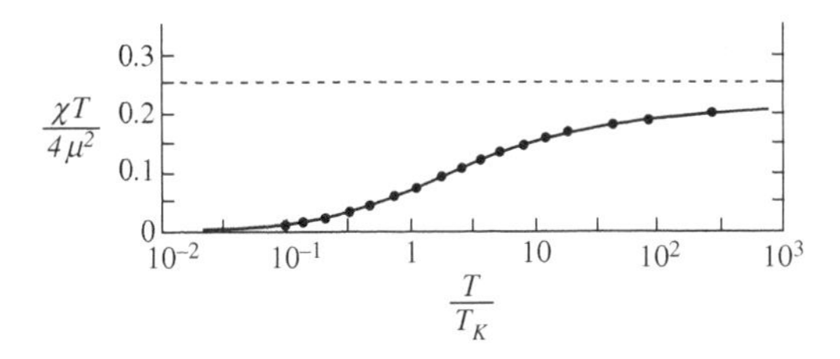

Figure 8.6 χT vs T. χT for the free localized spin is equal to μ^2 (dashed line). However, if s–d interaction is present, it decreases at low temperature and tends to 0 as $T \to 0$. This corresponds to the situation shown in Fig. 5.13, namely, when U is increased, $|J|$ decreases. Hence T_K also decreases, and the temperature at which χ starts to deviate from the Curie law decreases. The closed circles correspond to the results calculated by Krishna-Murthy *et al.* (1980) using a renormalization-group method based on the Anderson model. From N. Andrei, K. Furuya and J. H. Lowenstein, *Rev. Mod. Phys.* **55** (1983) 331.

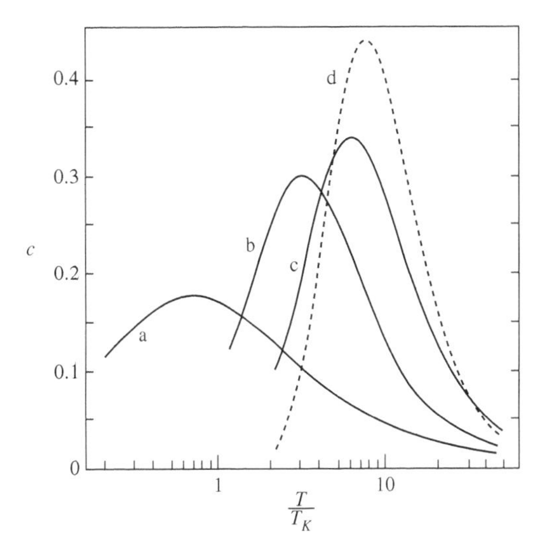

Figure 8.7 c vs T. The specific heat for (a) $\mu H/T_K = 0$, (b) $\mu H/T_K = 4.76$, (c) $\mu H/T_K = 9.05$, and (d) the specific heat of a free spin for $\mu H/T_K = 9.05$. The fact that there is little difference between (d) and (c) shows that the effect of the s–d interaction is suppressed by the high magnetic field in (c).

dependence on temperature and the magnetic field of the physical quantities are independent of the three different models adopted in Chapters 5, 7 and 8, and are identical with one another if the parameter T_K is the same, where, of course, $\mu_B H, k_B T \ll D$ is assumed.

By solving eqs. (8.122)–(8.124) numerically, the temperature dependence of χ and c can be obtained. We show the results in Figs. 8.6 and 8.7.

References and further reading

Andrei, N. (1980) *Phys. Rev. Lett.* **45**: 379.
Andrei, N. and Lowenstein, J. H. (1981) *Phys. Rev. Lett.* **46**: 356.
Andrei, N., Furuya, K. and Lowenstein, J. H. (1983) *Rev. Mod. Phys.* **55**: 331.
Filyov, V. M., Tsvelick, A. M. and Wiegmann, P. B. (1981) *Phys. Lett.* **81A**: 175.
Krishna-Murthy, H. R., Wilkins, J. W. and Wilson, K. G. (1980) *Phys. Rev.* B**21**: 1003.
Schulz, H. (1982) *J. Phys.* **C15**: L37.
Takahashi, M. (1971) *Prog. Theor. Phys.* **46**: 401, 1388.
Wiegmann, P. B. (1981) *J. Phys.* **C14**: 1463.
Yang, C. N. (1967) *Phys. Rev. Lett.* **19**: 1312.

9
Recent developments

There have been a number of notable developments in this field after the publication of the Japanese-language version of this book. The purpose of this chapter is to discuss two such topics.

The first topic involves a series of intermetallic compounds, which usually contain Ce. These have been found to have electronic heat capacity that is several hundred to thousand times greater than the corresponding heat capacity of ordinary metals. Here, although a Ce atom is magnetic, it does not interact with the neighboring Ce atoms, and it behaves as if it is an isolated magnetic atom. This fact explains the large heat capacity.

The other topic is the quantum dot. This tiny 'artificial atom' can be connected to leads to give rise to a system which is analogous to a system of metals with impurities. When there are an odd number of electrons in the quantum dot, the quantum dot acquires a spin, and its behavior becomes similar to that of a magnetic atom in metals. In particular, the transmission probability of an electron through the leads becomes 1 at absolute zero, and this corresponds to the unitary limit of electrical resistivity in systems with magnetic impurity.

9.1 The spin-flip rate

A spin placed inside a metal undergoes an interaction with the conduction electrons of the form eq. (5.90), which causes the inversion of the direction of the spin. This spin-flip rate, in the lowest order of J, is found to be

$$\Gamma = \frac{2\pi}{\hbar} J^2 \rho^2 k_B T. \tag{9.1}$$

The inverse of this quantity is then the time τ for the spin-flip to take place, that is, the relaxation time:

$$\tau = \frac{\hbar}{2\pi J^2 \rho^2 k_B T}. \tag{9.2}$$

When Γ is very large, the flipping of spin between the up and down states becomes so frequent that one can think of the state as a half–half mixture of the up and down-spin states. On the other hand, when Γ is very small, one can consider the spin to be pointed in either of the two directions. The crossover between these two cases occurs near the temperature T_K. We would like to discuss this point below.

Let us start with an impurity spin that is at temperature T. This system then has a thermal energy of about $k_B T$. The energy cannot be exactly $k_B T$, and there is an uncertainty in the energy which is also about $k_B T$. It then follows, from the uncertainty relationship between energy and time, that we are looking at this system through a timescale of about $h/k_B T$. Let us compare this observation timescale with the relaxation time of eq. (9.2). We then find that, since $J^2\rho^2 \ll 1$, the relaxation time is much longer than the observation timescale. That is, the spin orientation almost remains constant during the observation. Spin is pointed either in the up or the down direction, and there is a two fold degeneracy.

When the temperature is less than T_K, this argument no longer applies, because the expression for Γ, or equivalently τ, becomes inappropriate. Equation (9.1) is the lowest-order term in J. There are higher-order terms which contain $\log T$, so eq. (9.1) needs to be modified at low temperatures.

The result of this modification will be as shown in Fig. 9.1. Γ is proportional to T at high temperature, and becomes of the order of $k_B T_K/h$ near T_K. Below $T \approx T_K$, it does not fall off too much, and tends to a constant towards absolute zero.

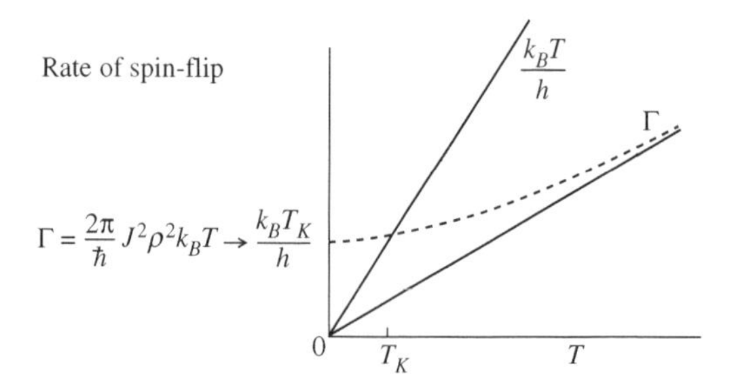

Figure 9.1 The dotted line represents the correct temperature dependence of Γ.

As can be seen from this figure, Γ crosses over with k_BT/h near $T = T_K$. $\Gamma > k_BT/h$ below T_K. The inverse of this inequality yields $\tau < h/k_BT$. In other words, the relaxation time becomes shorter than the observation timescale below $T = T_K$. That is, the spin-flip becomes frequent and the up and down orientations of spin appear equally. The twofold spin degeneracy is lost, and the behavior of the susceptibility changes from the Curie law to the Pauli susceptibility behavior. In other words, the spin is pointed in either the up or the down direction for $T > T_K$, whereas we have two spin orientations appearing equally for $T < T_K$.

Let us bring up a related point here, of the electrical resistance of metals with magnetic impurities. We first apply a theorem which is due to Friedel. According to Friedel, the electrical resistance due to an impurity is small when the number of electrons is zero or 1 in the outer shell of the impurity that causes electron scattering. The electrical resistance is large when the number of electrons is a half.

For temperatures greater than T_K, the spin is pointed in either of the two directions, so that either the number of up-spin electrons is 1 and the number of down-spin electrons is zero, or vice versa. In either case, the electrical resistance is small. On the other hand, when the temperature is below T_K, the up-spin and down-spin electrons appear equally, and the number of electrons in each state becomes 1/2. Hence the electrical resistance becomes large. It follows that when the temperature is lowered through T_K, the electrical resistance increases. The vanishing of the spin corresponds to the unitary limit of the electrical resistance, which behavior is heralded by the $\log T$ that appears at higher temperatures.

9.2 The heavy electrons

So far, we have considered the concentration of the impurity to be small, and neglected the mutual interaction in between the impurity spin. However, when the concentration goes up, one must start taking this mutual interaction into account. There is RKKY interaction[§] mediated by the conduction electrons in between the impurity spin. This fixes the spin orientations through the influence of the other spins. The impurity atoms are distributed randomly, and hence the spin orientations do not adopt a regular structure such as that of ferromagnetism

[§] A spin placed inside a metal causes the spin polarization of the conduction electrons via the interaction of eq. (5.90). Let us now say that we have another spin. This spin is then affected by the spin polarization, leading to an effective interaction with the first spin. This is called the RKKY (Ruderman–Kittel–Kasuya–Yosida) interaction. The strength of the interaction decreases proportionally with the cube of the distance between the spins, and oscillates with period π/k_F.

or antiferromagnetism. Rather, the spins are fixed in random orientation, and this is called the spin glass.

In the presence of such fixing of the spin orientations, the resistance minimum appears no longer. As we saw earlier, the $\log T$ term had emerged because of the spin-flip that occurs during the intermediate state in the scattering. This spin-flip cannot occur when the spin orientations are fixed.

In the typical magnetic alloys, such as Cu with Fe or Mn, the addition of a small percentage impurity brings about the spin-glass state. A stark contrast to this situation was provided by the discovery of a new compound which does not exhibit the spin-glass behavior regardless of the concentration of the magnetic impurity. This compound, $La_{1-x}Ce_xCu_6$, possesses the same crystal structure for $0 \leq x \leq 1$. Ce has one 4f electron, and is a magnetic atom. The electrical resistance of this compound is shown in Fig. 9.2.

There is a typical electrical resistance minimum behavior for $x = 0.094$. What is unusual here is that this behavior persists regardless of the value of x as x is increased, up to even $x = 1$. When $x = 1$, Ce is no longer an impurity and forms a regular lattice structure.

Since Ce atoms have spin, it would seem that there is interaction between neighboring spin which fixes the spin orientations in a certain order, but there is no indication of spin ordering in the measurement of magnetic susceptibility. Furthermore, the very appearance of the electrical resistance minimum phenomenon is evidence of the non-fixation of the spin orientations. Simply stated, each spin seems independent of the surrounding spin, and interacts with the

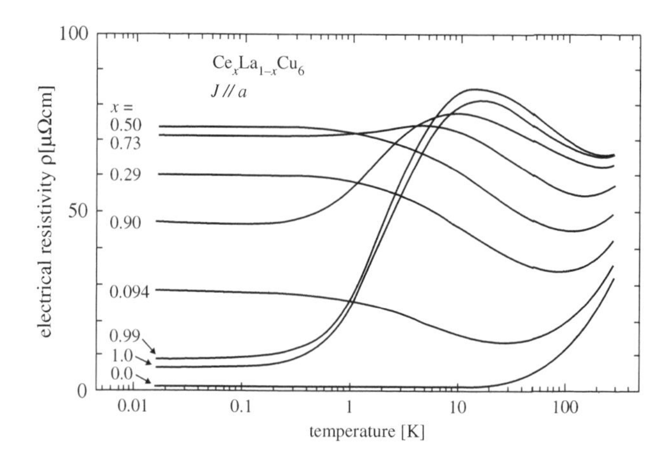

Figure 9.2 The temperature dependence of the electrical resistivity of $Ce_xLa_{1-x}Cu_6$. From A. Sumiyama *et al.*, *J. Phys. Soc. Japan* **55** (1986) 1294.

conduction electrons to give rise to the electrical resistance minimum. This is the case even when $x = 1$. Since this is the case, we can talk about T_K even when x is large. Taking T_K to be slightly above the region where the electrical resistivity approaches a constant at low temperature, we can say that T_K is around 5 K.

The electrical resistivity at $x = 1$ exhibits an interesting temperature dependence. Since Ce forms a regular lattice structure at $x = 1$, there is no impurity as such and, in this sense, we expect low electrical resistivity. However, the results in Fig. 9.2 show high resistivity at high temperatures. Temperatures of the order of 100 K being higher than T_K, it follows, as we have discussed before, that the spin orientation of each Ce atom is pointed in up or down directions in an irregular manner, and this irregularity gives rise to electrical resistance. Electrical resistance minimum occurs, furthermore.

On the other hand, at low temperatures near 1 K, we have $T < T_K$, and the spin orientations are changing all the time. Hence all spins behave in this way, there is no irregularity, and the electrons are not subject to scattering. As a result, the electrical resistance becomes quite small, as seen in Fig. 9.2.

Another marked characteristic of this compound is its electrical heat capacity. In eq. (7.67), we set $r = 2$, and use eq. (7.60) for χ. We then obtain, for the electrical heat capacity below $T = T_K$,

$$c = k_B T / T_K. \tag{9.3}$$

Here, we have neglected a numerical factor which is of order 1.

Even when we have a large concentration of magnetic impurities, so long as each impurity behaves in the same way as an isolated impurity, the heat capacity should be given by eq. (9.3) multiplied by the number of impurity atoms. We then take the case $x = 1$, that is, the case of $CeCu_6$, and consider 1 mol of the compound. The heat capacity is then given by RT/T_K, where R is the gas constant. The point is that this R is very large. In ordinary metals, the electrical heat capacity is of the order of $R(k_B T/\varepsilon_F)$. Here ε_F, being the Fermi energy of that metal, is about 5 eV for noble metals, and is several ten thousands in degrees K. On the other hand, T_K is at most a few tens of degrees K, and so RT/T_K is greater than the ordinary electrical heat capacity of metals by about one thousand times.

Since the electrical heat capacity of a metal is proportional to the effective mass of the conduction electrons of that metal, it follows that when the heat capacity becomes one thousand times that of ordinary metals, the effective mass of the electrons in this system may be interpreted to be one thousand times greater than that of the ordinary electrons. Many other compounds with such large heat capacity have been discovered, and these are collectively called the heavy electron systems.

One would ask, why is it that in $CeCu_6$ the spin of a Ce atom is unaffected by the surrounding Ce spin and behaves as if it is an isolated spin? The answer is that the RKKY interaction between the Ce spins is small. As we stated earlier, a Ce atom has one 4f electron, and its spin is $1/2$. The RKKY interaction is proportional to the square of the spin, and so the RKKY interaction is small for Ce and the spins are almost unaffected by one another.

We then ask, is it not possible that, although the interaction between the spins is small, such a small interaction triggers the ordering of spin at sufficiently low temperature, where the thermal fluctuation is reduced?

An important point here is that the orientation of spin is continuously altered through the interaction with the conduction electrons. Each spin then acquires an energy gain of the order of $k_B T_K$. If the spins were ordered, the flipping of spin would be stopped, and the energy gain would have disappeared. Therefore the ordering of spin does not take place.

9.3 Quantum dots

A quantum dot is a small 'puddle' of charge which contains a well-defined number of electrons. Usually one attaches two leads which are called the "source" and the "drain." The dot and the leads are connected weakly so that the dot is almost isolated. Specifically, one arranges the system so that the electrons in the leads can tunnel through the barrier to the dot, and vice versa.

A dot typically contains a few tens of electrons, and is spoken of as an "artificial atom." Each electron in the dot occupies a state with energy ε_i $(i = 1, 2 \cdots)$. Let us say that we have one electron in state i. Putting one more electron, of opposite spin, in this level requires an energy ε_i plus the extra energy due to the interaction with the other electrons. We may, as a first approximation, suppose that this energy is independent of i, and write this as E_C. That is, we require energy $\varepsilon_i + E_C$ to put the second electron in state i.

The lead behaves like a metal and is filled up to the Fermi level μ by electrons. The system comprising a dot connected weakly with leads is analogous to the system of an impurity atom embedded in a metal. What is different is that, unlike the impurity atom, this dot can be artificially tuned. For example, one can attach a gate electrode beside the dot, and modify the electrical potential of the dot by means of a gate electrical potential. This causes ε_i and $\varepsilon_i + E_C$ to change but, since μ remains constant, the electrons fall to the lead from state i when ε_i becomes greater than μ.

Let us now call the Fermi levels of the two leads μ_L and μ_R, respectively for the leads on the left and the right sides. By raising μ_L very slightly above μ_R, one is able to measure the conductance of the dot.

When ε_i coincides with μ, and state i is vacant, an electron at μ_L can go through ε_i as a resonance and reach μ_R. In the same case, but with an electron occupying state i, the electron in state i moves to state μ_R and the ε_i that has been vacated is then occupied by an electron from μ_L, again as a resonance process. The conductance becomes large in both cases.

The case where $\varepsilon_i + E_C$ coincides with μ is analogous and the conductance again becomes large. As one varies the gate voltage, the conductance should exhibit peaks whenever ε_i or $\varepsilon_i + E_C$ passes through μ. These peaks are called the Coulomb peaks, and the low regions in between these peaks are called the Coulomb valleys.

What will happen when μ is in between ε_i and $\varepsilon_i + E_C$? In this case, an electron can move from the lead on the left to the lead on the right (when μ_L is slightly higher than μ_R) through the process depicted in Fig. 9.3. The conductance then remains much smaller than in the case with resonance. In the same way, the conductance is small when μ is in between $\varepsilon_i + E_C$ and ε_{i+1}. However, unlike in this case, something extra happens in the case when μ is between ε_i and $\varepsilon_i + E_C$.

To see this point, let us make a note of the following. When μ is in between ε_i and $\varepsilon_i + E_C$, the state i is occupied by one electron, and the dot has a spin. This is analogous to the case of magnetic impurities in metals. We then compare our present case with the Anderson model, to see that ε_0 in eq. (5.81) corresponds to ε_i, and U corresponds to E_C. Furthermore, V_0 corresponds to the tunneling potential barrier between the leads and the dot.

Let us, in the same way as in the derivation of the s–d interaction of eq. (5.90) from the Anderson model, obtain the Hamiltonian for the present model. To do so, we treat the higher excitations as a perturbation. The state where an electron is moved from ε_i to the lead, and the state where an additional electron is put in ε_i from the lead, are at higher energies, and can be treated as a perturbation. This

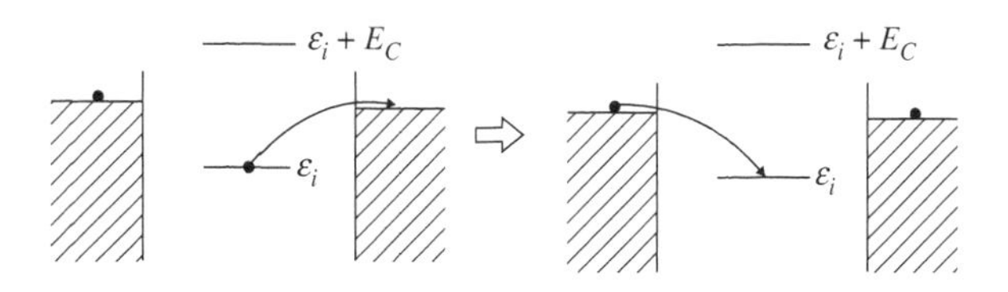

Figure 9.3 An illustration of the transition of an electron from the lead on the left to the lead on the right through a dot.

then gives us the Hamiltonian for the process of moving an electron from one lead to the other with the inversion of the spin of the dot. We also obtain, in the same way, the Hamiltonian for the process that does not involve the spin-flip. We also have the Hamiltonian for the process where an electron is scattered back into the same lead, giving rise to the inversion of the spin of the dot, or to no inversion of the spin of the dot, and these must also be taken into account.

Using these Hamiltonians, one can work out the transmission probability for the transition of an electron from one lead to the other. As in the case of magnetic impurities in metals, the transmission probability is independent of temperature in the lowest order. However, terms with logarithmic temperature dependence appear at higher orders.

Since the quantity that corresponds to J in eq. (5.90) is negative also in this case, the transmission probability increases as the temperature decreases, and approaches unity for temperatures going towards absolute zero (Glazman and Raikh, 1988; Ng and Lee, 1988; Kawabata, 1991). In terms of the conductance, this quantity tends to $2e^2/h$.

When μ is between ε_i+E_C and ε_{i+1}, the number of electrons in the dot is even, and the transmission probability remains small because the above processes do not take place. Since by varying the gate voltage, the system goes through these two regions alternately, it follows that, at absolute zero, the conductance alternates between small values and $2e^2/h$. This can be seen in Fig. 9.4.

We show the experimental result in Fig. 9.5. The horizontal axis gives the gate voltage, and the vertical axis gives the conductance. At high temperature of 800 mK, we have Coulomb peaks with nearly equal separation. When the gate

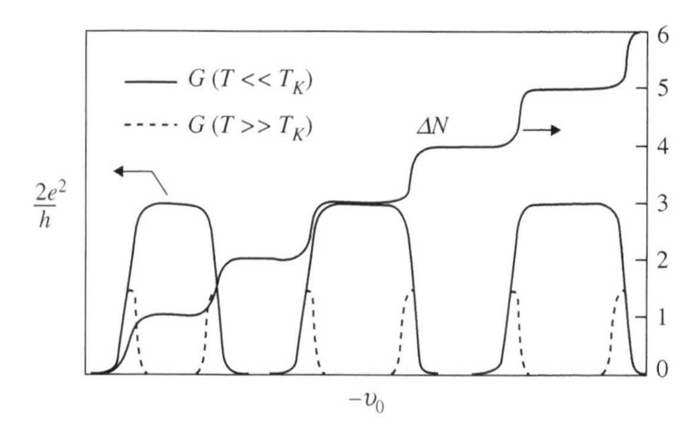

Figure 9.4 When the quantum dot has a spin, its conductance varies between $2e^2/h$ and zero as one varies the gate voltage (at absolute zero). From A. Kawabata, *J. Phys. Soc. Japan* **60** (1991) 3222.

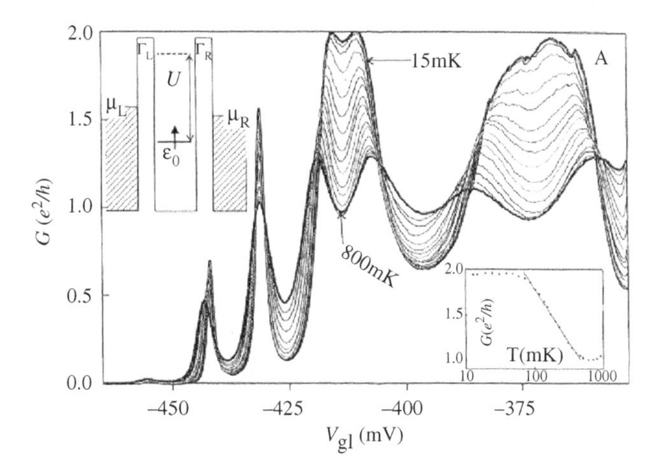

Figure 9.5 The gate voltage dependence of the conductance of the lateral quantum dot. We show the case of temperatures between 15 mK and 800 mK. From W. G. van der Wiel, *et al.*, *Science* **289** (2000) 2105.

voltage is about -413 mV, the conductance goes up as the temperature goes down, and we see that it approaches $2e^2/h$ at the lowest temperature of 15 mK.

For gate voltages near -400 mV and near -425 mV, the conductance increases as the temperature increases. We may infer that this is due to the thermal excitation of the dot electrons to the leads. When the gate voltage is near -413 mV, the separation distance between the two Coulomb peaks is decreasing as the temperature decreases. This point is not well understood. We can presumably say that this figure embodies the features of Fig. 9.4 in their most essential aspects.

References and further reading

Glazman, L. I. and Raikh, M. E. (1988) *JETP Lett.* **47**: 452.

Kawabata, A. (1991) *J. Phys. Soc. Japan* **60**: 3222.

Ng, T. K. and Lee, P. A. (1988) *Phys. Rev. Lett.* **61**: 1768.

Sumiyama, A. Oda, Y., Nagano, H., Onuki, Y., Shibutani, K. and Komatsubara, T. (1986) *J. Phys. Soc. Japan* **55**: 1294.

van der Wiel, W. G., De Franseschi, S., Fujisawa, T., Elzerman, J. M., Tarucha, S. and Kouwenhoven, L. P. (2000) *Science* **289**: 2105.

Appendices

Appendix A Matrix elements between Slater determinants

Let us say that we have two Z-electron Slater determinants of the form eq. (1.12). We abbreviate them as follows:

$$\Phi = \frac{1}{\sqrt{Z!}} \left| \phi_{k_1}(1)\, \phi_{k_2}(2) \, \cdots \right|, \tag{A. 1}$$

$$\Phi' = \frac{1}{\sqrt{Z!}} \left| \phi_{k'_1}(1)\, \phi_{k'_2}(2) \, \cdots \right|. \tag{A. 2}$$

When the Hamiltonian H is composed of a one-electron term denoted $h(i)$ and a two-electron term denoted $V(i,j)$ as in eq. (1.18), let us evaluate the quantity

$$D = \int \Phi'^{*} H \Phi \, d\tau_1 \cdots d\tau_Z = D_1 + D_2, \tag{A. 3}$$

where $d\tau_i$ corresponds to the integration over space and spin coordinates of the ith electron; D_1 and D_2 represent the contributions from the one-electron and two-electron terms in H, respectively. By expanding the determinants, we obtain

$$\Phi = (Z!)^{-1/2} \sum_P (-)^P P \left[\phi_{k_1}(1) \phi_{k_2}(2) \cdots \right], \tag{A. 4}$$

and an analogous equation for Φ'. P is an operator that permutes the coordinates of electrons $1, 2, \ldots$; $(-)^P$ is -1 for odd permutations and $+1$ for even permutations. Substituting these into eq. (A. 3), we obtain

$$D = (Z!)^{-1} \sum_{PQ} (-)^P (-)^Q \int Q \left[\phi_{k'_1}(1) \phi_{k'_2}(2) \cdots \right]^{*} HP \left[\phi_{k_1}(1) \phi_{k_2}(2) \cdots \right] d\tau.$$

Here $d\tau_1 \cdots d\tau_Z$ is abbreviated as $d\tau$. The integral is unaffected by the order of integration. If the variables of integration $1, 2, \ldots$ are permuted by P^{-1}, $P[\cdots]$ reduces to $\phi_{k_1}(1)\phi_{k_2}(2)\cdots$, whereas $Q[\cdots]^*$ becomes $P^{-1}Q[\phi_{k'_1}(1)\phi_{k'_2}(2)\cdots]^*$. H is unchanged. Therefore

$$D = (Z!)^{-1} \sum_{PQ} (-)^P (-)^Q \int P^{-1}Q[\phi_{k'_1}(1)\phi_{k'_2}(2)\cdots]^* H \phi_{k_1}(1)\phi_{k_2}(2)\cdots d\tau.$$

We first consider summing over the $Z!$ permutations Q while keeping P fixed. Here, all $Z!$ permutations appear in $P^{-1}Q$, in different order from Q (see §8.3). Moreover, $(-)^P \times (-)^Q$ equals $+1$ and -1 for even and odd permutations, respectively, under $P^{-1}Q$. Therefore the summation over Q is independent of P. Hence

$$D = \sum_{Q} (-)^Q \int Q\left[\phi_{k'_1}(1)\phi_{k'_2}(2)\cdots\right]^* H \phi_{k_1}(1)\phi_{k_2}(2)\cdots d\tau.$$

Now, D_1 is given by:

$$D_1 = \sum_{lQ} (-)^Q \int Q\left[\phi_{k'_1}(1)\phi_{k'_2}(2)\cdots\right]^* h(l)\phi_{k_1}(1)\phi_{k_2}(2)\cdots d\tau. \qquad \text{(A. 5)}$$

Let us analyze this result for a number of different cases.

Firstly, we consider case (A):

(A) $\qquad k_l = k'_l \quad (l = 1, \ldots, Z).$

For the identity element $Q = E$ in the summation over Q, because of the normalization of ϕ_{k_l}, we see that the sum reduces to

$$\sum_{l} h(k_l, k_l). \qquad \text{(A. 6)}$$

Here, $h(k_l, k_l)$ is as defined by eq. (1.20). All other permutations $Q \neq E$ yield zero because of the orthogonality of ϕ_{k_l}, as no two k_ls are the same. Hence, for case (A), eq. (A. 6) gives the value of D_1.

Secondly, we consider case (B):

(B) $\qquad k_i \neq k'_i, \quad k_l = k'_l \quad (l \neq i).$

Taking the $Q = E$ permutation to start with, all terms other than $h(k'_i, k_i)$ vanish owing to the orthogonality between ϕ_{k_i} and $\phi_{k'_i}$. All other permutations $Q \neq E$ yield zero.

For all cases other than these two, if they can be reduced to (A) or (B) by changing the order of k_l or k'_l, the above results are directly applicable after multiplying by a factor of -1 or $+1$ corresponding to an odd or even permutation, respectively. D_1 vanishes for all other cases, that is, when Φ is different from Φ' on two or more orbitals.

Next, let us calculate D_2:

$$D_2 = \sum_{l>m} \sum_{Q} (-)^Q \int Q\left[\phi_{k'_1}(1)\phi_{k'_2}(2)\cdots\right]^* V(l,m)\phi_{k_1}(1)\phi_{k_2}(2)\cdots d\tau. \tag{A.7}$$

For case (A), the $Q=E$ term is given by

$$\sum_{l>m} V(k_l, k_m; k_l, k_m).$$

V is defined by eq. (1.21). All other Qs vanish because of the orthogonality of ϕ_{k_l}, except the lm transposition term $Q=(lm)$. This is because the orthogonality between ϕ_{k_l} and ϕ_{k_m} does not owing due to the presence of $V(l,m)$. The contribution of this term is

$$-\sum_{l>m} V(k_m, k_l; k_l, k_m).$$

Next, for case (B), the $Q=E$ contribution is non-zero under the orthogonality between ϕ_{k_i} and $\phi_{k'_i}$ only when l or m is equal to i. The contribution of this term is therefore

$$\sum_{l\neq i} V(k'_i, k_l; k_i, k_l).$$

The only other non-vanishing contributions are due to the transposition of i with l or m. Taking $V(i,l)$ or $V(i,m)$ from $V(l,m)$, neither the integration over i nor that over l, m vanishes. The contribution of this term is given by

$$-\sum_{l\neq i} V(k_l, k'_i; k_i, k_l).$$

All other Qs give vanishing contributions.

Next, let us consider case (C):

$$\text{(C)} \qquad k_i \neq k'_i, \quad k_j \neq k'_j \quad (k_l = k'_l,\ l \neq i, j).$$

In the $Q=E$ term, out of the summation over l and m, only the $(l,m)=(i,j)$ term is non-zero. This term contributes

$$V(k'_i, k'_j; k_i, k_j).$$

The only other non-vanishing contribution is due to the permutation Q that exchanges i and j. Here, in the summation over l and m, again only the $(l,m)=(i,j)$ term remains, and this reads:

$$-V(k_j', k_i'; k_i, k_j).$$

As in the calculation of D_1, for those contributions that can be reduced to (A), (B) or (C) by changing the order of k_l, the above results are applicable, together with sign changes as appropriate. D_2 is zero in the other cases, that is, when Φ differs from Φ' on three or more orbitals.

Summarizing the above results, we have:

$$\text{(A)} \quad D=\sum_l h(k_l,k_l)+\sum_{l>m}[V(k_l,k_m;k_l,k_m)-V(k_m,k_l;k_l,k_m)], \quad \text{(A. 8)}$$

$$\text{(B)} \quad D=h(k_i',k_i)+\sum_{l\neq i}[V(k_i',k_l;k_i,k_l)-V(k_l,k_i';k_i,k_l)], \quad \text{(A. 9)}$$

$$\text{(C)} \quad D=V(k_i',k_j';k_i,k_j)-V(k_j',k_i';k_i,k_j). \quad \text{(A. 10)}$$

Appendix B Spin function for N-electron systems

We define functions $\alpha(\zeta)$ and $\beta(\zeta)$ to represent up- and down-spin respectively, where ζ is the spin coordinate of an electron. These functions satisfy $\alpha\left(\zeta=+\frac{1}{2}\right)=1, \alpha\left(\zeta=-\frac{1}{2}\right)=0, \beta\left(\zeta=+\frac{1}{2}\right)=0$, and $\beta\left(\zeta=-\frac{1}{2}\right)=1$, and are the eigenfunctions of both s_z and s^2, as follows:

$$s_z\alpha=\tfrac{1}{2}\alpha, \qquad s_z\beta=-\tfrac{1}{2}\beta, \quad \text{(B. 1)}$$

$$s^2\alpha=S(S+1)\alpha, \quad s^2\beta=S(S+1)\beta \quad \left(\text{where } S=\tfrac{1}{2}\right). \quad \text{(B. 2)}$$

We then define the spin ladder operators as

$$s_\pm=s_x\pm is_y,$$

and

$$s_+\alpha=0, \quad s_-\alpha=\beta, \quad \text{(B. 3)}$$

$$s_+\beta=\alpha, \quad s_-\beta=0, \quad \text{(B. 4)}$$

where s_+ raises the z-component of spin by 1, and s_- lowers it by 1.

When there are N electrons, there are 2^N independent spin functions. Our notation is such that, for example, $\alpha\beta\alpha\alpha\cdots$ represents $\alpha(\zeta_1)\beta(\zeta_2)\alpha(\zeta_3)\alpha(\zeta_4)\cdots$. That is, the leftmost part is a function of the spin coordinate of the first electron, the second from the left is a function of the spin coordinate of the second electron, and so on. In this way, there are 2^N functions, starting with $\alpha\alpha\alpha\cdots$ and ending with $\beta\beta\beta\cdots$. All of these are eigenfunctions of

$$S_z = \sum_{i=1}^{N} s_{iz}, \tag{B.5}$$

but, in general, are not eigenfunctions of

$$\boldsymbol{S}^2 = S_x{}^2 + S_y{}^2 + S_z{}^2. \tag{B.6}$$

Now, how should we go about forming a suitable linear combination of these functions that is an eigenfunction of both S_z and $\boldsymbol{S}^2$? The $\boldsymbol{S}^2$ eigenvalue of such an eigenfunction should be given by $S(S+1)$. Here, S is an integer or half-integer which represents the magnitude of the total spin of N electrons.

In particular, $\alpha\alpha\cdots\alpha$ is an eigenfunction whose S_z is given by $\frac{N}{2}$ and the magnitude of its spin S is also given by $\frac{N}{2}$. In general, applying the operator

$$S_- = \sum_{i=1}^{N} s_{i-} \tag{B.7}$$

to function Θ_{SM} whose magnitude of spin is given by S and S_z is given by M, we obtain

$$S_-\Theta_{SM} = \sqrt{S(S+1) - M(M-1)}\,\Theta_{S\,M-1}, \tag{B.8}$$

so that S_z is lowered by 1. By the actual application of S_- on $\alpha\alpha\cdots\alpha$ and using eq. (B.3), we obtain

$$\frac{1}{\sqrt{N}}(\beta\alpha\alpha\cdots\alpha + \alpha\beta\alpha\cdots\alpha + \cdots + \alpha\alpha\alpha\cdots\beta), \tag{B.9}$$

which is a function that satisfies $S = \frac{N}{2}$ and $S_z = \frac{N}{2} - 1$. Repeating this operation, we obtain functions with $S = \frac{N}{2}$ and $S_z = -\frac{N}{2}, \ldots, \frac{N}{2}$.

Next, let us consider products that involve only one β. In all, there are N such products, and S_z is given by $\frac{N}{2} - 1$. Out of the linear combinations of these products, we already know that eq. (B.9) has $S = \frac{N}{2}$. Writing down the $N - 1$ linear combinations that are orthogonal to eq. (B.9), we see they all have $S = \frac{N}{2} - 1$. The reason for this is that S cannot be less than $\frac{N}{2} - 1$ (as $S_z = \frac{N}{2} - 1$). Furthermore, S cannot be equal to $\frac{N}{2}$, because if S were equal to $\frac{N}{2}$, we could apply the operator

$$S_+ = \sum_i s_{i+}, \tag{B.10}$$

which raises S_z by 1. This would give rise to a function with $S = \frac{N}{2}$ and $S_z = \frac{N}{2}$. However, we have already counted this function, in eq. (B.9). Applying S_- to each of these $N - 1$ functions, which have $S = \frac{N}{2} - 1$ and $S_z = \frac{N}{2} - 1$, we obtain $S_z = \frac{N}{2} - 1, \ldots, -\left(\frac{N}{2} - 1\right)$.

Next, there are ${}_N\mathrm{C}_2$ products that involve two βs. Out of the linear combinations of these products, one combination has $S = \frac{N}{2}$, and $N - 1$ are those obtained by applying S_- to the above $N - 1$ functions that have $S = \frac{N}{2} - 1$ and $S_z = \frac{N}{2} - 1$. The remaining ${}_N\mathrm{C}_2 - N$ functions are defined to be orthogonal to these N functions, and they satisfy $S = \frac{N}{2} - 2$. By operating on each of them by S_-, we obtain $S_z = \frac{N}{2} - 2, \ldots, -\left(\frac{N}{2} - 2\right)$.

As for the products that involve M βs, there are ${}_N\mathrm{C}_M$ of them. Out of their linear combinations, ${}_N\mathrm{C}_{M-1}$ arise from applying S_- to the functions with $S > \frac{N}{2} - M$. The remaining

$$f_{NS} = {}_N\mathrm{C}_M - {}_N\mathrm{C}_{M-1} \tag{B.11}$$

functions, which are constructed to be orthogonal to these, have $S = \frac{N}{2} - M$ and $S_z = \frac{N}{2} - M$. One characteristic property of these functions is that operating on them by S_+ defined by eq. (B.10) makes them vanish, that is:

$$S_+ \Theta_{\frac{N}{2}-M\ \frac{N}{2}-M} = 0. \tag{B.12}$$

This is because S_z is already equal to S and so S_z cannot be raised any more by S_+, which increases S_z by 1. The converse is also true, namely:

> For an arbitrary linear combination of ${}_N\mathrm{C}_M$ spin functions that involve M βs, if operating on it by S_+ yields zero, then $S = \frac{N}{2} - M$ for this linear combination. (B.13)

We will show, later on, that the function that emerges as the solution to the secular equation in Appendix N possesses this feature.

As an example, when $N = 2$,

$$\alpha\alpha,$$

$$\frac{1}{\sqrt{2}}(\alpha\beta + \beta\alpha),$$

$$\beta\beta$$

are the functions with $S = 1$ and $S_z = 1, 0, -1$, and

$$\frac{1}{\sqrt{2}}(\alpha\beta - \beta\alpha)$$

is the function with $S = 0$.

Appendix C Fourier expansion of three-dimensional periodic functions

By defining oblique axes x, y, z to be along $\boldsymbol{a}_1, \boldsymbol{a}_2, \boldsymbol{a}_3$, we may write eq. (4.9) as $f(x, y, z) = f(x + n_1 a_1, y + n_2 a_2, z + n_3 a_3)$. f is hence a periodic function in x, y, z, and it can be expanded into a Fourier series, as follows:

$$f(x, y, z) = \sum_{m_1 m_2 m_3} f_{m_1 m_2 m_3} \exp 2\pi i \left(\frac{m_1 x}{a_1} + \frac{m_2 y}{a_2} + \frac{m_3 z}{a_3} \right). \tag{C. 1}$$

Because $\boldsymbol{r} = (\boldsymbol{a}_1 x/a_1) + (\boldsymbol{a}_2 y/a_2) + (\boldsymbol{a}_3 z/a_3)$, and using eq. (4.6), we obtain

$$\boldsymbol{K} \cdot \boldsymbol{r} = 2\pi \left(\frac{m_1 x}{a_1} + \frac{m_2 y}{a_2} + \frac{m_3 z}{a_3} \right).$$

Substituting this into eq. (C. 1) yields eq. (4.10).

Let us consider the following integral:

$$\int_{\Omega_0} e^{i\boldsymbol{K} \cdot \boldsymbol{r}} dv. \tag{C. 2}$$

Here, Ω_0 indicates integrating over one unit cell. Since the integrand is invariant under $\boldsymbol{r} \to \boldsymbol{r} + \boldsymbol{R}$, the choice of the unit cell is arbitrary. Let us define a unit cell $\Omega_0{}'$, which is Ω_0 spatially translated by a certain vector $\boldsymbol{d}$. Now, we have

$$\int_{\Omega_0} e^{i\boldsymbol{K} \cdot \boldsymbol{r}} dv = \int_{\Omega_0{}'} e^{i\boldsymbol{K} \cdot \boldsymbol{r}} dv.$$

As the right-hand side of this expression is equal to

$$\int_{\Omega_0} e^{i\boldsymbol{K}\cdot(\boldsymbol{r}+\boldsymbol{d})} dv,$$

we obtain

$$(e^{i\boldsymbol{K}\cdot\boldsymbol{d}} - 1) \int_{\Omega_0} e^{i\boldsymbol{K}\cdot\boldsymbol{r}} dv = 0.$$

When $\boldsymbol{K} \neq 0$, because the above equation must hold for arbitrary $\boldsymbol{d}$, we obtain

$$\int_{\Omega_0} e^{i\boldsymbol{K}\cdot\boldsymbol{r}} dv = 0 \quad (\boldsymbol{K} \neq 0). \tag{C. 3}$$

Let us consider multiplying eq. (4.10) by $e^{-i\boldsymbol{K}\cdot\boldsymbol{r}}$ and integrating it over a unit cell. Since the integrand is again invariant under $\boldsymbol{r} \to \boldsymbol{r} + \boldsymbol{R}$, the choice of the unit cell is arbitrary. By using eq. (C. 3), eq. (4.11) follows immediately.

Appendix D Proof of eq. (5.29)

Let us introduce $\psi_k = \alpha j_l(kr) + \beta n_l(kr)$. Because ψ_k satisfies eq. (5.4) for $r > r_0$, it also satisfies eq. (5.19). If α and β are independent of k, the derivative of ψ_k with respect to k can be written in terms of the derivative with respect to r. Equation (5.19) is then rewritten as

$$2kr^2\psi_k{}^2 = \frac{1}{k}\frac{d}{dr}\left[r^3\psi_k{}'^2 - r^2\psi_k\psi_k{}' - r^3\psi_k\psi_k{}''\right].$$

Here, $'$ denotes the derivative with respect to r. Now, because this equation only contains derivatives with respect to r, it also holds when α and β depend on k. In other words, it is valid when ψ_k is given by eq. (5.8). Let us integrate this equation from $r = r_0$ to R. Using eqs. (5.17) and (5.20) at $r = R$ on the right-hand side of this equation, we obtain

$$\int_{r_0}^{R} r^2 v_k{}^2 dr = \left[1 + \frac{1}{R}\frac{d\delta_l}{dk}\right]^{-1} - \frac{1}{2k^2}(r^3 v_k{}'^2 - r^2 v_k v_k{}' - r^3 v_k v_k'')_{r=r_0}.$$

Here, the second term on the right-hand side is of order r_0/R or $1/kR$, and therefore it is negligible compared with $R^{-1}d\delta_l/dk$ when δ_l varies vigorously, as is the case near ε_0 in Fig. 5.5, that is, when there is strong localization. We therefore omit this term, and integrate eq. (5.23) outside $r = r_0$. Using the above equation on the right-hand side of eq. (5.23), we obtain

$$\int_{r_0}^{R} n(r,k) 4\pi r^2 dr = \sum_{l} \frac{2(2l+1)R}{\pi}.$$

The right-hand side is the density of states $2\rho(k)$ when there is no impurity (see eq. (5.21)). In other words, it is equal to the integral of $n_0(r,k)$ over the whole space. Hence we obtain eq. (5.29).

Appendix E Relations between Green's functions

Using eq. (5.14), we find that eq. (5.34) satisfies

$$\left[-\frac{d^2}{dr^2} - \frac{2}{r}\frac{d}{dr} + \frac{l(l+1)}{r^2} \right] G_k(r,r') = k^2 G_k(r,r') - \frac{\delta(r-r')}{r^2}. \qquad \text{(E. 1)}$$

Integrating this equation from $r = r' - 0$ to $r' + 0$, we obtain

$$\frac{\partial}{\partial r} G_k(r, r-0) - \frac{\partial}{\partial r} G_k(r, r+0) = \frac{1}{r^2}. \qquad \text{(E. 2)}$$

Although r and r' are equivalent in G, the derivative of G is discontinuous at $r = r'$. Hence $G_k(r,r')$ of eq. (5.34) satisfies the following three conditions:

1. When $r \neq r'$,

$$\left[-\frac{d^2}{dr^2} - \frac{2}{r}\frac{d}{dr} + \frac{l(l+1)}{r^2} \right] G = k^2 G$$

   is satisfied, and so is the same equation with r replaced by r'.
2. The derivative is discontinuous at $r = r'$, and the magnitude of the discontinuity is given by eq. (E. 2).
3. It vanishes when either r or r' is equal to R, or when either r or r' is equal to 0.

Now, it is easy to see that $G_k(r,r')$ of eq. (5.35) also satisfies these conditions. As for condition Appendix E, this is verified by keeping eq. (5.37). Hence eq. (5.35) is equivalent to eq. (5.34).

We define the following functions from this $G_k(r, r')$:

$$A(k,r) = r^2 G_k(r,r) = kr^2 j_l(kr)[n_l(kr) - c(k)^{-1} j_l(kr)], \tag{E. 3}$$

$$B_1(k,r) = r^2 \frac{\partial}{\partial r} G_k(r, r+0) = kr^2 j_l'(kr)[n_l(kr) - c(k)^{-1} j_l(kr)], \tag{E. 4}$$

$$B_2(k,r) = r^2 \frac{\partial}{\partial r} G_k(r, r-0) = kr^2 j_l(kr)[n_l'(kr) - c(k)^{-1} j_l'(kr)], \tag{E. 5}$$

$$C(k,r) = r^2 \frac{\partial^2}{\partial r \partial r'} G_k(r,r')\big|_{r'=r} = kr^2 j_l'(kr)[n_l'(kr) - c(k)^{-1} j_l'(kr)], \tag{E. 6}$$

$$B(k,r) = r^2 \frac{\partial}{\partial r} G_k(r,r')\big|_{r'=r} = r^2 \sum_n \frac{\phi_n(r)\phi_n'(r)}{k^2 - \kappa_n{}^2}. \tag{E. 7}$$

Here, $j_l'(kr) = \frac{d}{dr} j_l(kr)$, etc. By using eq. (5.37), we obtain the following relations between these equations:

$$B_2 - B_1 = 1, \quad AC = B_1 B_2. \tag{E. 8}$$

Furthermore, by Dirichlet's theorem, the value of the Fourier series at the discontinuity is equal to the mean of the two limiting values. Hence

$$B_2 - B = B - B_1 = \tfrac{1}{2}. \tag{E. 9}$$

Then, as the first step in rewriting eq. (5.10), we note that eq. (5.10) can be written as

$$1 = kr_0{}^2 \left[n_l(kr_0) - c(k)^{-1} j_l(kr_0) \right] \cdot \left[L(k) j_l(kr_0) - j_l'(kr_0) \right]$$

using eq. (5.37). This is equivalent to

$$1 = L(k)A(k, r_0) - B_1(k, r_0) \tag{E. 10}$$

by eqs. (E. 3) and (E. 4). Multiplying this by $C(k, r_0)$, and using eq. (E. 8), we obtain

$$\begin{aligned} 0 &= LAC - B_1 C - C = LB_1 B_2 - (B_2 - 1)C - C = B_2(LB_1 - C) \\ &= B_2(LB_2 - C - L), \end{aligned}$$

and hence:

$$L(k)B_2(k, r_0) - C(k, r_0) = L(k). \tag{E. 11}$$

Combining eqs. (E. 10) and (E. 11), while taking care of eq. (E. 9), yields

$$L(k)^2A(k,r_0) - 2L(k)B(k,r_0) + C(k,r_0) = 0. \tag{E. 12}$$

Substituting the initial definition, eq. (5.34), of the Green's function and eq. (E. 7) into this, we obtain eq. (5.39).

Substituting $L = L_0 + \Delta L$, where $L_0 \equiv L(k_0)$, into eq. (E. 12) yields

$$L_0{}^2A - 2L_0B + C = -2\Delta L(LA - B) - (\Delta L)^2A.$$

Since $LA - B = \frac{1}{2}$ according to eqs. (E. 9) and (E. 10), we obtain

$$L_0{}^2A - 2L_0B + C = -\Delta L,$$

neglecting the term of order $(\Delta L)^2$. Using the defining equation for the Green's function and eq. (5.41), we obtain

$$r_0{}^2\sum_n \frac{[L(k_0)\phi_n(r_0) - \phi_n{}'(r_0)]^2}{k^2 - \kappa_n{}^2} = \frac{k^2 - k_0{}^2}{r_0{}^2u_{k_0}(r_0)^2},$$

which is precisely the equation preceding eq. (5.42).

Appendix F Expansion of free energy to order J^2

We define eq. (5.93) plus the spin Zeeman energy as H_0:

$$H_0 = \sum_{\boldsymbol{k}} \left\{(\varepsilon_{\boldsymbol{k}} - \Delta)a_{\boldsymbol{k}\uparrow}{}^\dagger a_{\boldsymbol{k}\uparrow} + (\varepsilon_{\boldsymbol{k}} + \Delta)a_{\boldsymbol{k}\downarrow}{}^\dagger a_{\boldsymbol{k}\downarrow}\right\} - 2\Delta S_z, \tag{F. 1}$$

where

$$\Delta = \mu_B H. \tag{F. 2}$$

Here, we set the g factor equal to 2 for both the conduction electrons and the localized electron. We choose H_{sd} of eq. (5.90) as the perturbation (where V is set equal to zero), expand the free energy up to the second order in J, and investigate its dependence on T and H.

In general, when $H = H_0 + H'$, we may use the following expansion ($\beta = 1/k_BT$):

$$e^{-\beta H} = e^{-\beta H_0}\left[1 - \int_0^\beta H'(u)du + \iint_{\beta>u>u'>0} H'(u)H'(u')dudu' - \cdots\right]. \tag{F. 3}$$

Here,

$$H'(u) = e^{uH_0} H' e^{-uH_0}$$

is called the interaction representation of H'. This is derived as follows. We first write

$$e^{-uH} = e^{-uH_0} S(u),$$

and then differentiate both sides with respect to u. This yields

$$-H' e^{-uH_0} S(u) = e^{-uH_0} \frac{dS}{du}.$$

Multiplying both sides by e^{uH_0} yields

$$-H'(u)S(u) = \frac{dS}{du}.$$

We then integrate both sides from $u = 0$ to β. Using $S(0) = 1$, we obtain

$$S(\beta) = 1 - \int_0^\beta H'(u)S(u)du.$$

Equation (F. 3) is obtained by expanding this expression with respect to H' using the successive approximation method.

Now, the free energy F is given in terms of the partition function Z as $Z = e^{-\beta F}$, and Z is represented as

$$Z = \mathrm{Tr}(e^{-\beta H}).$$

Here, for a physical quantity A, $\mathrm{Tr}(A)$ is defined as $\sum_n \langle n|A|n\rangle$ using an arbitrary complete set $|n\rangle$. We can easily show that the value of this expression is independent of the choice of the complete set. In the expression for Z, we can adopt the eigenstates of H as $|n\rangle$, and this leads to the ordinary definition of the sum over states. By substituting eq. (F. 3) into the right-hand side of the above equation, we obtain

$$\mathrm{Tr}(e^{-\beta H}) = \mathrm{Tr}(e^{-\beta H_0}) - \int_0^\beta \mathrm{Tr}(e^{-\beta H_0} H'(u))du$$
$$+ \iint_{\beta > u > u' > 0} \mathrm{Tr}(e^{-\beta H_0} H'(u)H'(u'))du du' - \cdots . \quad \text{(F. 4)}$$

In general, for an operator A,

$$\frac{\mathrm{Tr}(e^{-\beta H_0} A)}{\mathrm{Tr}(e^{-\beta H_0})}$$

is the thermal average of the physical quantity A in an unperturbed system H_0. This is denoted as $\langle A \rangle$. Furthermore, since the free energy F_0 of the unperturbed system is given by

$$e^{-\beta F_0} = \mathrm{Tr}(e^{-\beta H_0}),$$

eq. (F. 4) turns into

$$e^{-\beta F} = e^{-\beta F_0}\left[1 - \int_0^\beta \langle H'(u)\rangle du + \iint_{\beta>u>u'>0} \langle H'(u)H'(u')\rangle du du' - \cdots \right].$$

Now, making use of the relation $\mathrm{Tr}(AB) = \mathrm{Tr}(BA)$, we find that $\langle H'(u)\rangle \propto \mathrm{Tr}(e^{-(\beta-u)H_0}H'e^{-uH_0})$ is independent of u.

Moreover, since $\langle H'(u)H'(u')\rangle \propto \mathrm{Tr}(e^{-(\beta-u+u')H_0}H'e^{-(u-u')H_0}H')$, we see, by making use of the same relation, that $\langle H'(u)H'(u')\rangle$ depends only on $u-u'$, and it takes the same value when $u-u'$ is replaced by $\beta - u + u'$. Hence,

$$\iint_{\beta>u>u'>0} \langle H'(u)H'(u')\rangle du du' = \frac{1}{2}\beta \int_0^\beta \langle H'(u)H'\rangle du.$$

Therefore,

$$e^{-\beta F} = e^{-\beta F_0}\left[1 - \beta\langle H'\rangle + \frac{1}{2}\beta \int_0^\beta \langle H'(u)H'\rangle du - \cdots \right].$$

From this equation, we obtain

$$F = F_0 + \langle H'\rangle + \frac{1}{2}\beta\langle H'\rangle^2 - \frac{1}{2}\int_0^\beta \langle H'(u)H'\rangle du + \cdots . \qquad \text{(F. 5)}$$

This is the general second-order perturbative formula for the free energy.

Let us now adopt H_{sd} as H'. Then, $\langle a_{\boldsymbol{k}\uparrow}{}^\dagger a_{\boldsymbol{k}\uparrow}\rangle$, for example, is written as $f(\varepsilon_{\boldsymbol{k}} - \Delta)$ since it is the average of the number of electrons occupying $\boldsymbol{k}\uparrow$. Furthermore, $\langle a_{\boldsymbol{k}\uparrow}{}^\dagger a_{\boldsymbol{k}'\uparrow}\rangle$ $(\boldsymbol{k} \neq \boldsymbol{k}')$ vanishes. This is because $\boldsymbol{k}\uparrow$ and $\boldsymbol{k}'\uparrow$ are the eigenstates of the unperturbed system. Similarly, $\langle a_{\boldsymbol{k}\uparrow}{}^\dagger a_{\boldsymbol{k}'\downarrow}\rangle$ also vanishes. Hence,

$$\begin{aligned}\langle H_{\mathrm{sd}}\rangle &= -J\langle S_z\rangle \sum_{\boldsymbol{k}} [f(\varepsilon_{\boldsymbol{k}} - \Delta) - f(\varepsilon_{\boldsymbol{k}} + \Delta)] \\ &= -J\langle S_z\rangle(N_\uparrow - N_\downarrow).\end{aligned}$$

As seen in §3.2, if Δ is sufficiently smaller than $\varepsilon_F = D$, $N_\uparrow - N_\downarrow = 2\rho\Delta$, where, ρ is the density of states per spin at the Fermi level. Since $\langle S_z\rangle = (1/2)\tanh(\mu_B H/k_B T)$ $(S = 1/2)$, we obtain

$$\langle H_{\mathrm{sd}}\rangle = -J\rho\mu_B H \tanh\frac{\mu_B H}{k_B T}.$$

Next, we need to know the interaction representation, $H_{\rm sd}(u)$, of $H_{\rm sd}$. When $H_0 = \varepsilon a^\dagger a$, we have $e^{uH_0} a e^{-uH_0} = e^{-\varepsilon u} a$ and $e^{uH_0} a^\dagger e^{-uH_0} = e^{\varepsilon u} a^\dagger$. It is easy to confirm these relations by operating on the eigenstates of $a^\dagger a$ by them. Furthermore, the relations $S_+(u) = e^{-2\Delta u} S_+$, $S_-(u) = e^{2\Delta u} S_-$ and $S_z(u) = S_z$ are also easily confirmed by operating on the eigenfunctions of S_z by them. Now, since the interaction representation of the products is the same as the product of the interaction representations, we obtain

$$H_{\rm sd}(u) = -J \sum_{kk'} e^{(\varepsilon-\varepsilon')u} [(a_{k\uparrow}{}^\dagger a_{k'\uparrow} - a_{k\downarrow}{}^\dagger a_{k'\downarrow}) S_z + a_{k\uparrow}{}^\dagger a_{k'\downarrow} S_- + a_{k\downarrow}{}^\dagger a_{k'\uparrow} S_+].$$

Here $\varepsilon = \varepsilon_k$ and $\varepsilon' = \varepsilon_{k'}$.

When we evaluate $\langle H_{\rm sd}(u) H_{\rm sd}\rangle$, the $\langle\quad\rangle$ of the products of four as appear. These may, for instance, be evaluated as follows:

$$\langle a_{k\uparrow}{}^\dagger a_{k'\downarrow} a_{k''\downarrow}{}^\dagger a_{k'''\uparrow}\rangle = \langle a_{k\uparrow}{}^\dagger a_{k'''\uparrow}\rangle \langle a_{k'\downarrow} a_{k''\downarrow}{}^\dagger\rangle = \delta_{kk'''}\delta_{k'k''} f_\uparrow (1 - f'_\downarrow),$$

$$\langle a_{k\uparrow}{}^\dagger a_{k'\uparrow} a_{k''\uparrow}{}^\dagger a_{k'''\uparrow}\rangle = \langle a_{k\uparrow}{}^\dagger a_{k'\uparrow}\rangle \langle a_{k''\uparrow}{}^\dagger a_{k'''\uparrow}\rangle + \langle a_{k\uparrow}{}^\dagger a_{k'''\uparrow}\rangle \langle a_{k'\uparrow} a_{k''\uparrow}{}^\dagger\rangle$$
$$= \delta_{kk'}\delta_{k''k'''} f_\uparrow f''_\uparrow + \delta_{kk'''}\delta_{k'k''} f_\uparrow (1 - f'_\uparrow).$$

We adopted the notations $f'_\uparrow = f(\varepsilon_{k'} - \Delta)$, $f''_\downarrow = f(\varepsilon_{k''} + \Delta)$, etc. In this way, we obtain

$$\langle H_{\rm sd}(u) H_{\rm sd}\rangle = J^2 \langle S_z{}^2\rangle \sum_{kk'} (f_\uparrow - f_\downarrow)(f'_\uparrow - f'_\downarrow)$$
$$+ J^2 \sum_{kk'} e^{(\varepsilon-\varepsilon')u} \Big\{ \langle S_z\rangle^2 [f_\uparrow (1 - f'_\uparrow) + f_\downarrow (1 - f'_\downarrow)]$$
$$+ \langle S_- S_+\rangle f_\uparrow (1 - f'_\downarrow) + \langle S_+ S_-\rangle f_\downarrow (1 - f'_\uparrow) \Big\}.$$

Next, note that

$$\int_0^\beta e^{(\varepsilon-\varepsilon')u} du = \frac{1}{\varepsilon - \varepsilon'} [e^{\beta(\varepsilon-\varepsilon')} - 1].$$

Using the relations, $e^{(\varepsilon-\varepsilon')\beta} f_\uparrow (1 - f'_\uparrow) = f'_\uparrow (1 - f_\uparrow)$ etc., we obtain

$$\int_0^\beta e^{(\varepsilon-\varepsilon')u} du [f_\uparrow (1 - f'_\uparrow) + f_\downarrow (1 - f'_\downarrow)] = \frac{1}{\varepsilon - \varepsilon'} (f'_\uparrow - f_\uparrow + f'_\downarrow - f_\downarrow).$$

Hence,

$$
\begin{aligned}
R &\equiv \int_0^\beta \langle H_{\rm sd}(u) H_{\rm sd}\rangle du \\
&= 4J^2\rho^2\langle {S_z}^2\rangle \Delta^2\beta \\
&\quad + J^2 \sum_{kk'} \frac{1}{\varepsilon-\varepsilon'} \Big\{ \langle {S_z}^2\rangle (f'_\uparrow - f_\uparrow + f'_\downarrow - f_\downarrow) \\
&\qquad\qquad + \langle S_- S_+\rangle [e^{2\Delta\beta}(1-f_\uparrow)f'_\downarrow - f_\uparrow(1-f'_\downarrow)] \\
&\qquad\qquad + \langle S_+ S_-\rangle [e^{-2\Delta\beta}(1-f_\downarrow)f'_\uparrow - f_\downarrow(1-f'_\uparrow)] \Big\}.
\end{aligned}
$$

Taking the density of states ρ to be constant, the sum over $\boldsymbol{k}$ may be replaced by the integral between $-D$ to D. We now define $E(\Delta)$ as

$$
E(\Delta) = \iint_{-D}^{D} \frac{1}{\varepsilon-\varepsilon'} (f'_\uparrow - f_\uparrow + f'_\downarrow - f_\downarrow) d\varepsilon d\varepsilon', \tag{F.6}
$$

and $F(\Delta)$ as

$$
F(\Delta) = \iint_{-D}^{D} \frac{1}{\varepsilon-\varepsilon'} (1-f_\uparrow) f'_\downarrow d\varepsilon d\varepsilon'. \tag{F.7}
$$

The other integrals are then reduced to F by changing the variables of integration, $\varepsilon \to -\varepsilon$, $\varepsilon \to \varepsilon'$, etc. Hence,

$$
\begin{aligned}
R = 4J^2\rho^2\langle {S_z}^2\rangle \Delta^2\beta &+ J^2\rho^2\langle {S_z}^2\rangle E(\Delta) \\
&+ J^2\rho^2\langle S_- S_+\rangle [e^{2\Delta\beta}F(\Delta) + F(-\Delta)] \\
&+ J^2\rho^2\langle S_+ S_-\rangle [e^{-2\Delta\beta}F(-\Delta) + F(\Delta)].
\end{aligned}
$$

As for $\langle S_- S_+\rangle$, we have

$$
\langle S_- S_+\rangle = \frac{\sum_M e^{-2\Delta M\beta}[S(S+1) - M(M+1)]}{\sum_M e^{-2\Delta M\beta}}.
$$

For $S = 1/2$, $\langle S_- S_+\rangle = e^{\Delta\beta}/(e^{\Delta\beta} + e^{-\Delta\beta})$. Similarly, $\langle S_+ S_-\rangle = e^{-\Delta\beta}/(e^{\Delta\beta} + e^{-\Delta\beta})$. Furthermore, $\langle {S_z}^2\rangle = 1/4$.

After changing the variables of integration, we obtain

$$
F(\Delta) = \iint_{-D-\Delta}^{D-\Delta} \frac{1}{\varepsilon+\varepsilon'+2\Delta} [1-f(\varepsilon)][1-f(\varepsilon')] d\varepsilon d\varepsilon'.
$$

We then carry out partial integration twice, and let $f(D - \Delta) = 0$ and $f(-D - \Delta) = 1$ at the boundary. All factors involving $f(\varepsilon)$ are transformed into $df/d\varepsilon$. When $\Delta/D \ll 1$, this yields

$$F(\Delta) = D\,2\log 2 - 2\Delta(1 + \log \beta D) + TG(2\Delta\beta),$$

where

$$G(x) \equiv \iint (\varepsilon + \varepsilon' + x)\log|\varepsilon + \varepsilon' + x| \cdot \frac{1}{(e^{\varepsilon}+1)(e^{-\varepsilon}+1)(e^{\varepsilon'}+1)(e^{-\varepsilon'}+1)} d\varepsilon d\varepsilon'.$$

$G(x)$ defined here is an odd function, and has the following properties:

$$G(0) = G''(0) = 0,$$

$$G'(0) = 1 + \iint \log|\varepsilon + \varepsilon'| \frac{1}{(e^{\varepsilon}+1)(e^{-\varepsilon}+1)(e^{\varepsilon'}+1)(e^{-\varepsilon'}+1)} d\varepsilon d\varepsilon' = 1.26066,$$

$$G(x) \longrightarrow x\log x + O\left(\frac{1}{x}\right) \quad x \to \infty.$$

In exactly the same way, we obtain $E(\Delta) = D\,8\log 2$. Making use of all of these results, we obtain

$$\begin{aligned}
F &= F_0 + F_1 + F_2 + \cdots, \\
F_0 &= -k_B T \log(e^{\Delta\beta} + e^{-\Delta\beta}), \\
F_1 &= -J\rho\Delta \tanh \Delta\beta, \\
F_2 &= -\tfrac{1}{2}\beta J^2\rho^2\Delta^2 \mathrm{sech}^2\Delta\beta - 3\log 2 \cdot J^2\rho^2 D \\
&\quad + 2J^2\rho^2\Delta(1 + \log D\beta)\tanh\Delta\beta \\
&\quad - J^2\rho^2 k_B T \tanh \Delta\beta \cdot G(2\Delta\beta).
\end{aligned}$$

Therefore, firstly, the susceptibility is given by

$$\chi = -\left.\frac{\partial^2 F}{\partial H^2}\right|_{H=0} = \frac{{\mu_B}^2}{k_B T}\left[1 + 2J\rho + 4J^2\rho^2\left(G'(0) - \frac{3}{4} + \log\frac{k_B T}{D}\right)\right]. \tag{F. 8}$$

Next, the magnetization $M = -\partial F/\partial H$ at $T \to 0$ is given by

$$M = \mu_B\left(1 + J\rho - 2J^2\rho^2 \log\frac{D}{2\mu_B H}\right). \tag{F. 9}$$

In the above calculations, we adopted a model in which the density of states is constant and the integration over energy is restricted to between $-D$ and D. For this model, the above results are exact up to order J^2. These results are necessary when discussing universality.

Appendix G Calculation of $g_\pm$

Let us first define $S_\pm(\lambda)$ as follows:

$$S_\pm(\lambda) = \mathrm{Tu}\exp\left[\pm\lambda J\int_0^\beta U(u)c_\pm{}^\dagger(u)c_\pm(u)du\right]. \tag{G. 1}$$

This is equal to $S_\pm$ for $\lambda = 1$. Remember that the emergence of $S_\pm$ corresponds to the change of the Hamiltonian from $H_\pm$ to $H_\mp$ over the time region U. In the same way, it is easily found out that $S_\pm(\lambda)$ arises when the Hamiltonian changes from $H_\pm$ to $H_\mp(\lambda)$. Here,

$$H_\mp(\lambda) = H_\pm \mp \lambda J c^\dagger c = H_0 \pm \left(\frac{1}{2} - \lambda\right) J c^\dagger c. \tag{G. 2}$$

$H_\mp(\lambda)$ is equal to $H_\mp$ for $\lambda = 1$ and $H_\pm$ for $\lambda = 0$. Let us define the Green's function for the Hamiltonian $H_\pm(\lambda)$, as usual, as follows:

$$G_\pm^\lambda(u - v) = \langle \mathrm{Tu}\, c(u)c^\dagger(v)\rangle. \tag{G. 3}$$

Here, $c(u)$ is the Heisenberg representation of c (eq. (6.13)) when the Hamiltonian is $H_\pm(\lambda)$, and $\langle\ \ \rangle$ denotes the thermal average when the Hamiltonian is $H_\pm(\lambda)$. By eq. (6.31), $G_\pm^\lambda$ is equal to $G_\pm$ when $\lambda = 1$ and $G_\mp$ when $\lambda = 0$. $G_\pm^\lambda$ can be obtained in a similar form to eqs. (6.38) and (6.39), as follows. Denoting the phase shift at the Fermi surface for the Hamiltonian $H_\pm(\lambda)$ as $\delta_\pm(\lambda)$ ($\delta_\pm(\lambda)$ and $\delta_\pm(u)$ being totally different), we obtain, by eqs. (6.36) and (G. 2),

$$\tan[\theta - \delta_\pm(\lambda)] = \tan\theta \mp \left(\frac{1}{2} - \lambda\right)\frac{\pi J\rho}{\cos^2\theta}. \tag{G. 4}$$

$\delta_\pm(\lambda)$ is equal to $\delta_\pm$ for $\lambda = 1$ and $\delta_\mp$ for $\lambda = 0$. Now, let us replace $\delta_\pm$ by $\delta_\pm(\lambda)$ in eq. (6.38) and use eq. (6.39). This yields $G^\lambda_\pm$:

$$G^\lambda_\pm(u-v) = \frac{\pi T \rho \cos^2[\theta - \delta_\pm(\lambda)]}{\cos^2\theta} P\left(\frac{1}{\sin \pi T(u-v)}\right). \tag{G.5}$$

Next, let us define $\mathcal{G}^\lambda_\pm$ as follows:

$$\mathcal{G}^\lambda_\pm(u,v) = \frac{\langle \mathrm{Tu}\, c_\pm(u) c_\pm{}^\dagger(v) S_\pm(\lambda)\rangle_\pm}{\langle \mathrm{Tu}\, S_\pm(\lambda)\rangle_\pm}. \tag{G.6}$$

This corresponds to eq. (6.50). With regard to the Dyson equation for $\mathcal{G}^\lambda_\pm$, this is obtained by replacing J by λJ in eq. (6.52). When we transform it using eq. (6.40), we obtain the same expression as eq. (6.53). Here, $\delta_\pm(u)$ appearing there in (6.53) is not defined by eq. (6.54) but by

$$\delta_\pm(u) = \begin{cases} \delta_\mp(\lambda) & u \in U, \\ \delta_\pm & u \notin U. \end{cases} \tag{G.7}$$

Further, by using a similar procedure to eqs. (6.55) to (6.57), we obtain

$$\mathcal{G}^\lambda_\pm(u,v) = G^\lambda_\mp(u-v)\left(\frac{Y(u)}{Y(v)}\right)^{[\delta_\pm - \delta_\mp(\lambda)]/\pi}. \tag{G.8}$$

for $u \in U$ and $v \in U$.

$g_\pm(\lambda)$ is defined by

$$g_\pm(\lambda) = \langle \mathrm{Tu}\, S_\pm(\lambda)\rangle_\pm. \tag{G.9}$$

The $g_\pm$ that is required is $g_\pm(1)$. Now, we take the derivative of eq. (G. 1) with respect to λ, and take its thermal average. Noting that

$$\mathcal{G}^\lambda_\pm(u, u+\delta) = -\frac{\langle \mathrm{Tu}\, c_\pm{}^\dagger(u) c_\pm(u) S_\pm(\lambda)\rangle_\pm}{\langle \mathrm{Tu}\, S_\pm(\lambda)\rangle_\pm},$$

we obtain

$$\frac{1}{g_\pm(\lambda)}\frac{d}{d\lambda} g_\pm(\lambda) = \mp J \int_0^\beta U(u) \mathcal{G}^\lambda_\pm(u, u+\delta) du.$$

Here, δ is a positive infinitesimal. Integrating this between $\lambda = 0$ and 1 leads to

$$\log g_\pm = \mp J \int_0^1 d\lambda \int_0^\beta du U(u) \mathcal{G}^\lambda_\pm(u, u+\delta). \tag{G.10}$$

$g_\pm$ is thus given by $\mathcal{G}^\lambda_\pm(u, u+\delta)$. Let us substitute $v = u + \delta$ into eq. (G. 8) and expand it in terms of δ. Using eq. (G. 5), we obtain

$$\mathcal{G}^\lambda_\pm(u, u+\delta) = G^\lambda_\mp(-\delta) + \frac{\rho}{\pi \cos^2\theta}\frac{1}{Y}\frac{dY}{du}[\delta_\pm - \delta_\mp(\lambda)]\cos^2[\theta - \delta_\mp(\lambda)]. \tag{G. 11}$$

We multiply this by $U(u)$ and integrate it with respect to λ and u. This integral is separable in terms of these two variables of integration. First, let us consider $G^\lambda_\mp(-\delta)$. By eq. (G. 2), $H_\pm(\lambda)$ can be regarded as $H_\mp$ plus a perturbation given by $\pm\lambda J c^\dagger c$. Let us now compose

$$\int_0^1 \langle \pm J c^\dagger c\rangle d\lambda. \tag{G. 12}$$

Here, $\langle \quad \rangle$ is the average when the Hamiltonian is $H_\pm(\lambda)$. Then, by Feynman's theorem (see Appendix H), this is the free energy for $\lambda = 1$ minus that for $\lambda = 0$. In other words, it is the free energy for the Hamiltonian $H_\pm$ minus that for the Hamiltonian $H_\mp$. Let us denote this as $F_\pm - F_\mp$. This can be rewritten as $\pm(F_+ - F_-)$. Now, according to eq. (G. 3), $G^\lambda_\pm(-\delta) = -\langle c^\dagger c\rangle$, where $\langle \quad \rangle$ is the average when the Hamiltonian is $H_\pm(\lambda)$. Using this, eq. (G. 12) may be written as $\mp J\int_0^1 G^\lambda_\pm(-\delta)d\lambda$. As this is equal to $\pm(F_+ - F_-)$, we obtain

$$\int_0^1 [G^\lambda_+(-\delta) - G^\lambda_-(-\delta)]d\lambda = 0.$$

Second, let us consider the second term of eq. (G. 11). Here, the dependence on λ is through the last two factors. By eq. (G. 4),

$$\cos^2[\theta - \delta_\pm(\lambda)]d\lambda = \mp\frac{\cos^2\theta}{\pi J\rho}d\delta_\pm(\lambda).$$

Hence the integration over λ from 0 to 1 is transformed into that over $\delta_\pm(\lambda)$ from $\delta_\mp$ to $\delta_\pm$, and this can be evaluated. The integration with respect to u yields $\log Y(u)$. Summarizing the above results, we obtain

$$\log g_+ + \log g_- = \frac{(\delta_+ - \delta_-)^2}{\pi^2}\log\prod_i \frac{Y(u_i)}{Y(v_i)}. \tag{G. 13}$$

Appendix H Feynman's theorem

Let us consider a Hamiltonian which is linear in parameter λ:

$$H(\lambda) = H_0 + \lambda H'.$$

The eigenvalues and eigenfunctions are functions of λ:

$$H(\lambda)\psi_n(\lambda) = E_n(\lambda)\psi_n(\lambda). \tag{H. 1}$$

We differentiate this expression with respect to λ, multiply it by $\psi_n(\lambda)^*$, and then integrate it. Using the normalization of ψ_n and eq. (H. 1), we obtain

$$\langle\psi_n(\lambda)|H'|\psi_n(\lambda)\rangle = \frac{\partial E_n(\lambda)}{\partial\lambda}. \tag{H. 2}$$

The free energy of this system is also a function of λ:

$$F(\lambda) = -\beta^{-1}\log\left[\sum_n e^{-\beta E_n(\lambda)}\right].$$

The derivative with respect to λ is

$$\frac{\partial F(\lambda)}{\partial\lambda} = \frac{1}{Z(\lambda)}\sum_n \frac{\partial E_n(\lambda)}{\partial\lambda}e^{-\beta E_n(\lambda)}.$$

By substituting eq. (H. 2) into this, we obtain

$$\frac{\partial F(\lambda)}{\partial\lambda} = \langle H'\rangle_\lambda.$$

Here, $\langle\quad\rangle_\lambda$ is the thermal average in a system whose Hamiltonian is given by $H(\lambda)$. The integration of this with respect to λ from 0 to 1 yields

$$F(1) - F(0) = \int_0^1 \langle H'\rangle_\lambda d\lambda.$$

This result is called Feynman's theorem.

Appendix I Elimination of adjacent pairs

We divide the integral of eq. (6.64) into two parts. In the first part, none of the β_is is a distance of less than $\tau + d\tau$ from one another; and in the second part, a certain pair is separated by a distance which is between τ to $\tau + d\tau$, while all others are farther apart than $\tau + d\tau$. Combining the nth-order term of the former and the $(n+1)$th-order term of the latter yields

$$Z = \sum_n (J_\perp)^{2n} \iint d\beta_1 \cdots d\beta_{2n} e^{(2-\varepsilon)V} \times \left[1 + J_\perp{}^2 \sum_i \int d\beta' \int d\beta'' e^{(2-\varepsilon)V'} \right]. \tag{I. 1}$$

Here, $\beta_1, \ldots, \beta_{2n}$ are all more than $\tau + d\tau$ apart; β' and β'' are in between β_i and β_{i+1}, and the distance separating them is between τ and $\tau + d\tau$. V' is the interaction between β' or β'' and the other βs, the interaction between β' and β'' being identically 0. Thus,

$$e^{(2-\varepsilon)V'} = \left[\frac{\beta_{i+1} - \beta''}{\beta_{i+1} - \beta'} \cdot \frac{\beta_{i+2} - \beta'}{\beta_{i+2} - \beta''} \cdots \frac{\beta' - \beta_i}{\beta'' - \beta_i} \cdot \frac{\beta'' - \beta_{i-1}}{\beta' - \beta_{i-1}} \cdots \right]^{2-\varepsilon}.$$

We expanded all sine functions and took the leading terms in their expansion. $\beta' - \beta'' = \tau$, and this is much smaller than the mean distance among particles l_0 as long as η is much less than unity. The above equation is hence approximated by

$$\begin{aligned} e^{(2-\varepsilon)V'} &= \left[\left(1 + \frac{\tau}{\beta_{i+1} - \beta'}\right) \cdots \left(1 + \frac{\tau}{\beta'' - \beta_i}\right) \cdots \right]^{2-\varepsilon} \\ &\cong 1 + (2-\varepsilon)\tau \left(\frac{1}{\beta_{i+1} - \beta'} - \frac{1}{\beta_{i+2} - \beta'} + \cdots \right. \\ &\qquad \left. + \frac{1}{\beta'' - \beta_i} - \frac{1}{\beta'' - \beta_{i-1}} + \cdots \right). \end{aligned}$$

Integrating the last expression in brackets ( ) with respect to β' and β'', when the number of particles is large, we obtain

$$\sum_i \int d\beta' \int d\beta'' (\quad) = -4 d\tau V.$$

Since we may write $\sum_i \int d\beta' \int d\beta'' = \beta d\tau$, substituting these results into eq. (I. 1) and exponentiating it yields

$$Z = e^{J_\perp{}^2 \beta d\tau} \sum_n (J_\perp)^{2n} \int \cdots \int d\beta_1 \cdots d\beta_{2n} e^{(2-\varepsilon)(1-4J_\perp{}^2 \tau d\tau)V},$$

as $d\tau$ is small. Here, β_i are more than $\tau + d\tau$ apart from one another. As seen from the definition in eq. (6.65), V contains τ. Letting $\tilde{V}$ denote V with τ

replaced by $\tau + d\tau$, we have

$$\tilde{V} = V + \frac{n d\tau}{\tau}.$$

Introducing $2 - \tilde{\varepsilon} = (2 - \varepsilon)(1 - 4{J_\perp}^2 \tau d\tau)$, we obtain

$$Z = e^{{J_\perp}^2 \beta d\tau} \sum_n (J_\perp)^{2n} \left(\frac{\tilde{\tau}}{\tau}\right)^{(\varepsilon-2)n} \int \cdots \int d\beta_1 \cdots d\beta_{2n} e^{(2-\tilde{\varepsilon})\tilde{V}}.$$

Now defining

$$\tilde{J}_\perp = J_\perp \left(\frac{\tilde{\tau}}{\tau}\right)^{(\varepsilon-2)/2},$$

we obtain

$$Z(\tau T, \varepsilon, \eta) = e^{{J_\perp}^2 \beta d\tau} Z(\tilde{\tau} T, \tilde{\varepsilon}, \tilde{\eta}).$$

Appendix J Proof of eq. (6.80)

We expand the partition function of the Anderson model, given by eqs. (6.77) and (6.78), in terms of H', following eq. (6.43). We then take the trace with respect to the number of localized particles. For instance, when there are no localized particles, the trace yields

$$Z_0 \sum_n (\tilde{J})^{2n} \int \cdots \int e^{-\varepsilon_{\mathrm{d}}(u_1 - v_1 + \cdots)} \langle c^\dagger(u_1) c(v_1) \cdots c^\dagger(u_n) c(v_n) \rangle_0 du_1 \cdots dv_n.$$

Here, $\langle \cdots \rangle_0$ is given as a determinant composed of $G_0(u_i - v_j)$ in eq. (6.18) (see eq. (6.51)). Using Cauchy's identity, we can show that this becomes e^V. In other words, the partition function of this problem corresponds to setting $\varepsilon = 1$ in eq. (6.64). In particular, when ε_{d} is at the Fermi level ($= 0$), we can show that

$$\langle b^\dagger b \rangle_{\varepsilon_{\mathrm{d}}=0} = \left.\frac{\partial F}{\partial \varepsilon_{\mathrm{d}}}\right|_{\varepsilon_{\mathrm{d}}=0} = -\frac{T}{Z} \left.\frac{\partial Z}{\partial \varepsilon_{\mathrm{d}}}\right|_{\varepsilon_{\mathrm{d}}=0} = \frac{1}{2}. \tag{J. 1}$$

Taking a further derivative of this yields the "susceptibility" χ' of the number of localized particles,

$$\chi' = -\frac{d}{d\varepsilon_{\mathrm{d}}} \langle b^\dagger b \rangle \bigg|_{\varepsilon_{\mathrm{d}}=0} = T \left[\frac{\partial^2 Z}{\partial {\varepsilon_{\mathrm{d}}}^2} \frac{1}{Z} - \left(\frac{\partial Z}{\partial \varepsilon_{\mathrm{d}}}\right)^2 \frac{1}{Z^2} \right]_{\varepsilon_{\mathrm{d}}=0}.$$

Using the above representation of Z here and using eq. (J. 1), we obtain eq. (6.80).

Appendix K Transformation from plane-wave representation to spherical-wave representation

By eq. (5.13), the bases of the spherical-wave representation may be written as

$$\psi_{klm}(r) = \left(\frac{2k^2}{R}\right)^{\frac{1}{2}} j_l(kr) Y_{lm}(\theta, \varphi). \tag{K. 1}$$

The bases in the plane-wave representation are given by eq. (3.2). Using eq. (5.53), we immediately obtain the coefficients of the spherical-wave expansion of plane waves. From eqs. (2.51) and (2.54), we see that $a_{\boldsymbol{k}}$ can be expanded in terms of a_{klm} by using the same coefficients. In particular, summing eq. (5.53) over k leads to

$$\sum_{\boldsymbol{k}} \psi_{\boldsymbol{k}}(r) = \left(\frac{2}{3}\right)^{\frac{1}{2}} \frac{R^2}{\pi} \int k \psi_{k00}(r) dk. \tag{K. 2}$$

Here, on the right-hand side, the sum over $\boldsymbol{k}$ is replaced by the integral. Denoting a_{k00} as a_k, we obtain

$$\sum_{\boldsymbol{k}} a_{\boldsymbol{k}} = \left(\frac{2}{3}\right)^{\frac{1}{2}} \frac{R^2 k_F}{\pi} \int a_k dk, \tag{K. 3}$$

where we took k to be equal to k_F and took it out of the integral.

The part of the kinetic energy for $l = m = 0$ is given by

$$\sum_k \frac{\hbar^2 k^2}{2m} a_{k00}{}^\dagger a_{k00}.$$

By eq. (5.12), k is given by $n\pi/R$ ($n = 1, 2, \cdots$). The sum over k means the sum over n. We replace this sum by $(R/\pi)dk$ integration, redefine k as $k_F + k$, and linearize the integral with respect to this new k. The kinetic energy given above then takes the form

$$\frac{R\hbar^2 k_F}{\pi m} \int k a_k{}^\dagger a_k dk.$$

Now we write k/k_F as k, and represent a_k in terms of the new k. Equation (K. 3) is also rewritten in the same way, and we substitute it into eq. (5.90), taking into account the spin degree of freedom. This yields eq. (7.1).

Appendix L Derivation of eq. (7.33)

We write eq. (7.32) as

$$z_{N+2} = z_N + a{z_N}^2 + b{z_N}^3 + c{z_N}^4.$$

We expand z_{N+2} as a Taylor series around z_N and obtain

$$z_{N+2} = z_N + 2\frac{dz_N}{dN} + 2\frac{d^2 z_N}{dN^2} + \frac{4}{3}\frac{d^3 z_N}{dN^3}.$$

From these two equations, we obtain

$$2\frac{dz_N}{dN} = a{z_N}^2 + b{z_N}^3 + c{z_N}^4 - 2\frac{d^2 z_N}{dN^2} - \frac{4}{3}\frac{d^3 z_N}{dN^3}.$$

We differentiate this equation with respect to N to express $d^2 z_N/dN^2$ etc. as a power expansion in z_N, and then substitute the result into the right-hand side. This yields

$$2\frac{dz_N}{dN} = a{z_N}^2 + (b - a^2){z_N}^3 + \left(c - \frac{5}{2}ab + \frac{3}{2}a^3\right){z_N}^4.$$

Denoting the right-hand side as $\psi(z_N)$, we have

$$2\int_{z_{N'}}^{z_N} \frac{dz}{\psi(z)} = \int_{N'}^{N} dN.$$

Expanding $1/\psi(z)$ as a Laurent series in z, and integrating term by term yields eq. (7.33). Here,

$$\Psi(z) = \frac{1}{az} + \left(\frac{b}{a^2} - 1\right)\log az + \frac{\left(ac - \frac{1}{2}ba^2 + \frac{1}{2}a^4 - b^2\right)z}{a^3}.$$

Appendix M Derivation of eq. (7.35)

This section should be read after reading the first half of §7.4. An alternative expression of the first term of eq. (7.48) (in the $N \to \infty$ limit) is:

$$\frac{\mathrm{Tr}\, {S_\infty}^2 e^{-H/\tau}}{\mathrm{Tr}\, e^{-H/\tau}}.$$

Here, H is given by eq. (7.18). Omitting the $\tilde{J}$ term, we can diagonalize H using a linear transformation. The maximum of the eigenvalues of H is about 1. Below this, there are infinitely many eigenvalues near 0. For numerical calculation, it is possible to substitute the large-N solutions to H_N for solutions to H. Let us evaluate the above expression by expanding H of eq. (7.18) in terms of $\tilde{J}$. We take T to be equal to T_N, which is defined by

$$k_B T_N \equiv 2 \cdot 2^{-N/2} \Lambda. \tag{M. 1}$$

It is convenient to use $\tilde{\chi}$ instead of χ:

$$\tilde{\chi}(T) = \frac{k_B T \chi(T)}{2{\mu_B}^2} - 0.5. \tag{M. 2}$$

As discussed before, when $\chi(T_N)$ is expanded in terms of $\tilde{J}$, $\tilde{\chi}(T_N)$ takes the form of a series expansion, which begins at the first order in $J' = -{\alpha_1}^2 \tilde{J}$. The coefficient of the nth-order term of the series is of order N^{n-1}.

As discussed before, the expansion of z_N in terms of J' yields results such as eq. (7.31). The coefficient of the nth-order term is again of order N^{n-1}. By eliminating J' from these two series expansions, we obtain

$$z_N = -1.9406\tilde{\chi}(T_N) + 4.5998\tilde{\chi}(T_N)^2 + 10.2447\tilde{\chi}(T_N)^3,$$

when $\Lambda = 2$. These coefficients are of course independent of J', but so long as N is not too small, they are also independent of N. Substituting this z_N into $\Psi(z_N) + \frac{1}{2}N$ and expanding it yields

$$\left[\Psi(z_N) + \frac{1}{2}N\right] \cdot \log 2 = -\frac{0.5}{\tilde{\chi}(T_N)} - 0.30873 - 0.5 \log|2\tilde{\chi}(T_N)| - \log\frac{k_B T_N}{\Lambda} + 3.1648\tilde{\chi}(T_N), \tag{M. 3}$$

leaving the terms that can be considered to be meaningful. As described near eq. (7.33), the left-hand side is independent of N. Hence the right-hand side is independent of T_N.

Let us relate this result to the parameters used in Chapter 5. The series expansion of χ is given by eq. (5.105), according to which:

$$\tilde{\chi}(T) = J\rho + 2J^2\rho^2 \log\frac{k_B T}{0.6001D} + \cdots. \tag{M. 4}$$

Now substituting $\tilde{\chi}(T_N) = J\rho + 2J^2\rho^2 \log(k_B T_N/0.6001D)$ into the right-hand side of eq. (M. 3), we see that the dependence on T_N vanishes:

$$
\begin{aligned}
&-\frac{0.5}{\tilde{\chi}(T_N)} - 0.30873 - 0.5\log|2\tilde{\chi}(T_N)| - \log\frac{k_B T_N}{A} + 3.1648\tilde{\chi}(T_N) \\
&\quad = -\frac{0.5}{J\rho} - 0.30873 - 0.5\log|2J\rho| - \log\frac{0.6001D}{A} + 3.1648J\rho.
\end{aligned} \tag{M. 5}
$$

This is the right-hand side of eq. (7.35).

One may think that eq. (M. 5) contains no more information than eq. (M. 4) does, but this is not so. As stated before, the left-hand side of eq. (M. 5) is independent of T_N. The higher-order terms of eq. (M. 4) thus need to be of a form such that substituting them into the left-hand side of eq. (M. 5) makes the $\log k_B T_N$ terms vanish. To find out such higher-order terms, we simply need to solve eq. (M. 5) in terms of $\tilde{\chi}(T_N)$. Hence we obtain

$$
\begin{aligned}
\tilde{\chi}(T_N) &= J\rho + 2J^2\rho^2 \log\frac{k_B T_N}{\tilde{D}} \\
&\quad + J^3\rho^3\left[4\log^2\frac{k_B T_N}{\tilde{D}} + 2\log\frac{k_B T_N}{\tilde{D}}\right] \\
&\quad + J^4\rho^4\left[8\log^3\frac{k_B T_N}{\tilde{D}} + 10\log^2\frac{k_B T_N}{\tilde{D}} - 10.66\log\frac{k_B T_N}{\tilde{D}}\right] \\
&\quad + \cdots.
\end{aligned} \tag{M. 6}
$$

Of course, this equation is also correct for T other than T_N.

Appendix N Solution to the eigenvalue problem in §8.6

N. 1 Formulation of the problem

When N real constants $k_1 \cdots k_N$ and a positive constant c are given, solve the following eigenvalue problem:

$$X_{j+1j} \cdots X_{Nj} X_{1j} \cdots X_{j-1j}\phi = \varepsilon\phi. \tag{N. 1}$$

Here, the eigenfunction ϕ is the N-electron spin function $\phi(\zeta_1\zeta_2\cdots\zeta_N)$, ε is the eigenvalue, and

$$X_{ij} = \alpha_{ij} + \beta_{ij}P_{ij}.$$

P_{ij} is an operator that exchanges the spin coordinates ζ_i and ζ_j in the spin function that is located to the right of P_{ij}, and

$$\alpha_{ij} = \frac{k_i - k_j}{k_i - k_j + ic},$$
$$\beta_{ij} = \frac{ic}{k_i - k_j + ic}.$$

These satisfy

$$\alpha_{ij} + \beta_{ij} = 1. \tag{N. 2}$$

Before writing down the solution, let us define a number of coefficients.

N. 2 Definition of coefficients and relations among coefficients

We define

$$c' \equiv \frac{c}{2},$$

in terms of which we define the functions $f_i(\Lambda)$:

$$f_i(\Lambda) = \frac{k_i - \Lambda + ic'}{k_i - \Lambda - ic'} \quad (i = 1, \ldots, N).$$

Λ is to be determined in the following. Let us define $b_i(\Lambda)$:

$$\left. \begin{aligned} &b_1(\Lambda) = 1 \\ &b_i(\Lambda) = f_1(\Lambda) \cdots f_{i-1}(\Lambda) \quad (i = 2, \ldots, N+1) \end{aligned} \right\}.$$

In particular, we denote $b_{N+1}(\Lambda)$ as $\Pi(\Lambda)$:

$$\Pi(\Lambda) \equiv f_1(\Lambda) \cdots f_N(\Lambda) = b_{N+1}(\Lambda). \tag{N. 3}$$

The following relation holds:

$$b_{i+1}(\Lambda) = b_i(\Lambda) f_i(\Lambda). \tag{N. 4}$$

$d_i(\Lambda)$ and $a_i(\Lambda)$ are defined as follows:

$$\begin{aligned} d_i(\Lambda) &= \frac{1}{k_i - \Lambda - ic'}, \\ a_i(\Lambda) &= b_i(\Lambda) d_i(\Lambda). \end{aligned} \tag{N. 5}$$

There are some identities among the coefficients α_{ij} and β_{ij}, which are as follows:

$$\alpha_{ij} a_i(\Lambda) + \beta_{ij} b_{i+1}(\Lambda) d_j(\Lambda) = f_j(\Lambda) a_i(\Lambda), \tag{N.6}$$

$$\alpha_{ij} b_{i+1}(\Lambda) d_j(\Lambda) + \beta_{ij} a_i(\Lambda) = d_j(\Lambda) b_i(\Lambda), \tag{N.7}$$

$$a_1(\Lambda) + \cdots + a_j(\Lambda) = \frac{b_{j+1}(\Lambda) - 1}{ic}. \tag{N.8}$$

Equations (N. 6) and (N. 7) can be verified by dividing them by $a_i(\Lambda)$ and substituting the definitions into them. Equation (N. 8) is easily proved by induction.

When M numbers $\Lambda_1, \ldots, \Lambda_M$ are given, we define F and Q by the following equations:

$$F(\Lambda_\alpha, \Lambda_\beta) \equiv \frac{\Lambda_\alpha - \Lambda_\beta + ic}{\Lambda_\alpha - \Lambda_\beta - ic} \quad (\alpha, \beta = 1, \ldots, M),$$

$$Q_\alpha(\Lambda_1 \Lambda_2 \cdots) \equiv \prod_{\beta \neq \alpha}^{M} F(\Lambda_\beta, \Lambda_\alpha) \quad (\alpha = 1, \ldots, M). \tag{N.9}$$

Here, $F(\Lambda_\alpha, \Lambda_\beta) = F(\Lambda_\beta, \Lambda_\alpha)^{-1}$. We then define A by

$$\begin{aligned} A(\Lambda_1 \Lambda_2 \cdots) \equiv (\Lambda_1 - \Lambda_2 - ic)(\Lambda_1 - \Lambda_3 - ic) \cdots \\ \times (\Lambda_2 - \Lambda_3 - ic) \cdots \\ \times (\Lambda_{M-1} - \Lambda_M - ic). \end{aligned}$$

P is defined as the operator that permutes the M numbers $\Lambda_1, \ldots, \Lambda_M$. For a permutation P, the corresponding coefficient A_P is defined as

$$A_P(\Lambda_1 \Lambda_2 \cdots) \equiv (-)^P PA(\Lambda_1 \Lambda_2 \cdots).$$

When we choose $(\cdots \alpha\beta \cdots)$ and $(\cdots \beta\alpha \cdots)$ as P, in which all sequences other than α and β are the same, we can easily show that

$$A_{\cdots\beta\alpha\cdots} = F(\Lambda_\alpha, \Lambda_\beta) A_{\cdots\alpha\beta\cdots}. \tag{N.10}$$

Moreover, when we choose $(\alpha \cdots)$ and $(\cdots \alpha)$, in which the sequences other than α are the same, we can use eq. (N. 10) repeatedly and use eq. (N. 9) to show that

$$A_{\alpha\cdots} = A_{\cdots\alpha} Q_\alpha. \tag{N.11}$$

We have the following two relations which involve both A_P and $a_i(\Lambda)$ etc., where the latter has been introduced earlier. First, if a certain function $f(\Lambda_1\Lambda_2\cdots\Lambda_M)$ is invariant under the transposition of Λ_k and Λ_{k+1}, the following relation holds:

$$\sum_P A_P(\Lambda_1\Lambda_2\cdots)P[f(\Lambda_1\Lambda_2\cdots)a_l(\Lambda_k)b_{l+1}(\Lambda_{k+1})d_j(\Lambda_{k+1})]$$
$$= \sum_P A_P(\Lambda_1\Lambda_2\cdots)P[f(\Lambda_1\Lambda_2\cdots)b_l(\Lambda_k)d_j(\Lambda_k)f_j(\Lambda_{k+1})a_l(\Lambda_{k+1})]. \tag{N. 12}$$

Here, the summation is over the $M!$ Ps, and P applies this permutation on $\Lambda_1\cdots\Lambda_M$ which is to the right of it.

Second, if $f(\Lambda_1\Lambda_2\cdots\Lambda_M)$ is invariant under transposition of Λ_1 and Λ_k, the following relation holds:

$$\sum_P A_P(\Lambda_1\Lambda_2\cdots)P\left[\frac{f(\Lambda_1\Lambda_2\cdots)}{\Lambda_1-\Lambda_k-ic}F(\Lambda_1\Lambda_{k-1})\cdots F(\Lambda_1\Lambda_2)\right]=0. \tag{N. 13}$$

These relations may be proved with comparative ease by directly substituting the definitions into them.

N. 3 Solution to the problem

The eigenfunction of eq. (N. 1) is a function of the spins of N electrons. Some simple results concerning such a function are described in Appendix B. In all, there are ${}_N\mathrm{C}_M$ products composed of $N-M$ α functions and M β functions, and the solution to eq. (N. 1) can be represented as a linear combination of these products. It is easy to see that products that involve different M do not mix. Let us denote the product in which the i_1th, i_2th, …, and so on up to the i_Mth electrons are represented by spin function β and the rest by α as

$$\phi_{i_1i_2\cdots i_M}.$$

The order of $i_1\cdots i_M$ does not need to be specified.

Let us now write down the solution to eq. (N. 1). We define $\Lambda_1\cdots\Lambda_M$ as

$$\Pi(\Lambda_\alpha)=Q_\alpha(\Lambda_1\Lambda_2\cdots)\qquad(\alpha=1,\ldots,M), \tag{N. 14}$$

where

$$\Pi(\Lambda_1)=1 \tag{N. 15}$$

when $M = 1$. Using these, we may compose

$$\phi = \sum_{1 \leqq i_1 < i_2 < \cdots < i_M \leqq N} \sum_P A_P(\Lambda_1 \Lambda_2 \cdots) P a_{i_1}(\Lambda_1) \cdots a_{i_M}(\Lambda_M) \phi_{i_1 i_2 \cdots i_M}, \tag{N. 16}$$

which has the eigenvalue

$$\varepsilon = f_j(\Lambda_1) \cdots f_j(\Lambda_M) \tag{N. 17}$$

and satisfies eq. (N. 1). Here, the summation over P is for the $M!$ permutations, and P is the operator that permutes $\Lambda_1 \cdots \Lambda_M$, which is to the right of it. Furthermore, in addition to

$$S_z \phi = \left(\frac{N}{2} - M \right) \phi,$$

we may also show that

$$S_+ \phi = 0. \tag{N. 18}$$

Hence, by eq. (B. 13), the magnitude of spin S is also equal to $\frac{N}{2} - M$. In other words, eq. (N. 16) is one of, or a linear combination of, the f_{NS} functions, this number being given by eq. (B. 11).

N. 4 Proof of the solution

Let us now confirm that eq. (N. 16) is an eigenfunction, by operating on it by $X_{j-1\,j}$ etc. directly.

(i) $M = 1$

In this case, we have

$$\phi = \sum_{i=1}^{N} a_i(\Lambda_1) \phi_i, \tag{N. 19}$$

where the factors ϕ_i are all α functions except $\beta(\zeta_i)$. Therefore $P_{j-1\,j}\phi_i$ is equal to ϕ_i except when $i = j-1$ or j. As a result, by eq. (N. 2), $X_{j-1\,j}\phi_i$ is equal to ϕ_i, again, except when $i = j-1$ or j. Hence,

$$X_{j-1\,j}\phi = a_1\phi_1 + a_2\phi_2 + \cdots + X_{j-1\,j}(a_{j-1}\phi_{j-1} + a_j\phi_j) + \cdots + a_N\phi_N.$$

Here, the argument Λ_1 has been omitted; $X_{j-1\,j}\phi_{j-1} = \alpha_{j-1\,j}\phi_{j-1} + \beta_{j-1\,j}\phi_j$, etc. Using eqs. (N. 6) and (N. 7), we find that the terms that are multiplied by $X_{j-1\,j}$ in the above equation are

$$f_j a_{j-1}\phi_{j-1} + b_{j-1} d_j \phi_j.$$

When we operate on the above equation with X_{j-2j}, we need to consider the terms that include ϕ_{j-2} and ϕ_j. Then, similarly, $f_j a_{j-2}\phi_{j-2} + b_{j-2}d_j\phi_j$ appears. By carrying out these operations up to X_{1j}, we obtain

$$X_{1j}\cdots X_{j-1j}\phi = f_j(a_1\phi_1 + \cdots + a_{j-1}\phi_{j-1}) + b_1 d_j\phi_j + \cdots + a_N\phi_N.$$

Here, b_1, which appears in the coefficient of ϕ_j, is equal to 1 by definition, but it can be replaced by $b_{N+1}(\Lambda_1)$ using eq. (N. 3) if eq. (N. 15) holds. When we operate on the above equation with X_{Nj}, we should consider the terms that involve ϕ_j and ϕ_N. Here, if we replace b_1 by b_{N+1}, we can use eqs. (N. 6) and (N. 7) again, and the terms that involve ϕ_j and ϕ_N become

$$b_N d_j\phi_j + f_j a_N\phi_N.$$

Proceeding thus up to X_{j+1j}, all the terms apart from ϕ_j are multiplied by f_j, and the term that involves ϕ_j becomes $b_{j+1}d_j\phi_j$. This is equal to $f_j a_j\phi_j$ owing to eqs. (N. 4) and (N. 5), and thus, we find that all terms are multiplied by f_j.

Next, let us prove eq. (N. 18). By eq. (N. 19), we have

$$S_+\phi = \left(\sum_i^N a_i\right)\alpha\alpha\cdots\alpha,$$

where $\sum_i^N a_i = (b_{N+1}-1)/ic$ by eq. (N. 8). However, this vanishes by eq. (N. 15).

(ii) $M > 1$

Let us discuss the case of $M = 3$. It is easy to deduce the general case from this.

$$\phi = \sum_{i<i'<i''}\sum_P A_P P a_i(\Lambda_1)a_{i'}(\Lambda_2)a_{i''}(\Lambda_3)\phi_{ii'i''}. \qquad \text{(N. 20)}$$

Let us now see how the coefficient of $\phi_{ll'l''}$ changes when we operate on ϕ with $X_{j+1j}\cdots X_{j-1j}$. Here, $l < l' < l''$. If $l'' < j$, we immediately see that in the sum of eq. (N. 20) only the i, i' and i'' terms shown in Fig. N. 1 contribute to $\phi_{ll'l''}$. Firstly, when we operate on it with X_{j-1j}, we need to consider the terms

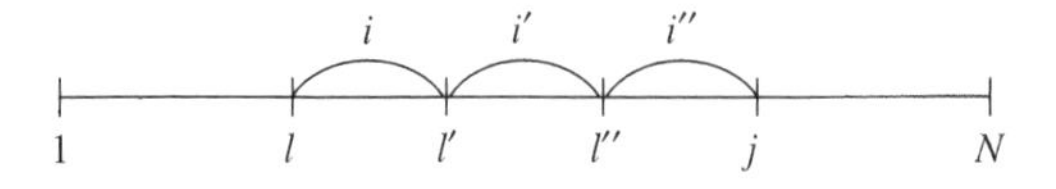

Figure N. 1

with $i'' = j$ or $j-1$. As $X_{j-1\,j}$ etc. commute with the operation P, as before, we obtain

$$X_{j-1\,j}(a_{j-1}(\Lambda_3)\phi_{ii'j-1} + a_j(\Lambda_3)\phi_{ii'j})$$
$$= f_j(\Lambda_3)a_{j-1}(\Lambda_3)\phi_{ii'j-1} + b_{j-1}(\Lambda_3)d_j(\Lambda_3)\phi_{ii'j}.$$

In the same way, when we carry out these operations up to $X_{l''j}$ (hereafter, we only show the terms that are necessary to obtain the coefficient of $\phi_{ll'l''}$), we obtain

$$\begin{aligned}&\sum_{i<i'}\sum_P A_P P a_i(\Lambda_1)a_{i'}(\Lambda_2)f_j(\Lambda_3)a_{l''}(\Lambda_3)\phi_{ii'l''}\\&\quad+\sum_i\sum_P A_P P a_i(\Lambda_1)a_{l''}(\Lambda_2)b_{l''+1}(\Lambda_3)d_j(\Lambda_3)\phi_{il''j}.\end{aligned} \tag{N. 21}$$

Here, we only kept $i'' = l''$ in the first term and $i' = l''$, $i'' = j$ in the second term, and omitted the others as they are not necessary for obtaining the coefficient of $\phi_{ll'l''}$. Then, using eq. (N. 12), we find that

$$\begin{aligned}&\sum_P A_P P a_i(\Lambda_1)a_{l''}(\Lambda_2)b_{l''+1}(\Lambda_3)d_j(\Lambda_3)\\&\quad=\sum_P A_P P a_i(\Lambda_1)b_{l''}(\Lambda_2)d_j(\Lambda_2)f_j(\Lambda_3)a_{l''}(\Lambda_3),\end{aligned}$$

and so we can write eq. (N. 21) as

$$\sum_P A_P P \sum_i a_i(\Lambda_1)\left[\sum_{i'} a_{i'}(\Lambda_2)\phi_{ii'l''} + b_{l''}(\Lambda_2)d_j(\Lambda_2)\phi_{ijl''}\right]f_j(\Lambda_3)a_{l''}(\Lambda_3).$$

When we operate on this with $X_{l''-1\,j}$, our operating targets are the term with $i' = l''-1$ and $\phi_{ijl''}$. Thus, in the same way as before, we obtain $f_j(\Lambda_2)a_{l''-1}(\Lambda_2)\phi_{il''-1l''} + b_{l''-1}(\Lambda_2)d_j(\Lambda_2)\phi_{ijl''}$. Proceeding up to $X_{l'j}$, we obtain

$$\begin{aligned}&\sum_P A_P P\left[\sum_i a_i(\Lambda_1)f_j(\Lambda_2)a_{l'}(\Lambda_2)\phi_{il'l''} + a_{l'}(\Lambda_1)b_{l'+1}(\Lambda_2)d_j(\Lambda_2)\phi_{l'jl''}\right]\\&\quad f_j(\Lambda_3)a_{l''}(\Lambda_3).\end{aligned}$$

Here, we only wrote the $i' = l'$ contribution in the first term inside square brackets, and the $i = l'$ contribution in the second term. Now, using eq. (N. 12) again, we find that

$$\sum_P A_P P a_{l'}(\Lambda_1) b_{l'+1}(\Lambda_2) d_j(\Lambda_2) f_j(\Lambda_3) a_{l''}(\Lambda_3)$$
$$= \sum_P A_P P b_{l'}(\Lambda_1) d_j(\Lambda_1) f_j(\Lambda_2) a_{l'}(\Lambda_2) f_j(\Lambda_3) a_{l''}(\Lambda_3),$$

and so we can write the above equation as

$$\sum_P A_P P \left[\sum_i a_i(\Lambda_1)\phi_{il'l''} + b_{l'}(\Lambda_1) d_j(\Lambda_1) \phi_{jl'l''} \right]$$
$$f_j(\Lambda_2) a_{l'}(\Lambda_2) f_j(\Lambda_3) a_{l''}(\Lambda_3).$$

When we operate on this with $X_{l'-1j}$, the targets are the $i = l' - 1$ term and $\phi_{jl'l''}$. Thus, in the same way as before, we obtain $f_j(\Lambda_1) a_{l'-1}(\Lambda_1) \phi_{l'-1l'l''} + b_{l'-1}(\Lambda_1) d_j(\Lambda_1) \phi_{jl'l''}$. When we carry out these operations up to X_{lj}, we see that $a_l(\Lambda_1) a_{l'}(\Lambda_2) a_{l''}(\Lambda_3) \phi_{ll'l''}$ is multiplied by $f_j(\Lambda_1) f_j(\Lambda_2) f_j(\Lambda_3)$. Since this factor can be taken out of P, we see that this factor is an eigenvalue.

It is easy to see that, if $j < l$, in the sum of eq. (N. 20) only the i, i' and i'' terms shown in Fig. N. 2 contribute to $\phi_{ll'l''}$. Out of the two cases in Fig. N. 2, we can use eq. (N. 20) for case (a), but, for case (b), it is more convenient to transform eq. (N. 20) as follows:

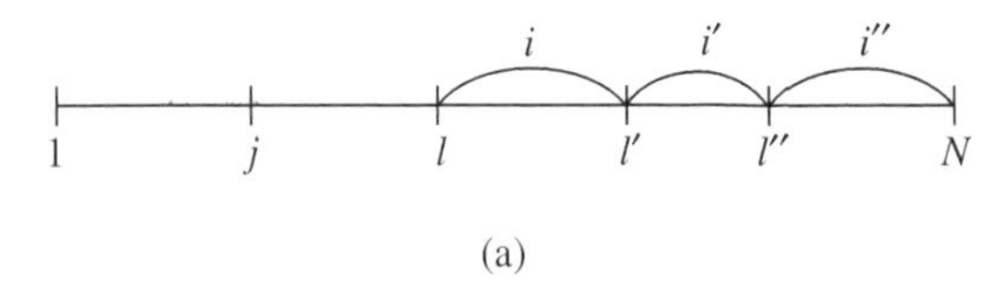

(a)

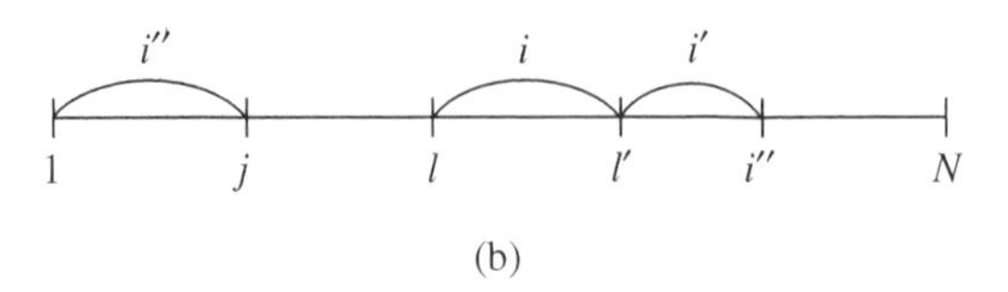

(b)

Figure N. 2

$$\sum_P A_P P \sum_{i''<i<i'} a_{i''}(\Lambda_1)a_i(\Lambda_2)a_{i'}(\Lambda_3)\phi_{ii'i''}$$
$$= \sum_P A_{PP_3} PP_3 \sum_{i''<i<i'} a_{i''}(\Lambda_1)a_i(\Lambda_2)a_{i'}(\Lambda_3)\phi_{ii'i''}$$
$$= \sum_P A_{PP_3} P \sum_{i''<i<i'} a_i(\Lambda_1)a_{i'}(\Lambda_2)a_{i''}(\Lambda_3)\phi_{ii'i''}.$$

Here we made use of eq. (8.27); P_3 denotes (3 1 2). Using eq. (N. 11), we can write the above as

$$= \sum_P A_P P \sum_{i''<i<i'} Q_3(\Lambda_1\Lambda_2\Lambda_3)a_i(\Lambda_1)a_{i'}(\Lambda_2)a_{i''}(\Lambda_3)\phi_{ii'i''}.$$

When we operate on this with $X_{j-1\,j}$, the $i''=j-1$ and j terms are affected. Carrying out these operations up to X_{1j} in the same way as before, we obtain

$$\sum_P A_P P \sum Q_3 a_i(\Lambda_1)a_{i'}(\Lambda_2)f_j(\Lambda_3)a_1(\Lambda_3)\phi_{ii'1}$$
$$+ \sum_P A_P P \sum Q_3 a_i(\Lambda_1)a_{i'}(\Lambda_2)b_1(\Lambda_3)d_j(\Lambda_3)\phi_{ii'j}.$$

We have only shown those terms that are needed for our discussion. Here, as $b_1=1$ and $Q_3=b_{N+1}(\Lambda_3)$ by eq. (N. 14), the second term in the above expression becomes

$$\sum_P A_P P \sum_{i<i'} a_i(\Lambda_1)a_{i'}(\Lambda_2)b_{N+1}(\Lambda_3)d_j(\Lambda_3)\phi_{ii'j}.$$

When this is combined with

$$\sum A_P P \sum a_i(\Lambda_1)a_{i'}(\Lambda_2)a_N(\Lambda_3)\phi_{ii'N},$$

coming from case (a), we can operate on it with X_{Nj} and proceed in the same way as above.

Using eq. (N. 14), the cases with other positional relationships among l, l', l'' and j can be worked out in the same way. Hence, we have proved that $f_j(\Lambda_1)f_j(\Lambda_2)f_j(\Lambda_3)$ is an eigenvalue.

Let us now move on to proving eq. (N. 18). $S_+\phi$ is represented as states with two spin-down electrons. The coefficient of $\phi_{ll'}(l<l')$ for these states can be

easily seen to be

$$\sum_P A_P P \left[\sum_{i<l} a_i(\Lambda_1) a_l(\Lambda_2) a_{l'}(\Lambda_3) \right.$$
$$+ \sum_{l<i<l'} a_l(\Lambda_1) a_i(\Lambda_2) a_{l'}(\Lambda_3)$$
$$\left. + \sum_{l'<i} a_l(\Lambda_1) a_{l'}(\Lambda_2) a_i(\Lambda_3) \right]. \tag{N. 22}$$

We then make use of eq. (8.27). In the brackets of eq. (N. 22), we replace P of the second term by PP_{12} and P of the third term by PP_{231}. Here, $P_{12} = (2\,1\,3)$ and $P_{231} = (2\,3\,1)$. This yields

$$= \sum_P A_P P \sum_{i<l} a_i(\Lambda_1) a_l(\Lambda_2) a_{l'}(\Lambda_3)$$
$$+ \sum_P A_{PP_{12}} P \sum_{l<i<l'} a_i(\Lambda_1) a_l(\Lambda_2) a_{l'}(\Lambda_3)$$
$$+ \sum_P A_{PP_{231}} P \sum_{l'<i} a_i(\Lambda_1) a_l(\Lambda_2) a_{l'}(\Lambda_3).$$

Now we use eq. (N. 10):

$$= \sum_P A_P P \left[\sum_{i<l} a_i(\Lambda_1) a_l(\Lambda_2) a_{l'}(\Lambda_3) \right.$$
$$+ \sum_{l<i<l'} F(\Lambda_1\, \Lambda_2) a_i(\Lambda_1) a_l(\Lambda_2) a_{l'}(\Lambda_3)$$
$$\left. + \sum_{l'<i} F(\Lambda_1\, \Lambda_2) F(\Lambda_1\, \Lambda_3) a_i(\Lambda_1) a_l(\Lambda_2) a_{l'}(\Lambda_3) \right].$$

We then carry out the summation over i using eq. (N. 8):

$$= (ic)^{-1} \sum A_P P a_l(\Lambda_2) a_{l'}(\Lambda_3) \{ b_l(\Lambda_1) - 1$$
$$+ F(\Lambda_1\, \Lambda_2)[b_{l'}(\Lambda_1) - b_{l+1}(\Lambda_1)]$$
$$+ F(\Lambda_1\, \Lambda_2) F(\Lambda_1\, \Lambda_3)[b_{N+1}(\Lambda_1) - b_{l'+1}(\Lambda_1)] \}.$$

Rearranging the expression inside { } leads to

$$\begin{aligned}&b_l(\Lambda_1) - b_{l+1}(\Lambda_1)F(\Lambda_1\,\Lambda_2)\\&+[b_{l'}(\Lambda_1) - b_{l'+1}(\Lambda_1)F(\Lambda_1\,\Lambda_3)]F(\Lambda_1\,\Lambda_2)\\&+F(\Lambda_1\,\Lambda_2)F(\Lambda_1\,\Lambda_3)b_{N+1}(\Lambda_1) - 1.\end{aligned}$$

Here, the third line can be written as

$$\frac{\Pi(\Lambda_1)}{Q_1} - 1$$

by eq. (N. 9), and so, according to eq. (N. 14), it vanishes. After some direct substitutions, the first line is found to be

$$a_l(\Lambda_1)\frac{\Lambda_1 + \Lambda_2 - 2k_l}{\Lambda_1 - \Lambda_2 - ic}.$$

Since

$$\sum_P A_P P \frac{\Lambda_1 + \Lambda_2 - 2k_l}{\Lambda_1 - \Lambda_2 - ic} a_l(\Lambda_1)a_l(\Lambda_2)a_{l'}(\Lambda_3)$$

is zero by eq. (N. 13), the contribution of the first line also vanishes. In the same way, the contribution of the second line also vanishes by eq. (N. 13).

Appendix O Wiener–Hopf integral equation

Let us consider the problem of solving

$$y(x) = \alpha e^{-\beta x} + \int_0^\infty R(x - x')y(x')dx' \quad (x > 0,\ \beta > 0), \tag{O. 1}$$

where R is as defined by eq. (8.64a). Say that eq. (O. 1) has been solved for $x > 0$. We then define $y_0(x)$ as follows for $x < 0$:

$$y_0(x) = \int_0^\infty R(x - x')y(x')dx' \quad (x < 0). \tag{O. 2}$$

The Fourier transforms are defined as follows:

$$Y(\omega) = \int_0^\infty y(x)e^{i\omega x}dx,$$

$$Y_0(\omega) = \int_{-\infty}^0 y_0(x)e^{i\omega x}dx.$$

Here, functions Y and Y_0 are holomorphic in the upper and lower half-planes, respectively. We multiply eq. (O. 1) by $e^{i\omega x}$ and integrate it from $\omega = 0$ to ∞. We also multiply eq. (O. 2) by $e^{i\omega x}$ and integrate it from $\omega = -\infty$ to 0. When we add the two together, the term that includes R becomes a convolution, and can be represented as a product of Fourier transformations. Keeping eq. (8.64a) in mind, we obtain

$$Y(\omega) + Y_0(\omega) = \frac{\alpha}{\beta - i\omega} + \frac{Y(\omega)}{1 + e^{c|\omega|}}.$$

Using this, we can derive

$$\frac{Y(\omega)}{1 + e^{-c|\omega|}} + Y_0(\omega) = \frac{\alpha}{\beta - i\omega}. \tag{O. 3}$$

Now, by using G_+ and G_- defined by eq. (8.74), we can show, for real ω, that

$$1 + e^{-c|\omega|} = G_+\left(\frac{c\omega}{2\pi}\right) G_-\left(\frac{c\omega}{2\pi}\right). \tag{O. 4}$$

Substituting this into eq. (O. 3) yields

$$\frac{Y(\omega)}{G_+\left(\frac{c\omega}{2\pi}\right)} + Y_0(\omega) G_-\left(\frac{c\omega}{2\pi}\right) = \frac{\alpha}{\beta - i\omega} G_-\left(\frac{c\omega}{2\pi}\right).$$

Here, when ω is extended to the complex plane, the first and second terms on the left-hand side are holomorphic in the upper and lower half-planes, respectively. If we transform the right-hand side into

$$\begin{aligned} \frac{\alpha}{\beta - i\omega} G_-\left(\frac{c\omega}{2\pi}\right) = {} & \frac{\alpha}{\beta - i\omega} G_-\left(\frac{c\beta}{2\pi i}\right) \\ & + \frac{\alpha}{\beta - i\omega}\left[G_-\left(\frac{c\omega}{2\pi}\right) - G_-\left(\frac{c\beta}{2\pi i}\right)\right], \end{aligned} \tag{O. 5}$$

the first and second terms on the right-hand side are holomorphic in the upper and lower half-planes, respectively. Thus, each of the two terms on the left-hand side should be equal to the corresponding term on the right-hand side. By equating the first terms of the two sides, we obtain eq. (8.76).

Appendix P Analytic continuation of eq. (8.82)

First of all, for real p, let us note the following two identities:

$$\frac{1}{2}\operatorname{sech}\frac{cp}{2} = \frac{i}{c}\sum_{n=0}^{\infty}(-)^n\left[\frac{1}{p+(2n+1)\dfrac{\pi i}{c}} - \frac{1}{p-(2n+1)\dfrac{\pi i}{c}}\right],$$

$$e^{-c|p|/2} = \frac{i}{\pi}\int_0^{\infty}\sin\frac{ct}{2}\left(\frac{1}{p+it} - \frac{1}{p-it}\right)dt.$$

When the latter is divided by the former, the left-hand side becomes $1+e^{-c|p|}$. Substituting eq. (O. 4) into this yields

$$G_+\left(\frac{cp}{2\pi}\right)e^{iBp}\sum_n(-)^n\left[\frac{1}{p+(2n+1)\dfrac{\pi i}{c}} - \frac{1}{p-(2n+1)\dfrac{\pi i}{c}}\right]$$
$$= \frac{c}{\pi}e^{iBp}\,G_-\left(\frac{cp}{2\pi}\right)^{-1}\int_0^{\infty}\sin\frac{ct}{2}\left(\frac{1}{p+it} - \frac{1}{p-it}\right)dt.$$

Here, we have multiplied both sides by e^{iBp}. Now, we extend p into the complex plane. For $B>0$, $G_+\left(\frac{cp}{2\pi}\right)e^{iBp}$ on the left-hand side is holomorphic in the upper half-plane. This expression multiplied by $(p+(2n+1)\pi i/c)^{-1}$ is also holomorphic in the upper half-plane. The same expression multiplied by $(p-(2n+1)\pi i/c)^{-1}$ can be written as the sum of two terms, of which the first is holomorphic in the upper half-plane and the second is holomorphic in the lower half-plane, in the same way as eq. (O. 5). Hence when $B>0$, the left-hand side can be written as the sum of a term that is holomorphic in the upper half-plane and another term that is holomorphic in the lower half-plane. Next, when $B<0$, $e^{iBp}G_-\left(\frac{cp}{2\pi}\right)^{-1}$ on the right-hand side is holomorphic in the lower half-plane. Hence, in the same way as before, the right-hand side can be written as the sum of a term that is holomorphic in the upper half-plane and a term that is holomorphic in the lower half-plane. The part that is holomorphic in the lower half-plane for $B>0$, for instance, needs to be the analytic continuation of the one for $B<0$ and vice versa, and therefore

$$-\sum_n G_+\left(i\left(n+\frac{1}{2}\right)\right)\frac{(-)^n e^{-(2n+1)(\pi/c)B}}{p-(2n+1)\dfrac{\pi i}{c}}$$

for $B > 0$ is analytically continued to

$$-\frac{c}{\pi}G_-\left(\frac{cp}{2\pi}\right)^{-1}e^{iBp}\int_0^\infty \sin\frac{ct}{2}\frac{dt}{p-it}$$
$$+\frac{c}{\pi}\int_0^\infty \sin\frac{ct}{2}\frac{1}{p+it}\left[G_-\left(\frac{cp}{2\pi}\right)^{-1}e^{iBp}-G_-\left(\frac{ct}{2\pi i}\right)^{-1}e^{Bt}\right]dt$$

for $B < 0$. By setting $p = 0$ and $e^{-\pi B/c} = \mu H/T_1$, we obtain eq. (8.85).

Appendix Q Rewriting eqs. (8.115) and (8.116)

Hereafter, we denote the Fourier transform of a certain function $A(\Lambda)$ by $A(\omega)$, so long as this is not misleading. For example, from eq. (8.110a), we can derive

$$F_n(\omega) = \int F_n(\Lambda)e^{i\omega\Lambda}d\Lambda = e^{-n|\omega c'|}. \tag{Q.1}$$

From this result and eq. (8.110b), we can derive

$$A_{nm}(\omega) = \coth|\omega c'|\left(e^{-|n-m||\omega c'|} - e^{-(n+m)|\omega c'|}\right), \tag{Q.2}$$

and easily arrive at the following relation:

$$A_{n-1\,m}(\omega) + A_{n+1\,m}(\omega) = 2\cosh\omega c'\left[A_{nm}(\omega) - \delta_{nm}\right], \tag{Q.3}$$

(with $A_{0m}(\omega) \equiv 0$).

$$F_n(\omega)A_{n+1\,m}(\omega) - F_{n+1}(\omega)A_{nm}(\omega) = \begin{cases} F_{m+1}(\omega) + F_{m-1}(\omega) & (m > n) \\ 0 & (m \leqq n). \end{cases} \tag{Q.4}$$

Furthermore, by defining $s(\Lambda)$ by eq. (8.117), we obtain

$$s(\omega) = \tfrac{1}{2}\mathrm{sech}\,\omega c'. \tag{Q.5}$$

Thus, if

$$B(\omega) = \tfrac{1}{2}\mathrm{sech}\,\omega c' \cdot A(\omega),$$

the following holds:

$$B(\Lambda) = s * A(\Lambda). \tag{Q.6}$$

Here, $*$ denotes convolution (see eq. (8.113a)).

Returning to eq. (8.115), we define the following functions to save space:

$$p_n(\Lambda) \equiv \log(1+\eta_n),$$
$$q_n(\Lambda) \equiv \log(1+\eta_n{}^{-1}).$$

By eq. (8.115), we obtain

$$Dg_n(\omega) + Tp_n(\omega) = T\sum_m A_{nm}(\omega)q_m(\omega). \tag{Q. 7}$$

Here, by eq. (8.104), we obtain

$$g_n(\omega) = -\frac{2\pi i}{\omega}e^{-n|\omega c'|} + \left(\pi - \frac{2\mu H}{D}n\right)2\pi\delta(\omega). \tag{Q. 8}$$

We add together eq. (Q. 7) multiplied by $-2\cosh\omega c'$, eq. (Q. 7) with $n-1$, and the same with $n+1$. Using eq. (Q. 3), we obtain

$$p_n(\omega) - q_n(\omega) = \tfrac{1}{2}\mathrm{sech}\,\omega c'\,[p_{n+1}(\omega) + p_{n-1}(\omega)] \quad (n \geqq 2), \tag{Q. 9}$$

$$p_1(\omega) - q_1(\omega) = \tfrac{1}{2}\mathrm{sech}\,\omega c'\left[p_2(\omega) + \frac{2\pi i D}{T}\left(\frac{1}{\omega} + \pi i\delta(\omega)\right)\right]. \tag{Q. 10}$$

Hence the infinite summation over m in eq. (Q. 7) disappears. This corresponds to inverting the matrix A. Now, using eq. (Q. 6), we can carry out inverse Fourier transformation. From eq. (Q. 9), we obtain

$$p_n(\Lambda) - q_n(\Lambda) = s * (p_{n+1} + p_{n-1}) \quad (n \geqq 2).$$

This is nothing other than eq. (8.119). We note that $\omega^{-1} + \pi i\delta(\omega)$ is the Fourier transform of the step function, and obtain, by eq. (Q. 10),

$$p_1(\Lambda) - q_1(\Lambda) = s * p_2 - \frac{2D}{T}\tan^{-1} e^{-\pi\Lambda/c}.$$

This is eq. (8.118).

Next, we multiply eq. (Q. 7) by $F_{n+1}(\omega)$, and subtract from it the product of eq. (Q. 7), with $n+1$, and $F_n(\omega)$. Using eqs. (Q. 4) and (Q. 8), we obtain

$$\frac{2\mu H}{T}2\pi\delta(\omega) + p_n(\omega)F_{n+1}(\omega) - p_{n+1}(\omega)F_n(\omega)$$
$$= -\sum_{m>n}[F_{m+1}(\omega) + F_{m-1}(\omega)]\,q_m(\omega).$$

Reverting to Λ space, this is expressed as

$$\frac{2\mu H}{T} + p_n * F_{n+1} - p_{n+1} * F_n = -\sum_{m>n} (F_{m+1} + F_{m-1}) * q_m.$$

Let us now consider the case of extremely large n. When $\Lambda_\alpha{}^n$ with extremely large n occurs, large excitation energy will be needed, and, for such n, only holes will be present, regardless of the value of Λ. Thus, $\eta_n{}^{-1}$ tends to zero, and $q_n(\Lambda)$ also tends to zero. Hence the right-hand side of the above equation tends to zero. This is represented by eq. (8.120).

We now discuss eq. (8.116). We first note the following identity:

$$NF_n(\Lambda) + F_n(\Lambda - 1) = NA_{1n} * s(\Lambda) + A_{1n} * s(\Lambda - 1). \tag{Q. 11}$$

Here, the second term on the right-hand side represents the convolution of $A_{1n}(\Lambda)$ and $s(\Lambda - 1)$. When we compare the Fourier transforms of the two sides, by eqs. (Q. 1), (Q. 2) and (Q. 5), we can confirm that they are the same. We then substitute this into eq. (8.116), and, using eq. (8.114), we obtain

$$F_\Lambda = -T \sum_n \int [Ns(\Lambda) + s(\Lambda - 1)] A_{1n} * \log(1 + \eta_n{}^{-1}) d\Lambda.$$

Moreover, making use of eq. (8.115) for $n = 1$, we obtain

$$\begin{aligned} F_\Lambda = &- D \int [Ns(\Lambda) + s(\Lambda - 1)]\, g_1(\Lambda) d\Lambda \\ &- T \int [Ns(\Lambda) + s(\Lambda - 1)] \log(1 + \eta_1) d\Lambda. \end{aligned}$$

The first term is a constant, and so this equation gives eq. (8.121).

Index